The Ebb and Flow of Chinese Petroleum

Ideas, History, and Modern China

Edited by

Ban Wang (*Stanford University*)
Wang Hui (*Tsinghua University*)

VOLUME 21

The titles published in this series are listed at *brill.com/ihmc*

The author, Mao Huahe

The Ebb and Flow of Chinese Petroleum

A Story Told by a Witness

By

MAO Huahe

Translated by

MAO Yiran and Thomas SEAY

BRILL

LEIDEN | BOSTON

The Library of Congress Cataloging-in-Publication Data is available online at http://catalog.loc.gov
LC record available at http://lccn.loc.gov/2019018993

Typeface for the Latin, Greek, and Cyrillic scripts: "Brill". See and download: brill.com/brill-typeface.

ISSN 1875-9394
ISBN 978-90-04-40272-0 (hardback)
ISBN 978-90-04-40273-7 (e-book)

This book is printed on acid-free paper and produced in a sustainable manner.

Contents

PART 3
The Daqing Era (1960–1969): China No Longer "Poor in Oil"

PART 4
The Jianghan Era (1970–1979): Against the Current, China Becomes a Big Oil-Producing Country

PART 5
Shengli Era (1980–1989): When the Gods Fight, Mortals Suffer: Rolling a Boulder Uphill

Foreword

Petroleum in China has a long history. The founding of the People's Republic of China on October 1, 1949 ushered in a new epoch in the development of the petroleum industry. After oil was discovered in Daqing in 1960, the development of this oilfield was basically completed after four years of hard work. This feat was then publicized and reported throughout the nation. In December 1963, Mao Zedong issued his call, "In Industry, Learn from Daqing." In the next thirty-plus years until 2000, Chinese petroleum forged ahead, following the path blazed in Daqing. With the advent of the new century, in the midst of rapid reform and opening, the slogan, "In Industry, Learn from Daqing," gradually faded from the petroleum industry, becoming a not-so-distant memory.

In June 1960, I retired from the People's Liberation Army and was assigned to Daqing Oilfield. Not long after that, due to a strange combination of circumstances, I became Daqing's "scribe." During my first four years there, I wrote most of the monthly and annual work reports from Daqing Oilfield. In November 1963, under the auspices of the State Economic and Trade Commission, a national meeting was convened in Beijing on industrial and transportation work. Kang Shi'en 康世恩, the Deputy Minister of Petroleum and director of the Daqing Oilfield, presented the "Daqing Oil Campaign Report." This report for the first time disseminated throughout the nation "The Daqing Man" and "The Daqing Experience," which became models for Chinese industry through the next few decades, as encapsulated by Mao's declaration, "In Industry, Learn from Daqing." The thinking and viewpoints of this report were entirely Kang Shi'en's. He dictated the original draft, and a colleague and I wrote it down word-for-word and arranged it slightly. In February 1964, I was transferred from Daqing to the Ministry of Petroleum Industry in Beijing, where I was still engaged in propagating the achievements of the oil industry. In 1969, I went to Jianghan Oilfield in Hubei Province, where I worked for four years. In 1975, I was transferred to Huabei Oilfield in Hebei Province, where I worked for eight years, and at the end of 1982, I was transferred back to the Ministry of Petroleum Industry in Beijing. Up to 2000, I was paying very close attention to all aspects of the Chinese oil industry.

Needless to say, I was a participant, witness, and steadfast guardian of the development of China's oil industry. Oil was my entire career. Oil has followed me like a shadow my whole life, and I have shared in both its glories and humiliations. To the extent that I affirm or deny the Chinese oil industry, I am really affirming or denying myself.

In 1993, I retired. In the more than five years from March 1999 to May 2004, I traveled more than 40,000 km visiting all the oil and natural gas fields on land and offshore in China (excluding Taiwan), sometimes making two or three return visits. I interviewed more than 700 persons in the oil industry and took part in conferences with more than 200 oil experts and bureau-level or higher directors. I gathered more than 100 kg of books and information. I have carefully read hundreds of historical documents and books on oil, and I have filled ten notebooks with interviews.

I joined the petroleum industry in 1960. After more than fifty years of solitary research, I am finally free of the shackles of utilitarian commitments and interference. The constraints normally placed on one who writes about Chinese oil history no longer constrain me. Using my own thinking and viewpoints to comb through memories of the Chinese oil industry, I have arranged and compiled valuable historical data and my personal stories so that they will not be hidden from history. For those who were my friends, I will not silence criticism; and for those who were my foes, I will not fail to mention their contributions. I want to clear the historical record of distortions about the Chinese petroleum industry and present the facts. After twenty years, including twelve drafts, I have completed this *Ebb and Flow of Chinese Petroleum.*

Mao Huahe
September 2018
Palo Alto, California

A Lingering Sentiment for Chinese Petroleum

After half a century of experience and five years of visits and interviews, I pour out my heart. I write this in my golden years, not as a celebration, but to warn posterity.

Without accepting the tutelage of the two Simas[1] or walking in the footsteps of any ancestor, I retell the tale of the Emperor's New Clothes. How Chinese petroleum, a child born in a tumultuous era, worries me so!

1 Referring to Sima Qian 司马迁 (c. 145/135–86 BCE), influential Chinese historian of the Han Dynasty whose *Shiji* 史记 (*Records of the Grand Historian*) covers more than two thousand years of Chinese history up to his own time, and Sima Guang 司马光 (1019–1086), high-ranking official and historian of the Song dynasty, whose *Zizhi tongjian* 资治通鉴 (*Comprehensive Mirror to Aid in Governance*) is a highly influential history.

Maps

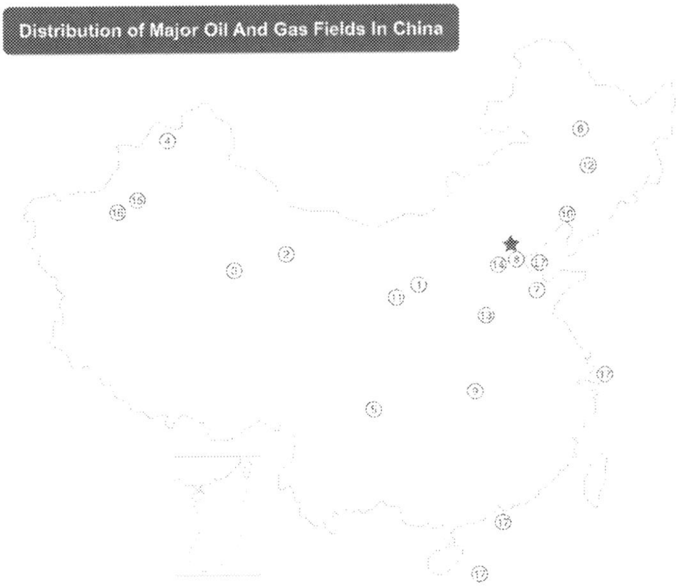

Legend

No.	Name	Founding Date	2014 Production (1,000 t)	Company Location
①	Yanchang	1905	13,240	Yan'an, Shaanxi
②	Yumen	1939	300	Jiuquan, Gansu
③	Qinghai	1955	6800	Dunhuang, Gansu
④	Xinjiang	1956	14,380	Karamay, Xinjiang
⑤	Sichuan	1958	11,100	Chengdu, Sichuan
⑥	Daqing	1960	42,280	Daqing, Heilongjiang
⑦	Shengli	1963	26,260	Dongying, Shandong
⑧	Dagang	1964	5,000	Tianjin
⑨	Jianghan	1969	1,800	Qianjiang, Hubei
⑩	Liaohe	1970	10,780	Panjin, Liaoning
⑪	Changqing	1971	55,440	Xi'an, Shaanxi
⑫	Jilin	1972	6,500	Songyuan, Jilin
⑬	Zhongyuan	1975	8,930	Puyang, Henan
⑭	Huabei	1976	5,100	Renqiu, Hebei
⑮	Tahe	1985	7,520	Wulumuqi, Xinjiang
⑯	Tarim	1986	24,660	Korla, Xinjiang
⑰	Offshore	1963	50,000 and up	Beijing

MAP 1 Distribution of major oil and gas fields in China

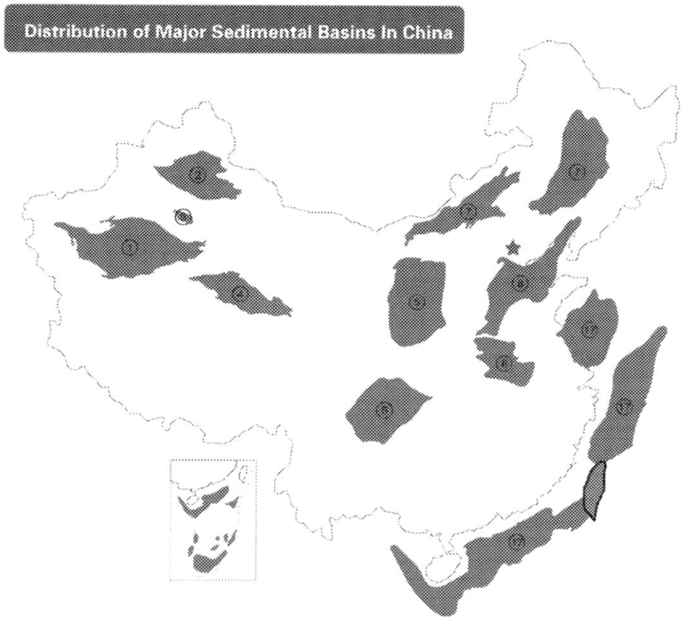

Legend

No.	Name	Area (1,000 km^2)	Location
①	Tarim	560	Xinjiang
②	Junggar	160	Xinjiang
③	Turpan	50	Xinjiang
④	Qaidam	120	Qinghai
⑤	Ordos	370	Shaanxi, Gansu, Ningxia, Shanxi, Inner-Mongolia
⑥	Sichuan	260	Sichuan
⑦	Songliao	260	Heilongjiang, Jilin, Inner-Mongolia, Liaoning
⑧	Bohai Sea Bay	180	Tianjin, Shandong, Hebei, Hanan, Liaoning
⑰	Offshore	1,000	Bohai Sea, Yellow Sea, East China Sea, South China Sea

MAP 2 Distribution of major sedimental basins in China

Notes on Author, Translators, and Others

Author: Mao Huahe

worked in China's petroleum industry for 33 years, from 1960 to 1993, before his retirement. He worked on the front line at oil campaign sites and at the Ministry of Petroleum in Beijing. Throughout much of his career, he worked as a speechwriter for the Ministry of Petroleum's leadership and as an executive in the oil industry, so he has broad, direct experience with China's petroleum industry. He started writing this book in 1999 and finished it only in 2018. This book is the summation of more than fifty years of his work and observations.

Translator: Mao Yiran

Besides visiting many of the oilfields discussed in this book, Mao Yiran lived in Daqing Oilfield and Jianghan Oilfield as a child and worked in Huabei Oilfield for a year (1977). After receiving her B.A. in English from Beijing Foreign Languages Institute, she worked for the Chinese Ministry of Education for a year (1982). She then pursued graduate study at the University of Iowa, where she received two M.A. degrees, in comparative literature (1985) and art history (1988). Her translation of Qian Zhongshu's novella *Cat* was published by Joint Publishing (H.K.) Co., Ltd. in 2001 and included in *Humans, Beasts, Ghosts* by Columbia University Press in 2011. She is a specialist in US–China trade and lives in Palo Alto, California.

Translator: Thomas Seay

is a proud native of Welch, West Virginia, presently exiled in California.

Chart Design: Selena Durst Mao

received her B.S. in product design from Stanford University (2014) and is now working towards a graduate degree in Urban Planning at the University of California, Los Angeles.

Illustrations: Arianna Seay

is a student at Palo Alto High School, Palo Alto, California.

Editor: Thomas E. Smith

received his Ph.D. in Chinese from the University of Michigan in 1992 and worked for Taiwan's Bureau of Foreign Trade as a translator and editor. He currently resides in Ann Arbor, Michigan.

PART 1

The Yanchang Era (1877–1949): Advanced in Antiquity, Backward in Modern Times

China is one of the countries that first discovered and used petroleum and natural gas, but in modern times, due to protracted feudal rule and wars, it fell behind. Challenging this backward situation was both destiny and opportunity for Chinese oilmen.

CHAPTER 1

Chinese Advances with Petroleum in Antiquity

1 China Was One of the First Countries to Discover Petroleum

In the first or second century BCE, our ancestors found petroleum in northern Shaanxi Province. The earliest record on petroleum in China appears in the Eastern Han era (25–220 CE) historian Ban Gu's 班固 (32–92 CE) *History of the Han Dynasty* (*Han shu* 《汉书》). In that book's "Geographical Treatise" he wrote: "In Gaonu, there is the Wei River, which is combustible." Gaonu 高奴 was the name of a county in the Qin dynasty (221–207 BCE), in the region of present-day Yanchang County 延长县 in Shaanxi Province. Wei River 洧水 is a tributary of present-day Yan River 延河. From then on, for two thousand years, there were records of oil discoveries in Yanchang in Shaanxi Province, in Jiuquan 酒泉 in Gansu Province, and in the Kuqa 库车 region and at the southern border of the Junggar Basin in Xinjiang Uyghur Autonomous Region.

The discovery of natural gas in our country may be traced further back to more than three thousand years ago. In the *Book of Changes*, written in the Western Zhou dynasty (eleventh century to 771 BCE), there is the statement, "There is fire in the lake," which is a realistic description of natural gas burning on the surface of pools, ponds, and lakes. Later on there are other records describing natural gas (*tianranqi* 天然气) in Sichuan, Yunnan, Guizhou, Hebei, Henan, Gansu, Shaanxi, as well as areas along the southeast coast and in Taiwan.

2 China's First Petroleum Geologist: Shen Kuo of the Song Dynasty

Shen Kuo 沈括 (1031–1095), the first known figure to use the term for petroleum *shiyou* 石油 (literally, "rock oil"), made special contributions in the natural sciences in Chinese history. As a scholar and a member of the imperial Hanlin Academy, he did research in astronomy, geography, physics, calendrics, music, medicine, and military affairs. In 1080, the Song dynasty Emperor Shenzong sent him to Luyan 鄜延 (present day Yan'an 延安) to be in charge of the military defense of northern Shaanxi Province. During his tenure there, he personally explored and carried out research on the petroleum in the Yanchang area. He held that unlike other types of oil, this type of oil was "born from the sand and rocks on the water's edge," that it "mixed with spring water and mysteriously

 | DOI:10.1163/9789004402737_002

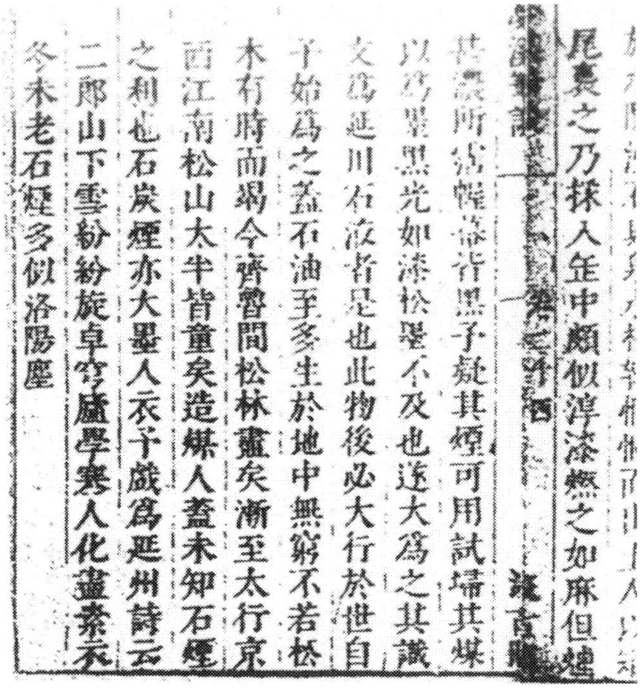
尾裛之乃採入缶中頗似淳漆燃之如麻但煙
甚濃所霑幄幕皆黑予疑其煙可用試埽其煤
以爲墨黑光如漆松墨不及也遂大爲之其識
文爲延川石液者是也此物後必大行於世自
予始爲之盖石油至多生於地中無窮不若松
木有時而竭今齊魯間松林盡矣漸至太行京
西江南松山太半皆童矣造煤人盖未知石煙
之利也石炭煙亦大墨人衣予戲爲延州詩云
二郎山下雪紛紛旋卓穹廬學塞人化盡素衣
冬未老石煙多似洛陽塵

FIGURE 1.1 Shen Kuo, *Mengxi bitan: Essays from the Brook of Dreams*

gushed forth." He called it *shiyou*. In his encyclopedia-like *Essays from the Brook of Dreams* (*Mengxi bitan* 《梦溪笔谈》), he proposed that, "The vast quantity of *shiyou* arises from the endless depths within the earth." Shen Kuo's in-depth research and scientific records on petroleum were vivid, concrete, and original. He may be regarded as China's first petroleum geologist.

According to recent studies carried out by Dunhuang scholars,[1] usage of the term *shiyou* had actually originated in documents dating to 908 CE, during the Later Liang dynasty (907–923). These documents were stored in one of the "Buddhist sutra caves" among the Mogao Caves of Dunhuang, Gansu Province. This predates Shen Kuo's usage of *shiyou* by more than a hundred fifty years.

1 See the report in *Shijie ribao* 《世界日报》, North American edition (July 28, 2000).

3 China's First Drilling Engineer: Governor of Shu Commandery Li Bing

Our country is also one of the first countries that put petroleum and natural gas to use. At the same time that our ancestors discovered petroleum and natural gas, they gathered, extracted, and utilized them. For thousands of years, they were primarily used for lighting, lubrication, antiseptics, ink-making, insect repellent, pesticides, treatments for sores, as well as boiling brine for salt. Oil was also used to make "fireballs" in wars. Our country pioneered the key technology for extracting oil and natural gas-drilling. According to *Chronicles of Huayang* (*Huayang guozhi* 《华阳国志》), after the demise of the Zhou dynasty, the King Xiaowen of Qin made Li Bing 李冰 the Governor of Sichuan (then known as Shu 蜀) around 256 BCE. "Not only did Li Bing know astronomy and geography, he also knew geology and drilled salt wells in Guangdu County …" As noted in *Commentary on the Waterways* (*Shuijingzhu* 《水经注》), "Li Bing knew the water arteries and drilled salt wells in Guangdu County, east of the river." That is to say, Li Bing, drawing from the experiences of local people on gathering brine and boiling it down for salt, was able to drill brine wells and boil the brine down for salt around present-day Shuangliu, near Chengdu, Sichuan Province. The fuel used to boil the brine was natural gas. At the time, people in Sichuan called natural gas wells *huojing* 火井 (fire wells). Salt wells and natural gas wells often existed in the same area and were developed simultaneously. In the Han dynasty (206 BCE–220 CE), the salt well production was already quite developed. Sichuan was known as the "Home of Fire Wells." It abounded in natural gas wells and gas seepage. The earliest and most famous was the natural gas well in Linqiong 临邛. According to *Basic Records on the Kings of Sichuan* (*Shuwang benji* 《蜀王本纪》), by the Western Han dynasty (206 BCE–9 CE) scholar Yang Xiong 杨雄, "There is a fire well in Linqiong, more than sixty *zhang* (= 3 1/3 m) deep." The *Chronicles of Huayang* notes, "The fire well in Linqiong County glows in the dark night. If people want fire from the well, they light it with a torch from their own homes, and instantly, with a loud noise like thunder, fire comes out, lighting up tens of *li*" (1 *li* = 500 m). In 67 BCE, during the Han dynasty Emperor Xuan's reign, there were twenty salt wells in Linqiong and Pujiang 蒲江. In the Eastern Han dynasty (25–220 CE), drilled salt wells covered 16 counties, including Linqiong, Chengdu, and Nanchong 南充, in Sichuan Province. During the Tang dynasty (618–907), there were 64 counties in Sichuan with drilled fire wells and salt wells. According to *Gazetteer of Renshou County* [Sichuan] (*Renshou xianzhi* 《仁寿县志》), the depth of the wells reached 550–888 *chi* (1 *chi* =

FIGURE 1.2 Ancient Chinese rig

1/3 m). Many historical records show that Li Bing was not only ancient China's most famous water conservation expert—he also led the people of Sichuan in developing drilling technology while they drilled salt wells and natural gas wells. He was undisputedly China's first drilling engineer.

4 The World's First Drilled Well: Shenhai-1

Chinese traditional drilling technology underwent two stages of development. In the first stage, prior to the Northern Song dynasty (960–1127), wells were usually shallow with large openings and were manually dug. In the second stage, starting with the Northern Song dynasty, wells were deeper with smaller openings. During the mid-Northern Song dynasty, around 1041, the drilling technology in Sichuan advanced from manual digging to mechanical punch drilling (*xiechong zuanzao* 械冲钻凿); this was a precursor of percussion drilling. This technology developed very quickly in Sichuan and was soon perfected. By the mid to late Qing dynasty (1644–1911), this kind of drilling technology had achieved a kind of miracle in the world history of drilling: in 1853, Shenhai-1 燊海一井, drilled in Zigong 自贡, Sichuan, reached the depth of 1001.42 m. This was the first man-made well that exceeded one thousand meters. For this reason, the well has the fine reputation of being the "great-granddaddy" of drilled oil wells around the world. This well produced mainly natural gas but also produced brine. To this day, it still produces 1000 m^3 of natural gas and 2,000 kg of salt every day. In 1988, the State Council listed it as a major historical and cultural site, protected at the national level.

The noted British sinologist Joseph Needham named deep-well mechanical drilling as one of the 26 important inventions that spread to Europe from China. He claimed that the well-drilling or cave-drilling technology used in today's oilfield exploration must have been invented by the Chinese. He also said that such technology had spread to other countries in the west by the eleventh century. According to Needham, all deep wells drilled in the world before 1900 were drilled using methods invented by the Chinese.[2]

Today, as we review this part of history when Chinese petroleum led the world and contemplate the achievements of Shen Kuo and Li Bing, we should endeavor to recreate the splendor of our ancestors.

2 Joseph Needham, *Science and Civilisation in China, Volume 1: Introductory Orientations* (Cambridge: Cambridge University Press, 1954), 242, 244–45.

CHAPTER 2

The Six Old Oilfields in Modern China: Small Scale and Weak Foundation

Long-term feudal rule and warfare severely delayed the development of our country's society and productivity. For nearly a thousand years after the emergence of mechanical punch drilling in the eleventh century (i.e., around 1041), the exploration, development, and usage of petroleum, excluding development in the application of natural gas in Sichuan, remained at its original primitive state. By the mid-nineteenth century, China was reduced to a semi-colonial and semi-feudal society. During the Qing dynasty's Self-Strengthening Movement (1861–1895), our country's modern petroleum industry finally started, with difficulty. From its inception, it encountered suppression and sabotage from invading foreign forces and tyrannical domestic feudal forces, and it was only with painstaking effort that operations were kept going. By the time the People's Republic of China was established in 1949, there were only six small oilfields.

1 The Chinese Petroleum Industry Begins: the 1877 Discovery of Taiwan's Miaoli Oilfield

In 1817, Miaoli County resident Wu Linfang 吴琳芳 found an oil slick on the banks of the Houlong River. In 1861, the last year of the Xianfeng Emperor's reign, Qiu Gou 邱苟, also of Miaoli County, manually dug a three-meter-deep well near Chuhuangkeng 出磺坑 (the name means, literally, "Sulfur-producing Pit") by Houlong River. The well produced 20 kg of oil daily, which was used for lamps. In 1877, Shen Baozhen 沈葆贞, the Qing dynasty Governor-general of Jiangsu, Anhui, and Jiangxi provinces, went to inspect Taiwan. He and Ding Richang 丁日昌, the Governor of Fujian Province, jointly presented a memorial to the Qing court requesting approval to drill for oil at the site. Approval for an officially managed well was granted. The following year, two American technicians, A. Port Karns and Robert D. Locke, were hired and percussion drilling equipment imported. Our nation's first modern drilling crew was formed. They drilled the first well in Miaoli, approximately 120 m deep, and it produced 750 kg of oil daily. Subsequently, after many trials, they found no new wells. In 1885, Taiwan was changed from a prefecture under the jurisdiction of Fujian Province to a province directly under the jurisdiction of

 | DOI:10.1163/9789004402737_003

the Qing court. Taiwan's first Governor, Liu Mingchuan 刘铭传, actively advocated drilling in Taiwan. The Qing court approved his proposals, and in 1887, the Miaoli Petroleum Bureau was founded, with Lin Zhaodong 林朝栋 appointed its director. This Bureau remained in operation for four years, during which time five wells were drilled, but only one produced oil, and at a low production rate. Due to financial problems, Taiwan's oil industry was in a precarious position. In 1895, Taiwan was occupied by the Japanese. During the fifty-year Japanese occupation (1895–1945), three relatively large-scale geological surveys were conducted in Taiwan, and five small oilfields were found in Jinshui 锦水, Chuhuangkeng, Zhudong 竹东, Niushan 牛山, and Zhutouqi 竹头崎, in that order. Two oil-containing structures were found by Liuchong River 六重溪 and at Dongzijiao 冻子脚. A total of 140 wells were producing oil and gas. Annual oil production peaked in 1927 at 19,000 tons. When the Japanese surrendered in 1945, the oil and gas fields were taken over by the Nationalist government-owned, Shanghai-based Chinese Petroleum Corporation (CPC Corp). From 1904, when the record begins, to 1948, Taiwan produced a total of 160,930 tons of oil.

2 China's Earliest On-Land Oilfield: Yanchang Oilfield

The first oil plant at Yanchang was founded in 1905. The first well was drilled in 1907, and refining also started, but for many years there was only one deep well producing oil. It was only after nearly forty years of effort that a certain scale of production was reached. In 1943, production at the Yanchang Oilfield reached its peak at 1,229 tons, and in 1949 it produced 820 tons. Yanchang will be discussed further in Chapter 3.

3 Dushanzi Oilfield, Xinjiang, 1909

In 1909, the regional government under the Qing court in Xinjiang drilled the first well in Dushanzi after it imported a rig and refining caldron from Russia, but the operation stopped in 1911. The price of imported Russian kerosene increased drastically. In 1916, the governor of Xinjiang, Yang Zengxin 杨增新 allocated more than 5,000 taels (= 156.25 kg) of silver to revive operations. Besides sending people to learn drilling and refining at Yanchang Oilfield, he ordered operations to resume at Dushanzi. A Xinjiang local, Pan Zuhuan 潘祖焕, also hired people for his oil business, and for a time he produced and sold his own oil products on the market. Later on, as imported oil from the former Soviet Union increased, oil prices decreased, and domestic oil operations

FIGURE 2.1
Yanchang Oilfield in Shaanxi Province

in Xinjiang were shut down, one after the other. In 1936, in cooperation with the Soviet Union, the local government in Xinjiang developed Dushanzi Oilfield. In 1942, its annual production peaked at 7,321 tons, and it became one of the three big on-land oilfields, together with Yumen 玉门 and Yanchang. Later, when Sino-Soviet collaboration ceased there, operations at the oilfield were intermittent. Until the peaceful liberation of Xinjiang in 1950, oil production in Dushanzi remained small. There were only 150 employees and two oil wells that produced only 3 to 5 tons of crude oil per day. From 1939 to 1950, a total of 14,152 m of wells were drilled in Dushanzi Oilfield, and 11,497 tons of crude oil were produced.

4 Natural Gas in Sichuan, 1936

In 1936, the National Resources Committee of the Kuomintang government formed the Sichuan Oilfield Exploration Division. A German technician was hired and a rotary drilling rig purchased. At Shiyougou 石油沟 in Ba County 巴县, Ba-1 was drilled to 1,402.2 m, and it produced 15,000 m^3 of natural gas per day. Ba-1 was the first natural gas well drilled with a rotary drilling rig in China. In December 1943, at Mt. Shengdeng 圣灯山 in Longchang 隆昌 County, another natural gas well was drilled. By 1949, five wells had been drilled in Ba County and Longchang County, but only two of them produced commercially viable natural gas flow.

5 Synthetic Oil Plants in the Northeast

After the Russo-Japanese War ended in 1905, the Japanese invaded the northeast of our country. Starting in 1928, using the oil shale discovered in Liaoning and Jilin Provinces, they built a few shale oil plants, retorts for processing oil shale, and coal oil refineries in Fushun 抚顺 and Jinzhou 锦州 in Liaoning and Huadian 桦甸 in Jilin. At its peak in 1942, the northeast produced 257,000 tons of oil from oil shale and 24,000 tons of oil from coal. In 1945, after Japan surrendered and the Kuomintang government took over, most of these plants were run down and ceased operations, until the liberation of the northeast in 1948.

6 China's First Modern Petroleum Industry Base—the Yumen Oilfield, 1939

From its discovery in 1939 up to 1949, Yumen Oilfield was China's largest and most productive oilfield, with an annual output of 80,000 tons.

From 1907 to 1948, apart from Taiwan and Dushanzi in Xinjiang, 169 wells with a total depth of 67,000 m were drilled nationwide. Some rough petroleum geological surveys were conducted in certain areas in Shaanxi, Gansu, Xinjiang, Sichuan, Heilongjiang, Liaoning, and Taiwan provinces. In 1948, there were only four small oilfields in Taiwan's Chuhuangkeng, northern Shaanxi's Yanchang, Xinjiang's Dushanzi, and Gansu's Yumen. There were only seven small gas fields in Sichuan's Ziliujing 自流井, Shiyou River 石油沟, and Mt. Shengdeng; there were also some small gas fields in Taiwan's Jinshui, Zhudong, Niushan, and Liuchong River. There were only a few old, shabby synthetic oil plants in the northeast. In 1949, there were only eight small rigs and proven petroleum reserves of 29 million tons in the entire country. Total on-land crude oil annual production was 120,000 tons. Such was the weak foundation of our national petroleum industry at the time.

CHAPTER 3

China's First On-Land Oilfield: Yanchang Oilfield

1 Yanchang Oilfield Established, 1905

Yanchang Oilfield is situated in the vast area east of Yan'an in northern Shaanxi Province and west of the Yellow River. Geologically, it belongs to the Ordos Basin, which covers the provinces of Shaanxi, Gansu, Ningxia, Shanxi, and Inner Mongolia. Its topography, with abundant gullies and ravines, is typical of the Loess Plateau. The existence of oil in northern Shaanxi has been well-documented in Chinese historical records since early times.

In 1905, the Qing dynasty Governor of Shaanxi Province, Cao Hongxun 曹洪勋, wrote a memorial to the throne, requesting permission to establish the Yanchang Oilfield. Although the Qing court granted his request, it did not give him any funding. He boldly used 81,000 taels (= 2,331.25 kg) of the local government's silver originally intended for wasteland reclamation to build the oilfield. He appointed Expectant District Magistrate Hong Yin 洪寅 to oversee the operation. After two years of preparation, they hired a Japanese technician, Satō Hisarō 佐藤弥朗, who started drilling at Yan-1 Well, in Qili Village 七里村, on June 5, 1907, using a percussion rig brought from Niigata, Japan. Drilling was completed at 81 m on September 10, and the well had an initial daily output of 1 to 1.5 tons, which it maintained for ten years. The crude oil from Yan-1 was processed in small copper pots and produced 12.5 kg of lamp oil per day. It was sent for analysis in Xi'an, and it was found that when burnt, it produced a small amount of white smoke. It was as good as imported kerosene. In October 1907, the refinery started production and filled 14 cases with lamp oil (approximately 344 kg), which were sent to Xi'an for sale. People liked the product, and the news spread. Yanchang Petroleum Plant had caught the attention of the Chinese people. Production at Yan-1 gradually dwindled until it dried up in 1934.

The 28 years of Yanchang Oilfield's existence spanned the last years of the Qing dynasty, the rule of warlords, and the Republic of China era. After it signed the "Agreement for the exploration and operation of petroleum oil fields in Chihli and Shensi Provinces" with the Republic of China government in February 1914, Standard Oil sent geologists and a drilling crew to northern Shaanxi. Since they had little success there, the agreement was rescinded in June 1917. Afterward, rumors of "oil-poor northern Shaanxi" began to spread. In response, Chinese geologists expressed their opinions on the issue, which we shall discuss below.

DOI:10.1163/9789004402737_004

2 The Geologists Who Argued That Northern Shaanxi Had Oil

The noted geologists Ding Wenjiang 丁文江, Weng Wenhao 翁文灏, and Xie Jiarong 谢家荣 sped up their survey in northern Shaanxi. In 1919, Ding Wenjiang was the first to point out: "Standard Oil's work in northern Shaanxi has proven to be a failure." In 1921, Ding Wenjiang (also known as V.K. Ting or Ting Wen-chiang) and Weng Wenhao (Wong Wen-hao) jointly published an article entitled "A Brief Record of Minerals in China." In it they pointed out that in the past, the oil industry in China had attracted the world's attention, but since the joint exploration with Standard Oil in 1914, the disappointing results had tarnished its image. They argued that the oil-producing region was vast, and that the exploration was not detailed and thorough. They went on to write that oil remained a promising industry in China. Weng Wenhao, China's first Ph.D. in geology and the director of the government's Central Geological Survey, sent the geologists Wang Zhuquan 王竹泉, Pan Zhongxiang 潘钟祥, Xie Jiarong, Yang Gongzhao 杨公兆, Hu Bosu 胡伯素, Zhou Zongjun 周宗俊, Yan Huimin 颜惠敏, and others to northern Shaanxi Province. Their many important discoveries proved the presumption of an "oil-poor northern Shaanxi" to be false.

3 Sun Yueqi Appointed Director of Northern Shaanxi Oilfield

In the spring of 1934, the Nationalist government appointed the noted businessman Sun Yueqi 孙越琦 to be the director of the Exploration Division of Northern Shaanxi Oilfield. Sun Yueqi organized and led personnel to move three percussion rigs as well as 99 tons of support equipment from Shanghai to Yanchang, in northern Shaanxi. In September of the same year, Yan Shuang

FIGURE 3.1
Geologist Weng Wenhao, Ph.D. (1889–1971)

严爽, an expert in mining and metallurgy, succeeded Sun Yueqi as director. Heeding the advice of Wang Zhuquan and Pan Zhongxiang, he oversaw the drilling of seven wells in the Yanchang and Yongping 永坪 region. Each well was approximately 100 m in depth. More than half produced oil, and some had a daily output of 1.5 tons, but the production setup did not achieve scale. In 1934, the oilfield produced only 44 tons.

4 Mao Zedong Praises the Yanchang Oilfield

On April 28, 1935, the Red Army led by Liu Zhidan 刘志丹 liberated Yanchang County. Mao Zemin 毛泽民 (Mao Zedong's younger brother), the Minister of Economics in the Northwest Office of the Chinese Soviet Republic, appointed Yan Shuang as director of Yanchang Oilfield and Gao Dengbang 高登榜 as the Party's representative. They organized employees and soon resumed production. On January 18, 1936, Mao Zedong, leading the Anti-Japanese Vanguard on its eastern expedition, passed through Yanchang County. He stayed at the oil worker He Yannian's 何延年 cave dwelling for four days. During this time, in addition to holding meetings to discuss military and political plans, Mao Zedong toured the oilfield and talked with the managers, showing much interest in the operation.

On July 7, 1937, the autonomous Communist government of Shaan-Gan-Ning Border Region was established in Yan'an. In the winter of 1940, the Qinghua University-educated geologist Wang Peng 汪鹏 took charge of the geological work in Yanchang Oilfield. Starting in 1941, he and his fellow geologists further explored the Qili Village Structure. The first well, Qi-1, initially gushed 96 tons per day and created a sensation in the region. Yanchang Oilfield was later developed, and in 1943, it achieved an annual oil production of 1,279 tons. The oil of Yanchang Oilfield and the salt of Dingbian County became the "two treasures" of the region. Yanchang Oilfield contributed to the War of Resistance Against Japan and improved people's lives in the region. Chen Zhenxia 陈振夏, the oilfield's director, twice received the autonomous government's "Exceptional Model Industrial Worker" award and Mao Zedong's calligraphy "*maitou kugan* 埋头苦干 [quietly put one's shoulder to the wheel]." After the victory of the War of Resistance Against Japan in 1945, the civil war broke out. The oil workers under the organization of director Zhang Jun 张俊 and the Party political commissar Wan Pinshan 万品山 sealed up the wells and equipment. When Yanchang was liberated in April 1948, oil production resumed immediately. In 1949, Yanchang Oilfield produced 820 tons of crude oil and refined 176 tons of natural gas. This aided the War of Liberation in the northwest.

In 2014, Shaanxi Yanchang Petroleum (Group) Corp. Ltd. (Yanchang Oilfield), China's oldest on-land oilfield, produced 12.76 million tons of crude oil. It is China's seventh largest oilfield, and the company is on the Fortune Global 500 list. Yanchang Oilfield's success proved the theory that, "There is oil in northern Shaanxi." It is a tribute to the Chinese petroleum predecessors such as Shen Kuo, Cao Hongxun, and Weng Wenhao.

CHAPTER 4

Yumen Oilfield, 1939: China's First Modern Petroleum Base

1 Looking for Oil at the Foot of the Qilian Mountains

First, let us talk about the origin of the name of Yumen, or "Jade Gate." In ancient times, the area now known as Hetian or Hotan 和田 in Xinjiang produced what is traditionally known as "Khotan jade." When jade was transported to the interior, it had to pass this place where the imperial officials set up check points for inspection, hence the name Yumen. Yumen Oilfield is situated at the northern slope and foot of the Qilian Mountains 祁连山, approximately 100 km from Yumen and 80 km from Jiuquan. On a clear day, one can see Yumen Oilfield from Jiayuguan, the last western pass of the Great Wall, which is 90 km away. The elevation is around 2,500 m. As early as the Western Jin dynasty (265–316), our ancestors had discovered oil here. Li Daoyuan 郦道元 (470–527) of the Northern Wei dynasty (386–534) noted in his *Commentary on the Waterways*: "The mountain in the south of Yanshou County in Jiuquan produces spring water … The water has fat like meat sauce that, when put in a container, first appears yellow and then black. It does not congeal, and it burns extremely bright; it is not unlike grease, but is inedible … it is called mineral paint (*shiqi* 石漆)." There are records throughout subsequent dynasties of the oil in this area. When he sat in command of Jiuquan, the famous Qing dynasty general Zuo Zongtang 左宗棠 (1812–1885) sent for someone to fetch an oil sample in Yumen to be sent to France for chemical analysis. The quality of the oil was found to be ideal, but he did not have the right equipment at the time for developing and using this resource.

2 Weng Wenhao, the First Geologist to Survey Yumen

Based on their findings, they wrote China's first petroleum geology report. In 1928, the Gansu government sent the geologist Zhang Renjian 张人鉴 to Yumen to investigate oil, and he submitted a detailed proposal to build an oilfield and refinery.

 | DOI:10.1163/9789004402737_005

3 Geologist Sun Jianchu Thrice Investigates Yumen

In the 1930s, a geologist from the Central Geological Survey, Sun Jianchu 孙健初 and his group went to Yumen three times to investigate oil, the first time in 1934 and 1935. He and Zhou Zongjun and others traveled to the mountainous region in the western part of Gansu and the northeastern part of Qinghai, but due to unrest, they could not penetrate it deeply. The second time was in June 1937, when the preparatory office of the China Kerosene Exploration Company organized an expedition to the northwest that was led by Shi Youming 史悠明 and geologist Sun Jianchu; they were accompanied by two Americans, the noted geologist Dr. James Marvin Weller and the engineer Dr. Frederic A. Sutton. In October 1937, while traveling westward from Jiuquan, they found oil in Baiyang River 白杨河 in Yumen County and Shiyou River 石油河 in Laojunmiao 老君庙. During their voyage they saw three peasants skimming oil off the Shiyou River by the old temple to Taishang Laojun (the deified Laozi) south of Yumen (from which the name of the place, Laojunmiao, derives). After a detailed investigation, Weller and his team wrote, in their exploration report:

> In this district ... development costs will be very high, and if oil prospecting here is to be considered as strictly a business proposition, most careful consideration must be given to all of the factors involved in development, production, refining, and marketing aspects before a decision is made to proceed further. If, on the other hand, the development of an oilfield in the northwest part of China is required for national defense, to be accomplished at any cost, the Shih Yu Ho [Shiyou River] anticline should certainly be drilled.... [E]ven with the utmost dispatch and the best of luck, it is likely that at least two years would be required to drill the first wells and install a small refinery.[1]

On April 18 and 19, 1938, Weller and Sun Jianchu reported twice in Hankou to Weng Wenhao, who was then the Minister of Economic Affairs and the director of Central Geological Survey. Weller's geological writings fully reflect the scientific spirit of a geologist seeking truth from facts, and he was generous in recognizing the contributions of Chinese geologists, like Sun Jianchu, who had done exploratory work before him.

1 J. Marvin Weller, *Caravan Across China: An American Geologist Explores the Northwest, 1937–1938*, edited by Harriet Weller (San Francisco: March Hare Publishing, 1984), 235–36.

FIGURE 4.1
Geologist Sun Jianchu and American geologists James Marvin Weller (middle) and Frederic A. Sutton (left)

We should mention in passing that Sun Jianchu died of gas poisoning in Beijing in 1954, but his statue can still be found in Yumen Oilfield's park. Oil people still remember his achievements. Also, more than fifty years later, after China opened up and communications between Chinese and Americans became more frequent, Dr. Weller's daughter, the writer Harriet Weller, and Sun Jianchu's son, Prof. Sun Honglie met by chance, and so, miraculously, a two-generation Sino-American friendship continued. Later, Harriet Weller, tracing the footsteps of her father, visited Yumen many times, following her father's route, and praised the change and progress being made in Yumen. The oil people in Yumen and in China generally will forever remember the achievements of these Chinese and American geologists in the development of Yumen Oilfield.

4 The Nationalists and Communists Cooperate on the Joint Development of Yumen Oilfield

By 1938, the War of Resistance Against Japan had been fought for more than a year, and oil became extremely scarce in the unoccupied areas of China. Under such circumstances, the Ministry of Economic Affairs in the Nationalist government decided to form a Gansu Oilfield Preparatory Bureau (i.e., Yumen Oilfield) and appointed Zhang Xintian 张心田 as acting director in charge of exploring and developing Yumen Oilfield. Weng Wenhao dispatched a reconnaissance group to Shiyou River in the Laojunmiao area of Yumen. Headed by Yan Shuang, who had been director of Yanchang Oilfield, geologist Sun Jianchu, measuring technician Jin Xigeng 靳锡庚, and others rode camels to the area in December of 1938. The Nationalist government did not have any rigs, and Yan Shuang knew that Yanchang Oilfield, which was under the control of the Chinese Communist Shaan-Gan-Ning Border Region, had two percussion rigs. So Weng Wenhao went to the Wuhan office of the Eighteenth Army of the National Revolutionary Army (i.e., the Communist

Eighth Route Army) to meet with Zhou Enlai. Zhou Enlai responded by saying that he had no objection at all to working together to serve the country. He agreed on the spot to have the rigs disassembled and shipped over to Yumen. He appointed Director Qian Zhiguang 钱之光 to carry out the task. Later, Lin Boqu 林伯渠, the Chinese Communist Party representative in Shaanxi, personally made the arrangements. Yanchang Oilfield sent 15 technical people, including drilling engineers, with two percussion rigs, two furnaces and steam engines, 12 casings, drilling bits, and steel cables, weighing in all more than 30 tons. The Eighth Route Army escorted the party to the Laojunmiao area. Laojunmiao-1, which had previously been dug manually, showed oil at 20 m. Once the rigs arrived, they were put to work. On August 11, 1939, when the well was drilled to 115.51 m, it produced 10 tons of crude oil per day. Such was the beginning of Yumen Oilfield, a story of Nationalist and Communist cooperation.

I might add that in 1984, Jin Xigeng, then in his eighties, had long retired from the Scientific Research Institute of Petroleum Exploration and Development. When he asked to join the Communist Party, as the Secretary of the Institute's Party Committee, I visited him at his house and later endorsed him to become a member of the Party.

In the winter of 1961, the writer Xu Chi 徐迟 told the story of Yumen Oilfield's beginnings in his long report "At the Foot of the Qilian Mountains (Qilian shanxia 祁连山下)," published in the Beijing-based monthly *People's Literature* (*Renmin wenxue* 《人民文学》). At the time, I was in Daqing 大庆. I was deeply moved by his story, and it piqued my interest in the history of Chinese petroleum. In the winter of 1999, when I started writing this book, I made a trip to the Guzijian Library in Beijing, found the story again, and copied it. The above is mostly based on this essay.

5 The Policy of "Self-Reliance and Charging Ahead Despite Difficulties"

The geologist Weng Wenhao, as the Nationalist government's Minister of Economic Affairs, knew very well the importance of oil in the War of Resistance Against Japan and the difficulty of its extraction. After the first well in Yumen produced oil in 1939, the Nationalist government decided in 1940, with Chiang Kai-shek's agreement, to establish the Gansu Petroleum Bureau, which Weng Wenhao then planned and organized. Weng strongly recommended businessman Sun Yueqi as the general director and Yan Shuang as the manager of Yumen Oilfield, and Chin Kai-ying 金开英 as manager of Yumen Oil Refinery. During the conference at the founding of the Gansu Petroleum Bureau, Sun

Yueqi first proposed the Bureau's policy of "self-reliance and charging ahead despite difficulties." He set a goal of "producing 1.8 million gallons in 1942 to support the frontline." He declared that if the plan were not realized, he would resign. After that, Yumen Oilfield and Yumen Oil Refinery started production and development on a grand scale. The remoteness of Yumen Oilfield, however, created tremendous problems for production and development, workers' livelihoods, the transportation of goods, financing, and so on. In addition, this happened during the War of Resistance Against Japan, so equipment imported from abroad was either blocked or bombed by the Japanese and could not be transported inland. At the height of Yumen Oilfield's construction, the workers along with their family members numbered ten thousand people. Though faced with all kinds of difficulties, the Yumen employees worked hard and were self-reliant, giving their all to increase production. For example, ten rotary rigs, which had been purchased from the US, would have been able to drill down to 1,000 m. However, most of them were destroyed in transit by Japanese bombing. Once the shipment reached the oilfield, only three and half rigs could be put together, with a lot of missing parts, so they designed and manufactured the missing parts and finally put three rigs in production. Also, when the imported refining equipment was destroyed by bombing, they dispatched people to places thousands of kilometers away to collect machine tools, oil and water pumps, steel pipes, valves, oil refining kettles, and spare parts. The employees eventually built 24 kettle refineries and installed four groups of three-stage continuous refineries, which increased refining capability, drawing gas, kerosene, and diesel from 18% of the crude oil. By 1942, the oilfield had produced 1.8 million gallons of gasoline, which played an important role in supplying friendly territories where fuel was extremely scarce.

In the summer of 1992, ten old oil men from Taiwan who used to work in Yumen Oilfield visited Sun Yueqi in Beijing and presented him with a refined silver platter of gratitude. On it were engraved the words *yinshui siyuan* 饮水思源 (drink the water and remember the source), to express their respect.

6 The Best and the Brightest

The news of large-scale oil extraction in Yumen spread throughout China. During the War of Resistance Against Japan, large groups of patriotic young people, especially intellectuals who had come back from overseas to take part in the war and graduates from colleges and trade schools, climbed mountains, waded rivers, and crossed the Gobi Desert in order to throw themselves into the embrace of Yumen Oilfield. Among their numbers we may count the following: Weng Wenhao's cousin Weng Wenbo 翁文波, fellow of the British Royal Society and grand master of geophysics as well as chief creator of earthquake monitoring in China; Weng Wenhao's son Weng Xinyuan 翁心源, the pioneer of oil transfer in China; Chin Kai-ying, who made great contributions to Taiwan's economy and was lauded as China's oil refining pioneer; the noted oil refining expert Lee Ta-hai 李达海, who later became the Minister of Economic Affairs in Taiwan; American-educated Tong Xianzhang 童宪章, the son of a famous Nationalist general, a chief designer of Daqing Oilfield and fellow of the Chinese Academy of Sciences; the American-educated oil engineering expert Shi Jiuguang 史久光, son of the noted diplomat Shi Youming 史悠明; American-educated oil geologist Chen Ben 陈贲; geologist Bian Meinian 卞美年; chemist Tan Shifan 谭世藩; there are many more. In only a few short years, a couple of hundred experts, scholars, and educated young people gathered in Yumen Oilfield. By creating China's first modern oilfield in the remote and desolate foot of the Qilian Mountains, they made remarkable contributions.

From 1983 to 1986, as the secretary of Party Committee of the Scientific Research Institute of Petroleum Exploration and Development, I became acquainted with Weng Wenbo, visiting him numerous times in his home and helping him organize his memoirs.

Weng Wenhao's son Weng Xinyuan suffered as a result of his father's position in the Nationalist government. He was tortured and died during the Cultural Revolution.

7 Chiang Kai-shek Visits Yumen Oilfield

The achievements and importance of Yumen Oilfield attracted the Nationalist government's close attention. In August 1942, accompanied by important officials such as Hu Tsung-nan 胡宗南, Chiang Kai-shek visited Yumen Oilfield. He listened to Sun Yueqi's report on the oilfield and visited a few wells and the refinery. As he toured the oilfield, he commented: "Not easy, not easy." After lunch, he heard reports from various departments of the oilfield. He proposed that in the future, Yumen Oilfield should renovate the city gate tower of Jiayu Pass, the western terminus of the Great Wall, plant trees, reclaim wasteland, raise livestock, use oil residue to pave roads, and enhance personnel training. Before leaving the oilfield, he commented to Hu Zongnan that it seemed that the difficulties of building an oilfield here were significant, and that Hu must support them. He then said to Sun Yueqi: "Should you have any problems later, tell me any time, I'll support you." He also expressed his intention to have his two sons visit Yumen. Soon afterwards, in order to solve the problem of transporting finished oil products inland, Sun Yueqi sent a special telegraph to Chiang, requesting oil drums for emergency usage. Chiang Kai-shek immediately ordered the military to transport 30,000 oil drums to Yumen Oilfield. Later, Chiang agreed to allocate 5 million US dollars and 120 million French francs to purchase cars and equipment. Chiang Kai-shek's support played a key role in Yumen Oilfield's development. In May 1943, Chiang Kai-shek's sons Chiang Ching-kuo and Chiang Wei-kuo, accompanied by Hu Tsung-nan's confidential secretary Xiong Xianghui 熊向晖, came to Yumen Oilfield to visit and learn. In 1951, soon after the founding of the new China, at a meeting of the Chinese People's Political Consultative Conference, Sun Yueqi and Xiong Xianghui met again. Only then did Sun know that Xiong had been an underground Communist at the time. There seems to have been a great deal of cloak-and-dagger and double spy activity around this time, making this an interesting oil-related anecdote.

8 Contributions to the War of Resistance against Japan

Yumen Oilfield was born amidst the turmoil of the War of Resistance Against Japan and played an important role in supporting the Chinese military effort. By 1949, after 11 years of development, its crude oil production had reached 80,000 tons, and it was capable of refining 100,000 tons annually. At the time, Yumen was the largest, most productive, and most technologically advanced oilfield in China. Meanwhile, employee living quarters were built at the foot of the desolate Qilian Mountains.

At its peak in 1942, the oilfield employed as many as 6,800 people; by 1949 this figure was adjusted to more than 4,000. After many years of technical training, a specialized and knowledgeable oil workforce emerged. People doing 53 different kinds of work comprised the modern, specialized petroleum team. The oilfield recruited college graduates and foreign-educated scientific and technical personnel and assigned them important tasks. These became China's first generation of oil experts and the backbone of China's oil industry.

PART 2

The Yumen Era (1950–1959): Laying the Foundation, Setting Petroleum Development Policy

The most meaningful achievements of this decade came as a result of numerous geological surveys and the practice of sending groups overseas to study. These achievements could also be attributed to the discussions that went on between the highest level of leadership in China, the nation's best geologists, as well as Soviet geologists in China. The whole nation attached importance to the petroleum industry, and believed that since "China has oil," we should "lose no time in oil exploration." Meanwhile, during this decade, Yumen Oilfield underwent rapid development. After 1956, with the successive discoveries of Lenghu Oilfield in Qinghai, Karamay Oilfield in Xinjiang, and Chuanzhong Oilfield in Sichuan, several sizable oil bases were built. All of these discoveries laid the foundation and prepared for the development of China's oil industry. By 1959, more than 50,000 people were employed in the oil industry nationwide, and annual oil production was 2.5 million tons. However, the oil industry was negatively impacted by the Anti-Rightist Movement of 1957 and the Great Leap Forward Movement of 1958.

CHAPTER 5

China's Petroleum Industry Takes Off

1 "Without Oil, Airplanes, Tanks, and Cannons Are no Better than a Dog Paddle"

Petroleum is an important strategic material. Just before the founding of the People's Republic of China in 1949, Stalin sent a verbal message through his special envoy, Anastas Ivanovich Mikoyan (1895–1978), saying that the experience of World War II had proved the importance of two things: one was rubber, the other petroleum. As soon as the new China was founded, the country's leaders put petroleum industry development high on their agenda. In 1951, the General Petroleum Administration Bureau, under the Ministry of Fuel Industries, held a petroleum industry exhibition in Beijing. Mao Zedong, Zhou Enlai, and Zhu De 朱德 (1886–1976), the commander-in-chief of the People's Liberation Army (PLA) and chairman of the Standing Committee of the People's Congress, came to see the exhibition. Zhu De personally cut the ribbon at the opening ceremony and said to Kang Shi'en, "Petroleum is very important! Modern wars are fought with steel and petroleum. Without petroleum, airplanes, tanks, and cannons are no better than a dog paddle."

As a country with a vast territory, China has rich petroleum and natural gas resources, so the key was (and still is) to be tirelessly persistent with oil exploration nationwide. At any time, and under any circumstances, the oil industry needs to give top priority to oil and natural gas exploration. It must first select exploration emphasis and penetration points, and it needs to adopt correct exploration methods and show a pioneering spirit with a strong work ethic.

2 Mao Zedong and Zhou Enlai Confer with the Geologist Li Siguang

In 1953, our country started its first Five-Year Plan. In late December 1953, Mao Zedong, Zhou Enlai and other leaders met with new China's first geology minister, the famed geologist Li Siguang 李四光, who had just returned from overseas. They sought out his advice on developing the oil industry. Mao Zedong and Zhou Enlai invited Li Siguang to the Chrysanthemum Library (Mao's personal residence at the time) in Zhongnanhai, the headquarters of the central government. Mao Zedong expressed the urgent need for more

 | DOI:10.1163/9789004402737_006

petroleum: "Mr. Li, none of our things that fly in the sky and run on the ground can run without oil!" Li Siguang, based on his profound geological expertise, thoroughly described the grand prospects for China's oil resources, and Mao Zedong and Zhou Enlai beamed with joy. Soon afterward, at a State Council bureau-chief meeting, Zhou Enlai said, "The geology minister is very optimistic! He told us that there are vast oil reserves underground in our country, and there is great hope."

During this time the country's top geologists, such as Li Siguang, Huang Jiqing 黄汲清, Xie Jiarong 谢家荣, Hou Defeng 侯德封, Zhang Wenyou 张文佑, and people in the Ministry of Petroleum Industry, such as Weng Wenbo, Sun Jianchu, Chen Ben, Wang Shangwen 王尚文, Tong Xianzhang, as well as some noted Soviet petroleum geologists, participated in discussions regarding China's oil prospects. Everyone was very confident about the future of the oil industry in China.

3 Establishing the Leading Governmental Institutions of the Petroleum Industry

The People's Republic of China was founded on October 1, 1949, and already on the 19th of the same month, the central government decided to form the Ministry of Fuel Industries for managing the coal, petroleum, and electric power industries. The senior revolutionary Chen Yu 陈郁 was appointed Minister. At the time, oil production was scant, and what resources there were was unclear. With 90% of oil coming from imports, China was truly an "oil-deficient nation." Nationwide, industries were in shambles, the economy was weak, and there was little money to invest in petroleum. Faced with such a situation, in April 1950, Minister Chen Yu held the first national oil industry conference in Beijing. More than 30 experts and professors, including Yuan Fuli 袁复礼, Pan Zhongxiang 潘忠祥, and Wang Jiayin 王嘉荫, went to the conference. A target of "restoring the existing foundation within three years" for the oil industry was set at the conference. After the conference, the General Petroleum Administration Bureau was established under the Ministry of Fuel Industries, to be in charge of the restoration and development of the oilfields in the northwest, gas fields in Sichuan, and synthetic oil plants in the northeast. Xu Jinqiang 徐今强, who had been the military representative in the Nationalist government's Chinese Petroleum Corporation (CPC), became the acting bureau director. Mining and metallurgy expert Yan Shuang was named deputy director, and the petroleum geologist Sun Jianchu was appointed head of the bureau's Exploration Division. Under this bureau, there was the Northwest Petroleum Administration with Kang Shi'en as chief and Yang Zhengmin

杨拯民, the elder son of the Nationalist General Yang Hucheng 杨虎城,[1] and Zou Ming 邹明, former manager of the Nationalist government's Yumen Oilfield, as deputies; there was also the Northeast Petroleum Administration with Tang Ke 唐克 as chief.

4 The PLA Contributes Greatly to the Establishment of a Petroleum Workforce

In 1949, the petroleum workforce nationwide (excluding Taiwan) amounted to 16,000 people, of whom approximately 6,000 worked in the oilfields, and 623 were technical staff. To develop the petroleum industry, the first thing needed was to build a larger workforce and train more technicians and professionals. In 1952, Kang Shi'en, the chief of the Northwest Petroleum Administration, made a request to this effect to the Fuel Minister Chen Yu and Defense Minister Peng Dehuai 彭德怀.[2] Acting upon this, Mao Zedong, in his capacity as chairman of the Central Military Commission, issued an order to discharge more than 8,000 persons from the 57th Division of the 19th Army of the PLA. The Division Commander Zhang Fuzhen 张复振 and Political Commissar Zhang Wenbin 张文彬 led these veterans to the petroleum industry. In the next few years, these people became the backbone of our nation's oil industry, and this would continue to be the case in the following decades. It became a regular practice, a tradition, for Mao Zedong, Zhou Enlai, and other leaders to dispatch large groups of people from the PLA to become oil workers. Consequently, the PLA has made great contributions to the development of the country's oil industry.

As the country's first Five-Year-Plan unfolded in 1953, the need for petroleum in all spheres increased. As the PLA continued to form various branches, the number of its military vehicles also increased. However, the oil shortage had become a big problem. In 1955 alone, the PLA needed as much as 500,000 tons of petroleum products.

---

1 Yang Hucheng (1893–1949) was a famous general during the War of Resistance Against Japan. On December 12, 1936, he teamed up with General Zhang Xueliang 张学良 and staged a coup d'etat, detaining Chiang Kai-shek and forcing him to join forces with the Communists to fight the Japanese. Afterwards, Chiang arrested and imprisoned Yang. In 1949, before fleeing to Taiwan, Chiang ordered Yang and his family killed. Few escaped.

2 Peng Dehuai (1894–1974) was one of the PLA's key leaders, one of the Ten Marshals, and the People's Republic of China's first defense minister. In 1959, after he criticized Mao Zedong's proposal of the Great Leap Forward, he was removed from his position and criticized. During the Cultural Revolution, he was tortured. He died in 1974. He was rehabilitated posthumously in 1978.

In February 2001, at the age of 93, the Red Army veteran and former Deputy Minister of Petroleum Zhou Wenlong 周文龙 excitedly recounted for me the following anecdote from more than forty years ago: "At the time, Chairman Mao and Premier Zhou were very concerned about the oil shortage in the military. Peng Dehuai, who was in charge of the army, was even more worried. He would often tell Premier Zhou that the military lacked oil. Premier Zhou spread his hands apart in annoyance, telling Peng Dehuai: 'I don't have any oil on hand either. If you need it in a hurry, you should choose some military officers and go find the oil yourself.'" This illustrates how the oil industry of the new China got its operations started by the PLA, and how General Peng played an instrumental role.

Meanwhile, large groups of cadres were being transferred from the old liberated areas to work in the oilfields. Technical personnel who worked in the oil industry in the old China and intellectuals educated abroad who came back after the founding of the new China formed the business and technical backbone of new China's oil industry. Furthermore, relevant central government agencies and local governments saw to it that large groups of technical school graduates and young urban dwellers continued to join the oil workforce. Starting in 1951, petroleum technical schools were built in Beijing, Xi'an, Yumen, Chongqing, Fushun, Lanzhou, and elsewhere, and so was Beijing Petroleum Institute. A number of colleges also offered majors in petroleum geology and petroleum engineering. Petroleum personnel were trained through these measures. In 1958, the petroleum workforce nationwide reached 141,000 people, of whom 46,000 were oil workers, cadres, and technical personnel dedicated exclusively to oil exploration. These efforts laid the foundation for future development.

5 The Country Adopts Three Measures for Petroleum Development

After repeated discussions had developed the consensus that "petroleum was both necessary and possible to find," the state adopted three specific measures for the petroleum industry's development. First, it gave top priority to looking for petroleum resources. From 1953 to 1959, the state's investment in oil exploration made up 71% of its total investment in the oil industry, resulting in significant increases in oil exploration and development. Second, in December 1954, the State Council decided that starting in 1955, while the General Petroleum Administration Bureau of Ministry of Fuel Industries would continue oil exploration and development, the Ministry of Geology and the Chinese Academy of Sciences respectively would be responsible for oil surveys and scientific research. In the decades since then, these

FIGURE 5.1 Coal gas rubber container on top of a bus in Beijing in 1961

agencies have continued to work together and complement each other; together, they have created a golden age of Chinese petroleum industry. Third, in July 1955, the State Council decided to form the Ministry of Petroleum Industry (usually called "Ministry of Petroleum" for short). On the recommendation of Peng Dehuai of the Central Military Commission, General Li Jukui 李聚奎, the political commissar of the PLA Logistics Academy, was appointed Minister.

6 Kang Shi'en and Soviet Experts Conduct Geological Surveys

After the founding of the new China, the Ministry of Petroleum undertook two vital and effective learning activities. The first occurred in October 1953, not long after Kang Shi'en assumed the position of chief of the General Petroleum Administration Bureau, when the Soviet Union sent a group of experts to help our country to conduct petroleum geological surveys in the northwest region and devise the petroleum industry's first Five-Year Plan. The group was led by the eminent petroleum geologist Andrei Alekseevich Trofimuk. Minister Chen Yu considered it a great opportunity and arranged for Kang

Shi'en and Chief Geologist Chen Ben, geologist Jiang Fuzhi 姜辅志, and others to accompany the group. The expedition lasted 156 days, and they surveyed Yumen, the Hexi Corridor, northern Shaanxi, Yanchang, the Ordos Basin, as well as eight provinces and regions, including Sichuan, Yunnan, Guizhou, Guangxi, and Guangdong. This was the first major petroleum geological survey in China after 1949. During the expedition, Soviet experts gave patient guidance and interacted closely with the Chinese geologists. In 1954, this expert group completed a monograph entitled *China's Oil and Gas Fields* (*Zhongguo youqitian* 《中国油气田》). Based on the fundamentals of petroleum geology, the book evaluated China's major sedimentary basins and appraised China's oil prospects. The writers expressed their belief that "China is extremely rich in oil resources. Because of the short history of the oil industry, little work has been done, and at present the extent of exploration is insufficient. We think that with more investment and more work in the oil industry, China could become self-sufficient in oil." This survey played an important role in the later development of the oil industry. No wonder decades later the elderly Kang Shi'en still said, "This national petroleum survey laid the foundation for my geological understanding."

7 Kang Shi'en Leads a Petroleum Delegation to Visit the Soviet Union

The Ministry of Petroleum went into operation on September 1, 1955, and on September 5, Minister Li Jukui, during the ministry's first meeting, decided that Deputy Minister Zhou Wenlong would lead a petroleum delegation to visit Czechoslovakia and East Germany, and Assistant Minister Kang Shi'en would lead a petroleum delegation to visit the Soviet Union. They hoped to find feasible solutions for developing China's oil industry. These were undoubtedly information-gathering trips, demonstrating once more the country's expectations for developing its oil industry. The petroleum delegation led by Zhou Wenlong, after visiting Czechoslovakian and East German synthetic oil production facilities, brought back the important suggestion that "Producing synthetic oil is technically complex and expensive. Our country is a big country, so in the long run, in order to solve the oil problem fundamentally, we must rely on extensively developing natural petroleum." The petroleum delegation headed by Kang Shi'en, which went to the Soviet Union, included: Zhang Jun 张俊, the chief of Petroleum Geology Bureau; Chief Geologist Wang Shangwen 王尚文; Yang Haipeng 杨海鹏, chief of the Planning Bureau; Yang Zhengmin, chief of Yumen Petroleum Administration Bureau; and Zhang Dingdi 张定一, chief of the Northeast Petroleum Administration Bureau. They arrived in Moscow in October 1955 and returned in early February 1956, having visited major Soviet oilfields, petroleum and scientific institutions and receiving advice from experts, scholars, and leaders in the oil industry. They were greatly rewarded by this visit. There was a qualitative leap in their understanding of where to look for oilfields. They decided to forego their previous practice of drilling oil at local structures and seepage sites on the edge of basins at the bases of mountains. Now they began to consider whole basins, searching for second-order structure zones and then drilling at the most promising structures. This proved to be a new guiding principle in oil exploration, and it had major ramifications for future oil industry development. The discoveries of Karamay Oilfield and Daqing Oilfield were both achieved by dissecting second-order structure zones. Twenty years later, Kang Shi'en recalled: "While the national petroleum geological survey of 1953 made us realize our country's rich oil and gas resources and gave us confidence for developing oil industry, the visit to the Soviet Union prepared us theoretically and methodologically."

8 Mao Zedong Presides over Petroleum Policy Discussions

In early 1956, Mao Zedong, Zhou Enlai, and other state leaders spent more than a month taking part in hearings with 34 different economic agencies. On February 16, the newly appointed Petroleum Minister Li Jukui was the first to report but was embarrassed at failing to answer some questions. Mao Zedong smiled and said, "You spoke very well. You have only been there for five months, so you should not be blamed for not being familiar with some situations." At the end of the meeting, Mao proposed having another talk. Li Jukui asked if Assistant Minister Kang Shi'en could make the report next time. Mao Zedong gladly accepted.

On February 26, after thorough preparation, Kang Shi'en and Li Jukui arrived at the Hall of Diligent Governance in Zhongnanhai. The meeting was attended by Mao Zedong, Liu Shaoqi 刘少奇, Zhou Enlai, Deng Xiaoping, Li Fuchun 李富春, Li Xiannian 李先念, Bo Yibo 薄一波, and so on. Kang Shi'en started by talking about petroleum's chemical composition and molecular structure and how it was formed, how it moved, and where it concentrates and covers. Besides introducing basic technical information on petroleum, he also reported on methodology and plans for exploration. As shown below, Mao Zedong occasionally chimed in as he was trying to learn more about petroleum, while he expressed his high expectations for the industry and gave his opinions on its development. Some of his main remarks are as follows.

First, when Kang Shi'en reported that Yumen was the key petroleum base, but that recently a well in Xinjiang had also begun to produce oil, Mao Zedong said, "The Americans said that the Chinese strata are too old, that there is no oil. It looks like at least places like Xinjiang and Gansu have it, right? Your Ministry of Petroleum gives us some hope!"

Second, when Kang Shi'en talked about how hard it was to look for oil in Qinghai and Xinjiang, Mao Zedong replied, "It seems that developing the oil industry requires a revolutionary and death-defying spirit."

Third, when Kang Shi'en talked about the three stages of oil exploration "from geological surveys, to drilling and exploring with rigs, to developing the discovered oilfields," Mao Zedong said, "It seems that you should give each county a rig and ask them to poke around!"

Fourth, when Kang Shi'en mentioned that the Soviet Union drilled many wells annually, and that its petroleum investment was 12% of its total investment, Mao Zedong asked Li Fuchun, the Vice Premier and Director of the State Planning Commission, what percentage our petroleum investment was. Li Fuchun replied, "Only a few percent." Mao Zedong responded, "We need to drill one million meters annually in our country." In 1956, only 500,000 m were drilled. Meanwhile, Mao Zedong also said that more cadres should be transferred to the Ministry of Petroleum.

Indeed, this meeting was a discussion on petroleum development policy at the highest level of leadership in China.

9 "Giving Each County a Rig"

Mao Zedong had said, "You should give each county a rig and ask them to poke around." Stalin too had once said something to the effect that one who had not seen as many rigs as trees in a forest knew nothing about petroleum. In the beginning of the twentieth century, when people in the American oil industry argued over whether their resources would dry up, an insightful geologist predicted that the day the US had rigs everywhere would be the day oil would be plentiful in the US.

Mao Zedong's statement, like Stalin's and the American geologist's, are equally to the point! The more wells you drill, the more oil you have. Or as Chinese drillers say, "Where rigs don't go, the oil won't flow." At the height of American production in the 1970s, the US had more than 4,000 rigs and 500 million tons of production annually. In China, when the number of rigs reached its maximum around 1984, there were only 1,000 rigs, and now more than half of them are idle. There are 2,861 counties in China. If some day each county were truly to have a rig and use it to drill continuously, I am sure the Chinese petroleum industry would find itself in a whole new situation!

If Mao Zedong's advice of "giving each county a rig" were followed, it would lead China from an "oil-deficient country" to the world's leading oil producer.

CHAPTER 6

Yumen Oilfield: Cradle of the Chinese Petroleum Industry

1 United to Welcome Liberation

In 1949, as the gunfire of the War of Liberation drew near to one of the largest industrial operations in the northwest—Yumen Oilfield—the Nationalist government's executive office in the northwest pressured Zou Ming, the oilfield's director, requesting him to submit a plan for destroying the oilfield. Instead, he and other patriots such as Sun Yueqi and the oilfield employees united to protect the oilfield. Meanwhile, the First Field Army of the PLA under the command of Peng Dehuai, sped up its advance. With lightning speed, the mechanized armor regiment led by General Huang Xinting 黄新廷 of the Third Army arrived at the oilfield on the morning of September 25. Yumen Oilfield was spared from destruction.

Three days later, on September 28, with the recommendation of General Wang Zhen 王震, General Peng Dehuai appointed Kang Shi'en, who was the Third Army's 9th Division's political commissar, as Yumen Oilfield's chief military representative.

Fifty years later when recounting this period, Kang Shi'en said that when Yumen was liberated, Zou Ming as its director did a good job protecting the oilfield. First, when the Kuomintang rule verged on collapse, plundering and sabotage occurred everywhere. Zou gathered gold bars and silver coins worth more than 100,000 yuan from Lanzhou, Guangzhou, Hong Kong, and so on and stored them at the oilfield. He put away enough to support oilfield operations for months. This guaranteed the financial stability of Yumen Oilfield. Meanwhile, he transported more than 10,000 *dan* (= 500,000 kg) of grain from Wuhan and Dunhuang, enough supplies to last the oilfield employees and their families four or five months. Second, he increased pay for the members of the oilfield police, asking them to deal with harassment and sabotage from Kuomintang stragglers and disbanded soldiers. Third, he formed and led an oilfield protecting force made up of oilfield employees.

Ten days or so after Yumen was liberated, Peng Dehuai traveled from Jiuquan to Yumen, in order to meet with Kang Shi'en and Zou Ming. Peng Dehuai praised Zou Ming for protecting the oilfield. Zou said modestly that the fact that Yumen was intact and undamaged was due to the PLA's speedy advance.

 | DOI:10.1163/9789004402737_007

More than fifty years later, Yumen Oilfield invited Zou Ming to visit. He was then in his eighties and retired comfortably in Beijing. His memories were clear when he talked about that period of history. He said in earnest: "For a long time, people have been saying that Yumen Oilfield restored production after Liberation. That's not correct. How can you talk about restoration when Yumen Oilfield had never stopped production? It should be that Yumen Oilfield continued to increase production …" He was absolutely correct.

2 General Peng Dehuai Says Yumen Oilfield Should Become the "Cradle"

In October 1949, when the smoke from the War of Liberation still had not fully dissipated from the northwest battlefields, Peng Dehuai traveled from Jiuquan to visit the Yumen Oilfield. At a meeting attended by eight hundred people, he urged the workers to unite, learn, strive, increase production, and support the PLA's march into Xinjiang.

Peng Dehuai, from his high, strategic position, issued a call to turn Yumen Oilfield into the "cradle" of the nation's oil industry!

This was a prediction, an expectation, and also a historical task assigned to the Yumen oil workers. For fifty years, generation after generation of oil workers in Yumen have fought wholeheartedly to carry out this mission for the development of the country's oil industry.

But in early 1950, Yumen Oilfield hit hard times. The silver and grain that Zou Ming had stored up were depleted, and more than four thousand employees had not been paid for months. Kang Shi'en reported the emergency to the Northwest Military and Political Committee, i.e., the Headquarters of the PLA's First Field Army. General Peng Dehuai immediately directed that his plane, then in Jiuquan, be flown out so that people from Yumen could take it to see him in Lanzhou, so as to save time. Jiao Liren 焦力人, the deputy military representative, Zou Ming, the manager of Yumen Oilfield, and Shi Jiuguang, the American-educated geologist, went. Once they made their report to Peng Dehuai, Peng immediately summoned General Zhang Nansheng 张南生, who was in charge of the Army Logistics. Peng asked, "How much money do you have on hand?" Zhang replied, "50,000 silver yuan." Peng said: "Give it all to save Yumen!" Because there were basically no highways at the time, it would take three to four days to walk from Lanzhou to Yumen, which would include crossing the 3,000–3,500 m tall Wushao Mountain. Bandits were rampant along the way. General Peng sent a fully armored squad to escort the money and the visitors back to Yumen. They traveled in two Dodge trucks, which the Communist army had seized from the Nationalists. After the meeting, General Peng, who was known for being simple and frugal, asked his chef to cook up some stewed chicken, and during the meal, he personally dished out food to Zou Ming and Shi Jiuguang. These are the kinds of stories that oilmen have never forgotten.

3 Where There's Oil, There Are Yumen Men

Yumen Oilfield's function as a "cradle" became most prominent later on as a ground for training and supplying large numbers of teams for new oil and gas fields. It became the Chinese petroleum industry's first "big school." During the fifty years from 1950 to 2000, Yumen Oilfield supplied as many as 100,000 people and more than 4,000 sets of equipment to different oilfields in the nation. Every year, without exception, there were transfer assignments. In 16 of those years, Yumen Oilfield transferred more than 1,000 people, and in seven of those years more than 5,000. For fifty years, the people in Yumen Oilfield considered one of their responsibilities to be supporting new oilfields, and at times that even became the central work and focus of the whole oilfield.

There were four major transfers involving approximately 20,000 people each time. The first occurred during the four years from 1955 to 1958, when Yumen Oilfield transferred 21,555 people to other oilfields: 3,090 to Qinghai's Qaidam Oilfield in 1955, 6,255 to Xinjiang's Karamay Oilfield in 1956, 6,544 to Gansu's Lanzhou Oil Refinery in 1957, and 5,666 people to Sichuan's Chuanzhong Oil Campaign in 1958. These transfers resulted in the oil and gas industry's large-scale development in the western part of China, forming several new oil industrial bases with Yumen Oilfield as the "head of the dragon." This also prepared for the industry's impending strategic eastern shift.

The second took place in the three years from 1959 to 1961, when Yumen Oilfield sent 19,770 people to Daqing Oilfield: 3,768 in 1959, 8,371 in 1960, and 7,631 in 1961. That is, two-thirds of Yumen Oilfield's employees were transferred to Daqing Oilfield. At the time, of the roughly 40,000 people employed in the Daqing Oil Campaign, half had come from Yumen Oilfield. This was undoubtedly a magnificent accomplishment in the history of China's petroleum development. In fact, the Songji-3 well, which led to the discovery of Daqing Oilfield, was drilled by the 32118 Drilling Crew, which had transferred from Yumen Oilfield. Through the organization and deployment of the Ministry of Petroleum and the leadership of Yumen Oilfield Director Jiao Liren, personnel in all specialties—from geophysical exploration, drilling, oil extraction, oil refining, oilfield construction, and downhole operations to water and electricity supply, mechanical repair, transportation, communications, geological research, and oil and gas storage and transport planning and design—were transferred to Daqing, where they worked at all levels in various agencies, service departments, and construction sites.

The third large transfer occurred around 1970, when the people of Yumen again contributed significantly to the effort to support Changqing Oilfield 长庆油田. Changqing Oilfield is situated in Ordos Basin, which extends across provinces and regions such as Shaanxi, Gansu, Ningxia, Shanxi, and Inner Mongolia, with a total area of 370,000 km^2. It is the second largest sedimentary basin in our country, rich in oil and gas resources. Already during the 1950s, Yumen Oilfield had organized special teams to work in Changqing Oilfield. In 1963, the Ministry of Petroleum designated work in this area to Yumen Bureau. In 1969, the State Council instructed the Ministry of Petroleum to conduct a large-scale oil exploration campaign in the area and placed the heavy burden of manpower entirely on the shoulders of Yumen Oilfield. Since the campaign headquarters was located in the village of Changqingqiao 长庆桥村, Ning County, Gansu Province, the oilfield was named Changqing Oilfield.

Yumen people considered it their responsibility to support the new oilfield and proposed the slogan, "Mobilize the whole Bureau—run to Changqing!" The leadership moved to Changqing and from Changqing directed work at both Changqing and Yumen oilfields. Thus two-thirds of Yumen's manpower and equipment were transferred to Changqing. More than 20,000 people were transferred to Changqing around 1970, including 13,972 people in 1970 alone. According to a February 1970 Yumen Oilfield report on the personnel transfers:

> There were 185 people at the Yumen Bureau administration building, and the first group to go to Changqing amounted to 101 people. That included nine of the 14 people in the Organization Group, eight of 12 people in the Finance Group, five of eight people in the Equipment Group, and eight of nine secretaries from the Bureau office. The transferred personnel even took their desks and file cabinets with them. All of a sudden, many of the offices in the Bureau's administration building became empty.

On June 22, 1970, the Ministries of Petroleum, of Coal, and of Chemical Industry were merged into a single Ministry of Fuel and Chemical Industries.

In 1971, by order of this new Ministry, Changqing Oilfield became independent, separating from Yumen Oilfield, whereupon the Changqing and Yumen Oilfields became brother units in the industry. After the separation, if we take into consideration the workforce, equipment, assets, and scale of the two oilfields, the "older brother" Yumen had become in reality the "smaller brother."

When Yumen oil workers started to labor in earnest at Changqing Oilfield in 1970, there were no major breakthroughs initially. By the mid-nineties, they were achieving remarkable results but without much fanfare, because the Ministry of Petroleum was waging a big oil campaign in the Tarim Basin and paying little attention to other locations. In 2014, Changqing's annual production was 55.44 million tons, becoming in effect "another Daqing" and even surpassing it. It has now become the largest oilfield in our nation.

Ironically, while Wang Tao 王涛 and Zhou Yongkang 周永康 were conducting a well-funded, full-scale campaign in Tarim and ended up finding little oil, the Yumen people found a second Daqing under much more modest circumstances. As we say in Chinese, "When one makes great effort to plant a flower, it never blooms, but when one plants a willow and leaves it alone, it becomes shade." One really does not know whether to laugh or cry at heaven's capriciousness!

The fourth major personnel transfer occurred in the 1980s and 1990s, when Yumen Oilfield worked hard to support Tuha Oilfield 吐哈油田. "Tuha" is short for Turpan and Hami. Tuha Oilfield is situated in the Turpan Basin in southeastern Xinjiang, a few hundred kilometers from Yumen Oilfield. Starting in the 1950s, the oil exploration work in Turpan Basin was carried out by Yumen Oilfield, except for two short periods when explorations were conducted by the Xinjiang Petroleum Administration Bureau, i.e., Karamay Oilfield. In 1959 and 1964, Yumen Oilfield completed much pioneering work in Turpan Basin, setting the stage for the eventual discovery of Tuha Oilfield.

In 1983, the Ministry of Petroleum decided to have Yumen Oilfield resume work in Tuha. Answering the battle cry of "Stage a comeback, show our vigor again," Yumen oil workers went all out for the Tuha Oil Campaign. Most of the current 20,000 employees and equipment at Tuha Oilfield came from Yumen Oilfield. Even some of the office furniture came from Yumen Oilfield. Thus, by the end of the twentieth century, the oil industry people from Yumen who had transferred to China's northwest had built another large modern oil and gas field.

The Yumen oil workers, called "Yumen men" (*yumen ren* 玉门人), are known for their hard work, self-sacrifice, and dedication, and so the poet Li Ji 李季 said, "Where there's oil, there are Yumen men."

4 Yumen Oilfield's "Three Bigs and Four Exports" Achievements

In August 1989, in a commemorative article on the 50th Anniversary of Yumen Oilfield, Kang Shi'en wrote, "In 1958, the Ministry of Petroleum set the policy of 'Three Bigs and Four Exports' for Yumen Oilfield (that is, a big school, big lab, big research institute, and export products, export experience, export technology, and export talent).”

For decades, Yumen Oilfield firmly carried out this “Three Bigs and Four Exports” policy as it made outstanding contributions to the development of the oil industry nationwide. Setting off from the banks of Shiyou River at the foot of the Qilian Mountains, Yumen men formed an elite group in building other oil and gas fields. This group included many accomplished oil experts, tens of thousands of oil managers on all levels, as well as model workers such as the much-lauded “Ironman” Wang Jinxi 王进喜.

Due to its unique position in the development of the oil industry, Yumen Oilfield ushered in a new era for new China’s oil technology. In the 1940s, Yumen Oilfield was the first to import advanced American and European oil exploration and development technology. In the fifties, it was the first in the country to create a series of technologies targeted at water-flooding development, mechanical oil recovery, downhole diagnostics, oil well paraffin prevention and removal. Yumen Oilfield made large-scale improvements to low permeable and low yield beds, in order to increase oil reserves and oil production. In the 1960s, it was the first oilfield in the country to test the technology of tertiary oil recovery. Yumen Oilfield provided a wealth of experience with regard to geological research, oil exploration, oilfield construction, petroleum chemistry, and oil machinery. Such experience proved valuable to oil and gas fields throughout the country.

5 Yumen, the “Cradle”

In late fall of 1999, strolling in front of Sun Jianchu’s Memorial Statue in Yumen Park, I pondered why Yumen Oilfield had always put consideration of the overall situation in the oil industry first and did its utmost to support new oil areas. I came to the following conclusions:

First, Yumen men considered developing the oil industry nationwide to be their own duty. An old oilman, Chen Zhongyong 陈忠勇, told me this story while I was standing in front of Sun Jianchu’s statue:

> In the 1930s, the three old Yumen pioneers Yan Shuang, Sun Jianchu, and Jin Xigeng arrived at the banks of Shiyou River at Laojunmiao in Yumen. During the day, they’d investigate oil seepage and do surveys. At night, they’d sleep in a yurt three meters in diameter. The yurt had a hole on top, so they’d lie in there, looking at the moon and counting stars. Once, on a whim, Sun Jianchu blurted out: “If we found an oilfield here, with

> one well producing 10 gallons per day and ten wells 100 gallons … and later, starting off from here, we go elsewhere in the country, finding more oilfields, then the oilfields throughout the country will be as countless as the stars in the sky."

Chen Zhongyong had heard this story from Jin Xigeng himself. Ever since Yumen Oilfield was founded, if any oilfield from the country asked for help, Yumen men would give it unconditionally. It was like this before Liberation in 1949, and it was even more so after Liberation. This quality of Yumen men, of counting the development of the Chinese oil industry to be their own responsibility, was their most outstanding feature.

Second, the people of Yumen Oilfield would willingly endure hardships, impoverish themselves, and give all they had to support new oil areas. They transferred the best personnel and most advanced and complete equipment to new areas. According to a January 28, 1961 inter-oilfield memo issued by Yumen Oilfield titled "Arrangement on Transferring Employees and Families to Other Locations (关于外调职工及家属调迁工作的安排)":

> 1. The Ministry of Petroleum directs the transfer of 3,035 persons (1,021 cadres) people from Yumen Bureau to Heilongjiang … 2. The following people shall not be transferred: a. sick employees; b. old and weak employees; c. employees without winter clothing and quilts and temporary workers of Gansu Province; d. employees who are significantly politically suspect; e. childbearing women within one to two months before or after delivery.

Third, Yumen people were highly organized and disciplined. They obeyed orders. Once the order was given, they would immediately act. They would give whatever the Ministry wanted and transfer whoever was needed. They would give whenever required. They would volunteer before requirements were made. In the same document mentioned above, directions were given to the 3,035 people who were being transferred to Daqing Oilfield:

> 3. Time arrangement: from February 15 to 25, [the 3,035 people are to be] divided into 11 groups and sent off. 4. Organization and leadership: set up a transfer headquarters for Yumen employees and families, with relevant plants also setting up corresponding organizations. At railroad stations in (Yumen) East Station, Lanzhou, Xi'an, and Zhengzhou, set up transfer supply booths, approximately 500 persons for each group, with five escort cadres (including medical doctors). Each transferring employee and

> family member needs to take three to five days of dried food with them and 10 to 12 days of food coupons. Transferring employees should not ask for leave or separate from the group during the journey. Every 10 people shall form a squad, with a leader; every three squads shall form a platoon, with a platoon leader.

They were like troops from the PLA marching to a new battle ground.

Fourth, they developed themselves while they supported new areas. They revitalized Yumen time and again. In 1949, Yumen produced 80,000 tons of oil, but by 1959, it had increased production to 1.4 million tons, or 51% of the national oil production.

In 1970, Yumen Oilfield was given the task of supporting Changqing Oilfield. The Communist Party Committee of Yumen Bureau appointed Shi Zuohan 时作汉, the Secretary of the Party Committee of Yumen Oilfield Construction Company, to form Changqing Oilfield Construction Company. Sparing no effort, he chose the best people and successfully formed the company. But right on the eve of his departure, the Bureau suddenly announced that he would stay in Yumen Oilfield, and that someone else would lead the personnel of the new company to Changqing Oilfield. At the time, there was a rhyme circling among Yumen Oilfield employees: "Sick men, old men, Shi Zuohan, see what you can do, old man." But Secretary Shi had no complaints or regrets, he happily stayed in Yumen. He labored on. Within three years he rebuilt and developed the Yumen Oilfield Construction Company after it had been depleted of personnel and equipment.

6 The Immortal Spirit of Laojunmiao

Yumen Oilfield's support for China's entire oil industry antedates the founding of the new China in 1949. After China's victory in the War of Resistance Against Japan in 1945, some technical people went to work in Taiwan to help restore Kaohsiung Refinery, which had been sabotaged by the Japanese before they left. Others were sent off to restore synthetic oil plants in the northeast, and yet others went to work on restoring the refinery in Gaoqiao, Shanghai. In the 1990s, when groups of veteran oilmen returned from Taiwan to visit Yumen Oilfield, they said that when they restored the oil refinery in Kaohsiung, they raised the slogan, "Carry on Laojunmiao's spirit of hard work."

In 1992, Dr. Ye Jinlong 叶金龙, Vice-president of Taiwan's Chinese Petroleum Corporation (CPC), gathering dozens of petroleum and petrochemical experts, published *The Petroleum Industry—Protagonist of Taiwan's Economic Miracle* (*Shiyou gongye—Taiwan jingji qiji de zhujiao* 《石油工业—台湾经济奇绩的主角》). In the section titled "The Cradle of China's Petroleum Industry—The Story of Laojunmiao," the book documented Yumen Oilfield's history and its impact on Taiwan's petroleum and petrochemical industry. Ye Jinlong's account started with a poem:

> Revisiting Laojunmiao,
> The old temple had disappeared.
> Though the new temple is big,
> It is not as good as the old.
> Having parted for forty years,
> Veteran oilmen, revisiting their hometown,
> Expressed their pride about past times.[1]

FIGURE 6.1 Translator Yiran Mao at the Laojun Temple (Laojunmiao) in 2001

1 Ye Jinlong 叶金龙, *Shiyou gongye—Taiwan jingji qiji de zhujiao* 《石油工业—台湾经济奇迹的主角》 [*The Petroleum Industry—Protagonist of Taiwan's Economic Miracle*] (Taipei: Zhongguo shiyou xuehui, 1992), 6.

Ye Jinlong continued, writing with affection:

> Our country's petroleum industry, starting with the first well drilled in 1939 in Laojunmiao, germinated at the foot of the Qilian Mountains beyond the Jiayu Pass at the western end of the Great Wall. This was the origin of [Taiwan's] Chinese Petroleum Corporation.
>
> Everything from Jiayu Pass to the Yumen Oilfield was the Gobi Desert, and apart from small bushes of camel thorn, there were no trees or people. It was a barren land. Oil people reclaimed the wilderness, used water from Shiyou River, built food and flour factories, used adobe for walls, built their own food-processing plants and flour mills, and developed their own farms to grow produce. In the desolate Gobi Desert, they built a vibrant oilfield city.
>
> The establishment of Yumen Oilfield during the War of Resistance Against Japan was equally as important as the reconstruction of Kaohsiung Refinery from rubbles after the retrocession of Taiwan. The former supported the War of Resistance Against Japan, the latter supported Taiwan's financial and economic development. They were both national and economic miracles!
>
> Mr. Hu Hsin-nan, the former Chairman of the Board of CPC, believes that managing a state-run cause in order to make money for the national treasury benefits everyone. The idea, "If one should make a profit, he should do so for the benefit of everyone under the sun," was emblematic of the perseverance of veteran oil people. This also demonstrates that the "Spirit of Laojunmiao is immortal." "If one should make a profit, he should do so for the benefit of everyone under the sun" is part of a calligraphy presented by Yu Yu-jen to Chiang Ching-kuo in 1961. May the Chinese veteran oilmen's "Spirit of Laojunmiao" shine forever![2]

2 Ibid., 12.

CHAPTER 7

Pioneering a Great Cause at Qinghai Oilfield

1 The Harsh Environment of Qinghai Oilfield

Qinghai Oilfield is located in the Qaidam Basin 柴达木盆地, so it is also known as the Qaidam Oilfield. In Mongolian, the word *qaidam* means "salt marsh." This basin is in the northwest part of Qinghai Province and surrounded by the Kunlun Mountains 昆仑山, Qilian Mountains, and Altun Mountains 阿尔金山. The average elevation is 3,000 m, typical of a basin in the highlands. The sedimental area of the basin is 120,000 km^2. There are many records in history about oil seepage in this basin. It is one of the basins in our country with the best oil and gas prospects. Over the years, many petroleum geologists and scholars have set their eyes on the basin.

However, Qaidam Basin's natural environment and weather is very harsh. It is a desert area and more than a thousand km from any city. It is arid, with little precipitation. It is hard for people, animals, and vegetation to survive there long-term. The annual precipitation is not even 200 mm, while the annual evaporation rate is 2,000 to 3,000 mm, one of the world's highest. One year it counted 3,602 hours of sunshine, making it surpass the "Sunshine City" of Lhasa in Tibet as the sunniest place in the nation. At the same time, however, its frost-free period is only 40 days. No wonder someone has said of this place, "No grass grows on the ground, no birds fly in the sky, yellow sand covers the sky, and when the wind blows the rocks roll."

While finding oil is a relatively tough job, finding it here is tougher. If you have ever been to Qaidam, you would know what I mean.

2 The Origin of Qinghai Oilfield

As early as May 1947, Guan Zuoshu 关佐蜀, Zhou Zongjun, and other staff members of the Exploration Office of the Gansu and Gansu Qinghai Division of CPC, together with the Qaidam Basin Geological Survey Team from the Nationalist Government's Ministry of Economic Affairs, conducted petroleum surveys in Qaidam and found the Yousha Mountains 油砂山 (literally, "the oil sand mountains") in the west of the basin.

Large-scale petroleum exploration started only after the new China was founded. In March 1954, Kang Shi'en, the chief of the General Petroleum

 | DOI:10.1163/9789004402737_008

Administration Bureau, held the Fifth National Petroleum Exploration Conference in Xi'an, where the petroleum exploration work of China's First Five-Year Plan was devised, including the exploration of Qaidam Basin. In April of that year, the Petroleum Administration Bureau sent out a young geological crew from Xi'an, which reached Hongliuquan 红柳泉 ("Red Willow Spring") in the west of the basin at the end of May. In September of the same year, a work group consisting of dozens of people was formed. The group was headed by Kang Shi'en. Among its members were Petroleum Administration Bureau's Zhang Jun, Shen Chen 沈晨, Yang Wenbin 杨文彬, and geologists Chen Ben and Wang Shangwen, as well as two Soviet experts, Andrei A. Trofimuk and one known only in Chinese as Qiyaqikefu 契亚契克夫. They conducted topographical and geomorphological surveys at locations including Aral 阿拉尔, Yousha Hill, and Mangya 茫崖 ("Boundless Cliff"). This was the largest petroleum geological survey ever conducted in the basin, and they made plans for its extensive oil exploration.

On June 1, 1955, the Qinghai Petroleum Exploration Bureau was established. Zhang Jun, then the chief of the Petroleum Geology Bureau, was appointed Director as well as the Secretary of its Party Committee. Chen Shouhua 陈寿华, Yang Wenbin, and Guo Jiusheng 郭究圣 were appointed deputy directors. After that, large groups of people from all over the country were sent into the basin, and the Ministry of Geology sent teams of people, too. Tent towns went up at Mangya and Lenghu 冷湖 ("Cold Lake"), all places named by geological and oil workers. The hard oil work began. In December 1955, the no. 1 well at Youquanzi 油泉子 ("Oil Spring"), near Mangya, produced oil. On September 5, 1955, *People's Daily* published an editorial calling for support for the Karamay and Qaidam Oilfields. Suddenly all eyes were on the oil exploration in Qaidam Basin. At its peak, 26,072 people were employed there. In June 1958, the Ministry of Geology found oil strata at 300 to 400 m in the Lenghu-5 Structure. On September 13 of the same year, on the same structure, the Ministry of Petroleum's crew completed the drilling of Dizhong-4, a high producer. This is the well that lead to the discovery of Lenghu Oilfield. At its site, a memorial was erected with this inscription, "The good names of the heroes of Dizhong-4 have spread far and wide." To this day it stands in the midst of the boundless Gobi Desert, as a tribute to those geological and oil people who expended their sweat, tears, and even their lives there.

3 Story of the Old Oilman Aji

During my visits to Qinghai Oilfield, people told me all kinds of stories about the "Qaidam Spirit." Out of these hundreds of stories, I have chosen a few that illustrate that spirit. Yisha Aji 伊沙·阿吉 was born in 1882 in Qiemo County in south Xinjiang. His ancestors had come from Uzbekistan. His life became the stuff of legend. When he was small, he traveled with his father to the hinterland of Qaidam Basin, so he became very familiar with the area's mountains, water, sand, and rocks. He had a hard life before Liberation. In 1945, he was arrested and imprisoned for three years by the Nationalist government in Xinjiang on charges of collaborating with the Communists. He was finally rescued in an operation that had broad participation.

In 1949, when Qinghai was liberated, it was as if spring had come to the withered tree for Aji. From 1949 to 1954, old Aji led the PLA into the basin three times to wipe out the remaining troops of the Nationalist army and bandits. Though he went through hell and high water, he made several near-miraculous achievements. In 1954, when Qaidam oil exploration began, Aji became an ideal and most welcome guide. There was no place he could not locate, and he always knew where to find water in the vast Gobi. Aji became the "compass" for the oil teams in the desert. Once in 1955, Aji took an oil crew to the Yousha Mountains area in the western part of the basin. They had exhausted their drinking water, and both camels and men were so thirsty they could not budge. While everyone's life was in danger, Aji was carefully looking around. Pointing to one spot, he said, "There is water here, come dig with your spades!" People went over and looked—it looked no different from any other dry patch in the Gobi. Still doubting, people started digging. They had dug less than a meter when fresh water seeped out.

When Kang Shi'en led a group conducting reconnaissance at the basin in 1954, he held a managers' meeting at Mangya. He wanted to use it as a base but worried that fresh water could not be found. He sought advice from Aji. Aji told him that there were Mongolian gazelles (*huangyang* 黄羊; *Procapra gutturosa*) in this area, and that if he dug down 80 m, there would be fresh water. Reassured, Kang Shi'en decided to build a base in Mangya.

Aji loved Qaidam, loved the petroleum industry, and loved China all his life. He made special and valuable contributions to the development and construction of Qaidam. His name is known among all the oil workers in Qaidam Oilfield and PLA soldiers there. His descendants also worked in the oil industry. In 1961, after dying of sickness in Qaidam, he was buried by oil workers in a public grave in the remote Gasi 尕斯 Oilfield, and he has become the tutelary deity of Qinghai Oilfield!

4 Coexisting with Wolves

On August 1, 2001, Liang Zexiang 梁泽祥, who had retired and was living in Dunhuang, Gansu after working in Qaidam Basin for more than 40 years, told me this moving story. In 1958, he was a driller, working at Lenghu-3 Structure, where more than 30 workers lived in a dozen tents on the open and wild Gobi Desert. One morning, in the kitchen tent, they found two wolves lying there. Frightened, the whole crew tried to shoo them out. Strangely, while being chased, they would run when chased but stop whenever the people stopped, and they would constantly look back and cast "friendly" glances at them. In the next few days, the two wolves would approach the drilling crew's basketball court punctually at five or six o'clock in the afternoon. Workers would give them some leftovers and bone scraps, and they started calling them "Huanghuang." The two would respond to people's calls and act in various ways that made the workers laugh. They would always keep a few meters' distance from people, not allowing people to touch them, but never attacking those workers. After a while, coexisting with the wolves and playing with them became part of the oil workers' life, breaking up their boredom.

In the spring of 1959, this drilling crew was ordered to move. They loaded up everything the day prior to the move. On the day of the move, strangely enough, those two wolves broke with their usual habit and came to the basketball court in the morning, and their eyes filled with puzzlement as they kept circling around the truck. When the truck drove away, the two of them ran ahead, waiting for the truck. After the truck rolled past, they ran after the truck. The workers called out "Huanghuang," with tears in their eyes. The wolves obviously did not want the workers to leave. Liang Zeliang's story of the oil workers living with wolves under those special circumstances never failed to move those who heard it.

5 Risking Life to Look for Oil

The following is a veteran oilman's personal experience. During the Great Chinese Famine (1959–1961), the 3274 Drilling Crew was in Lenghu. In the barren Gobi Desert, they swore that they would be "happy to work in the wild and proud to endure hardship, looking for oil and gas, and not quit despite countless hardships." They reorganized the original four groups into three, freeing up one group to find food in Jiuquan and Dunhuang. Soon, their kitchen was "stocked." In addition to everyone's daily grain rations, there was one vegetable dish, one soup, and one congee free of charge. The dish was "stir-fried wild chives" (grass), the soup "chlorophyll soup" (from moss), and the congee "nutritious congee" (wild herbs). These things were barely edible, but this went on, day in and day out. Some got so hungry that they cut up their leather work shoes and made soup to drink. Later on, they ate all of their sheepskin coats, too. The seven or eight older workers couldn't get up. Even so, one younger driller exclaimed that they could not stop drilling and volunteered to work 16 hours per day for two weeks. The workers did not take even a day's rest, and the rig did not stop for a second. Before long the well gushed out a high-yield oil stream.

Coincidentally, in 1957, the same young driller had gone out into the wild to do a survey with a co-worker from Geology Crew 107. One day during the survey, their compass malfunctioned, the two of them got lost, and they could not find their way back to camp. To help people locate them, the driller used a surveyor's pole to write out a poem in gigantic characters high on a sand dune. The poem, which was also given a title, "March into Qaidam," read:

> With surveyor's pole as my pen,
> Sweat as my ink,
> And sand dune my paper,
> I express my pride.

But the wind came and erased their footsteps and the poem on the dune. So there they were in the midst of the boundless, silent Gobi Desert at sunset, and they couldn't find the road. The water in their canteens was gone, and the two of them fainted. After an unknown lapse of time, a hunting dog from a camel caravan found them and licked them awake. In the end, something good came out of their misfortune, because after they analyzed why their compass had malfunctioned, they determined that it was due to a big deposit of magnetite.

6 Dangjin Pass

On September 16, 1999, Li Ye 李晔, the former secretary of Yu Qiuli 余秋里, Lieutenant-governor of Shandong Province, told me the following story. In early June, 1959, Yu Qiuli travelled by car from Lenghu Oil Base to the Ministry of Petroleum's transportation company in Dunhuang, Gansu Province. Yu and his entourage traveled there by going through the high Dangjin Mountain Pass that separates Dunhuang County in Gansu Province in the east and Qaidam Basin in Qinghai Province in the west. The elevation at the pass is around 3,700 m. In those days, there were no stops for food or lodging along the way for truck drivers. All of the crude oil from Lenghu Oilfield had to be transported by large imported oil tanker trucks to Liuyuan on the Lanzhou–Xinjiang railway, by which it was shipped to Lanzhou Oil Refinery. They encountered a group of lean and haggard oil truck drivers, each holding a bowl of corn porridge which was so thin that one could see his own reflection in it. They surrounded the Minister's car, and one complained: "Minister! This big bowl of porridge looks like a half kilo but really isn't. Once I drink it, it'll be gone soon as I piss a couple of times. This big oil tanker weighs 18 tons and hauls a trailer, and when I go uphill or downhill, I can barely turn the steering wheel. I truly don't have the energy." Everyone chimed in in agreement, some with tears in their eyes. All present felt the agony.

As soon as he returned to Dunhuang, Yu Qiuli spoke with Zhang Fuzhen, the Manager of the Ministry's transportation company, about the plight of those truckers. They decided that they urgently needed to create a farm, just as Mao Zedong had proposed in Yan'an: "By working with our own hands, we get ample food and clothing." They were determined to have grain and vegetables planted to ensure that workers had enough food. Starting from this time, all units in the oil industry got engaged in food production on the side. In reality, their individual decisions to do so was the direct result of hunger.

7 Lenghu Oil Tomb Ground

One afternoon in mid-September 1987, Qaidam oil veteran Yang Haiping 杨海平 took me to visit the Oil Tomb Ground on the west side of Lenghu town. At the time, Qinghai Oilfield was streamlining and reducing its workforce. At first he said, "We came to Qaidam to find oil and decided that we'd 'dedicate our youth, our lives, and our children's lives' to it." He said it in such a stirring tone, I was in awe. When we got close to the cemetery, he told me, in a despairing tone: "Now everyone who could leave and had to leave has left, so the only ones remaining are those who cannot leave." Pointing to the tomb ground, he joked, "In Lenghu, this is the only 'expanding and developing' unit." This remains the largest public tomb ground I have ever seen. I tried counting how many graves there would be, but it was impossible. I wandered ahead, as a solemn, respectful feeling crept over me.

On September 17, 1999, at the Petroleum Hotel in Dunhuang, the chief of Qinghai Petroleum Administration Bureau spoke with me. With fervor and assurance, he spoke about Qaidam's past and future. He was both confident and modest, saying, "Qinghai Oilfield is an old oilfield, a good oilfield. It is also a small oilfield and a difficult oilfield …"

Difficult indeed! The difficulties that Qaidam oil workers encountered were loneliness, hardship, and the most bitter misery. But after all that bitterness, things did turn sweeter. In 2014, Qinghai Oilfield produced 6.8 billion m^3 of natural gas, which is equivalent to 6.8 million tons of oil, and became the country's fifth largest oilfield. It is no longer a small oilfield. Such is the history of Qinghai Oilfield.

FIGURE 7.1 Gasi Oilfield in Qinghai Province (2005)

CHAPTER 8

Karamay: New China's First Big Oilfield

1 Oil- and Gas-rich Xinjiang

Xinjiang Uyghur Autonomous Region, in western China, covers an area of 1,664,897 km^2, or about one-sixth of the nation's total area; it is China's largest province or region. It is surrounded by tall mountains, with the luxuriant Kunlun Mountains in the south, the beautiful Altai Mountains in the north, and the majestic Tianshan Mountains in the middle. Nestled among these mountains are the three big basins of Tarim, Junggar, and Turpan. These basins are like treasure bowls, full of oil and gas resources as well as a multitude of other minerals.

Xinjiang's abundance of oil and gas resources has long been noticed by experts and scholars, whether ancient or modern, Chinese or foreign. In ancient times, people knew petroleum from seepage and outcrops, then learned how to use it. There were many accounts of seepage on both the south and north sides of the Tianshan Mountains. The earliest records can be found in ancient books such as *History of the Northern Dynasties* (*Beishi* 《北史》) and *The New History of Tang* (*Xin Tang shu* 《新唐书》). During the Tang dynasty, between 643 to 659, Li Yanshou 李延寿 wrote in *History of the Northern Dynasties*: "The kingdom of Guizi 龟兹国 … In the big mountains in the country's northwest, there is something like grease that flows out and forms rivers, which go on for a number of *li* and then disappear into the ground. It looks like paste and stinks badly." Guizi is in the area of present day Kuqa in southern Xinjiang, and the grease is petroleum. *Xin Tang shu*, written by Ouyang Xiu 欧阳修 and others in the Northern Song dynasty, describes a trip in which someone "crossed the Shiqi River, passed over the Cheling mountains, and arrived at Gongyue City 弓月城" after passing Changji 昌吉, Manas 玛纳斯, and Wusu 乌苏 on the southern edge of the Junggar Basin. In ancient times, as I have mentioned in Part 1, people called petroleum *shiqi*. This Shiqi River is in present-day Jinghe County, not far from Dushanzi Oilfield.

Toward the end of the Qing dynasty, that is to say, the beginning of the twentieth century, accounts about petroleum in local chronicles became even more abundant and detailed. The mountains and enterprises sections of *Maps and Records of Xinjiang* (*Xinjiang Tuzhi* 《新疆图志》), published in Xinjiang in 1911, recorded discoveries and extractions at a dozen seepages in south and north Xinjiang, primarily at Dushanzi, Qingshi Gorge 青石峡 in Tacheng (i.e.,

 | DOI:10.1163/9789004402737_009

Karamay Oilfield's Heiyoushan 黑油山), and Kuqa. Later explorations proved that Dushanzi and Karamay were oilfields with commercial value, so they have been developed on a large scale. In 1958, Yiqike Oilfield 依奇克油田 was also discovered near Kuqa and was developed.

2 Weng Wenhao and Huang Jiqing's First Survey in Xinjiang

In the latter half of the nineteenth century, Tsarist Russia attempted to take advantage of its vicinity to Xinjiang and the privileges it had been granted by unjust treaties to meddle with petroleum extraction in Xinjiang. Chinese and foreign geologists alike had voiced their high opinion of the petroleum resources in this area. Some traveled there in person, others wrote about it and recommended it to relevant authorities. In November 1942, the noted geologist Weng Wenhao, the Minister of Economic Affairs in the Nationalist government, arranged for Huang Jiqing 黄汲清 (also known as Te-Kan Huang), the director and chief geologist of the Central Geological Survey, to lead a geological crew to do a survey in Xinjiang. The crew included Yang Zhongjian 杨钟健, Cheng Yuqi 程裕淇, and Zhou Zongjun 周宗俊 of the Central Geological Survey and Bian Meinian and Weng Wenbo of the Gansu Petroleum Bureau. Overcoming tremendous difficulties associated with field work, they first surveyed the Dushanzi area in northern Xinjiang. Later, Huang, Yang, Cheng and Zhou surveyed vast areas around the northern edge of Tarim Basin in southern Xinjiang. When they returned in Chongqing in May 1943, Huang Jiqing wrote the report "Xinjiang Oilfield Geological Survey (Xinjiang youtian dizhi baogao 新疆油田地质报告)." This report played a role in shaping the oil development in Xinjiang.

According to the Second National Oil Resources Assessment of 1993, Xinjiang's oil resources were estimated at 20.86 billion tons and natural gas resources at 1.03 trillion m^3, amounting respectively to 30% and 34% of China's on-land oil and natural gas resources. Some insiders also estimate that the actual amount, in the long run, may well be higher.

3 Karamay—Black Oil Mountain

Karamay in Uyghur means "black oil mountain." In Chinese, it is called Heiyoushan, meaning the same. It got that name because of a gushing oil well near the mountain, which led to the discovery of the oilfield. Based on a proposal by Saifuddin Azizi, the first Chairman of Xinjiang Uyghur

Autonomous Region, the oilfield was formally named Karamay Oilfield on May 1, 1965.

On January 1, 1955, the Soviet Union's shares of the Sino-Soviet Petroleum Corporation were transferred to China. After the transfer, Xinjiang Petroleum Company was set up, and it was later renamed the Xinjiang Petroleum Administration Bureau. In January 1955, the 6th National Petroleum Exploration Conference was held in Beijing to discuss exploration plans for Xinjiang Petroleum Company. The decision was made at the conference to drill two exploratory wells in the Karamay area. In March 1955, the Dushanzi Oilfield geologists Wang Ke'en 王克恩, Wang Qiuming 王秋明, Wang Lianbi 王连壁 produced the geological and technical design for Heiyoushan-1 Well. The well was then drilled by the 36 young members (from eight different ethnic groups) of 1219 Crew from the Drilling Division of Dushanzi Petroleum Bureau. Under the direction of the technicians Lu Mingbao 陆铭宝 and Ai Shan 艾山, they vowed to "settle down, stay rooted, and never give up" until the well produced oil. They started drilling on July 6, 1955 and finished drilling on October 29. This was the well that started the Karamay Oilfield, which opened the curtain for tremendous growth in the Xinjiang oil industry.

4 Petroleum Minister Li Jukui Takes Full Charge

In mid-December 1955, Petroleum Minister Li Jukui took Exploration Division Chief Tang Ke, Deputy Shen Chen, geophysical expert Weng Wenbo, and a Soviet expert (his name in Chinese: 安德列依柯) to sites in Xinjiang. On December 17, with the assistance of this work group, a meeting was held to discuss the overall planning of Heiyoushan. A design committee was formed by Zhang Wenbin, chief of Xinjiang Petroleum Company, chief geologist Du Bomin 杜博民 and his deputy Zhang Kai 张恺, and chief engineer Shi Jiuguang. They produced a document, "The Overall Exploration Plan for Karamay Area (Kelamayi diqu kantan zongti guihua 克拉玛依地区勘探总体规划)," which proposed a well distribution that combined drilling on local structures and drilling on big profiles, an expanded scope of drilling; it recommended determining as soon as possible the oil reserve situation in southern Heiyoushan and the Nanxiaoshi 南小石 oil structures. In February 1956, at the Ministry of Petroleum's First Petroleum Exploration Conference, requests were made to "hasten drilling and oil testing at Heiyoushan and attempt to find out the commercial value of Heiyoushan Structure during the first half of the year."

On April 19, 1956, Li Jukui sent Assistant Minister Kang Shi'en, along with a group of more than 20 petroleum experts, including Soviet experts, to Karamay. During their stay of over three weeks, they surveyed sites, read information, heard reports, and had heated debates as to whether this place had oil. Building upon "The Overall Exploration Plan for Karamay Area," they developed a better plan: "To catch big fish, cast a big net." They deployed 10 drilling sections and decided to drill 20 exploratory wells for the first round. They telegraphed their plan to Li Jukui in Beijing. Li was very supportive. He immediately drafted a telegraph to Kang Shi'en, to Zhang Wenbin, the head of the Xinjiang Petroleum Company, and to Wang Qiren 王其人, the company's Party Secretary, and approved the plan. Accordingly, Xinjiang Petroleum Company immediately dispatched more reconnaissance crews to travel 150 kilometers northwest from Dushanzi to the appointed site. There were no roads at the time. Overcoming all kinds of obstacles like strong winds, lack of water, and well blowouts, they soon started exploration work.

5 Premier Zhou Enlai Gives Full Support

In 1956, the oil exploration in Karamay had the full attention and support of the country. Premier Zhou instructed relevant government agencies, provinces, and regions to give Karamay full support and help them solve any problems. Vice Premiers Chen Yun 陈云, Li Fuchun, and Bo Yibo, who were in charge of the national economy, held many hearings pertaining to the support system for Karamay. On September 5, 1956, *The People's Daily* published an editorial, "Support the Karamay and Qaidam Petroleum Areas (支援克拉玛依和柴达木油区)." The State Council organized its 13 ministries and commissions and the 16 provinces and municipalities to support Karamay. Large amounts of equipment were imported from the Soviet Union and eastern Europe. With unified coordination from the State Council, the Ministry of Petroleum quickly transferred troops and equipment from old oilfields in Yumen and Yanchang.

6 Transporting Materials with 1,000 Camels

Xinjiang Uyghur Autonomous Region formed a special support apparatus, organizing thousands of laborers to make roads. Tacheng District also formed a transportation team made up of 1,000 camels to transport materials for the oil industry. Donated materials from all over the country soon arrived. Large groups of young people, as well as graduates from colleges and universities, all flocked there. As the noted newspaper reporter Chu Anping 储安平 wrote in a long feature report, “The cars from Wulumuqi [Urumqi] thrust straight into Karamay Oilfield like a sharp sword.”

7 A Stirring Myth

With the entire country’s backing and an all-out effort by the Ministry of Petroleum, ten thousand oil workers converged on the Karamay area. They lived in cellars and tents, enduring a lot of hardship together. The backbone of the group consisted of around a thousand veterans who had retired from the PLA in 1952. Not intimidated by hardship, they dared to fight tough battles. In fine PLA tradition, they contributed greatly to the development of the oil industry in Xinjiang. By September 1956, 23 exploratory wells had produced oil. In the same month, when Minister Li Jukui announced the news at the Eighth National Congress of the Communist Party of China, it drew the attention of the congress and the whole nation. In 1958, when the Karamay Oilfield was being developed, it produced 330,380 tons of crude oil. By 1960, further exploration had established that the oilfield measured 290 km^2 in area. It produced 1.64 million tons of crude oil that year, becoming our country’s largest oilfield before the discovery of Daqing Oilfield. Zhu De, the PLA Commander in chief, visited Karamay in September 1958 and praised it: “Within three years, an oil city of 40,000 people was built upon the desolate Gobi Desert. This is a great achievement, as well as a stirring myth.” Afterwards, Karamay continued this myth. In 2014, it produced 14.38 million tons of oil and gas, making it the nation’s sixth largest oilfield.

8 Karamay, an Exclamation Mark

Liu Yuxi 刘禹锡 (772–842), a Tang dynasty poet, wrote “Through heaven and earth, the spirit of heroes / Inspires awe for thousands of years.” Savoring this poem, I have always felt that it describes the virtues of people of Karamay. The birth and development of Karamay Oilfield is the result of decades of their labor. What, then, are these virtues? I believe first that they loved their country and felt it was their own duty to develop the country’s oil industry. Second, they possessed an adventurous and creative spirit. Third, they maintained the tradition of hard work. Fourth, they were practical and realistic. For example, on September 25, 1958, the geologist Yang Zhenglu 杨拯陆, daughter of the

Nationalist General Yang Hucheng (who had been executed by Chiang Kai-shek), was doing oil exploration work with a team-mate in the wild when they were hit by a desert snowstorm and froze to death. She was only 22 years old. Premier Zhou Enlai eulogized her as "a hero for all time." She carried on her father's spirit of self-sacrifice and dedication to her country's independence and prosperity, and provided a role model for Chinese young people.

On October 20, 1999, after chatting over three rounds of drinks at his home, Karamay oil veteran and oil photographer Gao Rui 高锐 looked at the Karamay map on the wall and suddenly blurted out, "Karamay is an exclamation mark!" What a perfect description, I thought. The thousands of complex thoughts and emotions I have for Karamay Oilfield could all be summarized in this exclamation mark!

CHAPTER 9

High Hopes for Sichuan Oil and Gas

1 Sichuan Basin: Rich in Oil and Gas Resources

Sichuan Basin includes Sichuan Province, Chongqing Municipality, and some areas in Yunnan, Guizhou, and Hubei provinces. It is surrounded by mountain ranges such as Longmen 龙门, Dalou 大娄, Emei 峨眉, Wu 巫, and Daba 大巴, and its total area is 260,000 km^2. Sichuan Basin has marine and continental beds 6,000–12,000 m thick. Rich oil and gas resources are found in Paleozoic and Mesozoic strata. Due to crustal movements, there are many faults in this basin, and the oil and gas reservoirs are fractured. In addition, its strata are steeply tilted and hard, the oil and gas are buried deep, and bottom hole temperature and pressure are high. All of this increases the difficulty of oil and gas extraction.

2 Zhuge Liang Makes a Fire Well Burn Brighter with a Glance

Utilization of oil and gas in Sichuan has a long history. As early as the Western Han dynasty (206 BCE–9 CE), people were using natural gas in Qionglai 邛崃, in western Sichuan. By 616 CE, Linqiong had become "a county of fire wells." After 1041, mechanical punch drilling was invented and a set of unique natural gas extracting methods were developed.

During the Liu-Song dynasty (420–479), Liu Jingshu 刘敬叔 in his *A Garden of the Extraordinary* (*Yiyuan* 《异苑》), wrote, "There are fire wells in Linqiong County, in Sichuan. When the Han dynasty was prosperous, they burned very brightly. By the time of Emperor Ling 灵帝 of Han [r. 168–189 CE] and Emperor Huan 桓 [r. 132–168], the fire became weaker. With one glance, Zhuge Liang[1] made it burn more vigorously." What a vivid description Liu Jingshu had for the natural gas wells in Linqiong, Sichuan! He associated the prosperity of the natural gas production with the prosperity of the country and to whether the authorities attached importance to it. The broad view of the past and present has much to savor.

1 Zhuge Liang 诸葛亮 (181–234), chancellor of the state of Shu, was a statesman and strategist in the Three Kingdoms period (220–265). He became a symbol of resourcefulness and wisdom in Chinese folklore.

 | DOI:10.1163/9789004402737_010

3 Natural Gas Was Used to Boil Salt in Zigong

The natural gas-emitting artesian wells of Zigong, Sichuan are also parts of an ancient gas field. As early as 600 BCE, natural gas was being used to boil salt here. In 1770, a 560 m natural gas well was drilled here. By the end of the nineteenth century, daily natural gas production from this gas field already exceeded 1 million m^3. By the end of 1949, this ancient gas field had cumulatively produced more than 30 billion m^3, and this played an important role in the economic development of this area.

4 Petroleum Bureau of Ministry of Fuel Industries Sets Up Chongqing Office

After 1949, Sichuan's oil and gas industry received more attention and was further developed. In December, 1949, the Chinese Communist Southwest Military Committee's Industrial Department took over the Sichuan Oilfield Exploration Division and its Chongqing sales office and immediately resumed production at two gas wells and the drilling of another exploratory well. Noted geologist Huang Jiqing was appointed to establish the Southwest Geological Survey Institute for this area's oil and gas exploration.

On July 1, 1950, the Sichuan Oilfield Exploration Division and the Chongqing sales office were consolidated into the Chongqing Office of the Petroleum Administration Bureau, under the Ministry of Fuel Industries, and PLA veteran Jiao Yiwen 焦益文 was appointed its director. Following the policy, "Focus on restoring the existing foundation, explore and develop at key locations," set at the First National Petroleum Conference of April 1950, they concentrated on the southern part of the Sichuan Basin. In particular, they focused on natural gas production and on building a carbon black plant using natural gas as raw material. They made remarkable achievements. In May 1955, this division was expanded into the Sichuan Petroleum Exploration Bureau and relocated to Chengdu, with Jiao Yiwen as its director.

5 Mao Zedong Calls Yu Qiuli a Member of the "Children's Corps"

When Yu Qiuli was nominated to replace Li Jukui as Petroleum Minister by Defense Minister Peng Dehuai and recommended by Zhou Enlai, Mao Zedong had a talk with Yu Qiuli in late January, 1958. As for the content of their conversation, Yu Qiuli recalled, "The Chairman asked, 'How old are you?' I said,

'Forty-three.' The Chairman laughed, 'The Children's Corps!' The Chairman also said, 'Li Jukui is a very good comrade, but he's getting on in age. You're young and full of energy, so the Central Committee has decided to let the two of you switch places.'" Then, in late February 1958, Lieutenant General Yu Qiuli, the political commissar of the PLA's General Logistics Department, became the second Minister of Petroleum.

6 Yu Qiuli Organizes the Chuanzhong Oil Campaign

Soon after Yu Qiuli became Minister of Petroleum, he received much good news of high-yield wells in Chuanzhong (i.e., central Sichuan). On March 10, 1958, the Longnüsi-2 well (龙女寺) gushed out more than 60 tons; on March 12, Nanchong-3 (南充) produced more than 300 tons; and on March 16, Penglaizhen-1 produced more than 100 tons. These three successive "gushers" were more than 200 km away from one another, and the area of their respective structures each over 100 km. The whole Ministry of Petroleum was jubilant over the news. A Soviet expert when running into Yu Qiuli told him happily, "Ah, Minister, you're so lucky! Soon as you arrive, a big oilfield is found right away!" and "This oilfield in Chuanzhong is like a second Baku in our Soviet Union." At the site, the situation was even more exciting. Longnüsi-2 was located by a stream. Due to a lack of preparation, technology and equipment, when the well had a blowout, that section of the stream turned into a river of crude oil. Nationally, the campaign called the Great Leap Forward was well underway in 1958. Strongly optimistic sentiments permeated the oil industry. Everyone, including oil experts, all thought a big oilfield had been found in Chuanzhong.

Later that month, Yu Qiuli and Kang Shi'en visited Chuanzhong together. This was also Yu Qiuli's first time visiting an oilfield after he took charge of the Ministry of Petroleum. They visited the site of the three wells in Chuanzhong, as well as many oil units. On April 20, they held an on-site meeting with oil experts in Nanchong. Yu Qiuli decided to organize the Chuanzhong Oil Campaign and appointed Kang Shi'en as commander. They decided to concentrate the main force of the Sichuan Petroleum Administration Bureau and transfer personnel from oilfields in Yumen, Qinghai, and Xinjiang. They also enlisted help from Sichuan Province and relevant divisions of the PLA. The Nanchong, Penglaizhen, and Longnüsi structures were the main targets, and twenty key wells were to be drilled at seven structures.

In October of the same year, the Ministry of Petroleum held an on-site meeting in Karamay in Xinjiang. News came in from the Chuanzhong front lines. Of the 20 key wells, 19 had been drilled; few produced oil and, in addition, there was a significant disparity in output. Production at some wells drastically fell off. Some seemed stable, but once shut and re-opened, they would not produce a drop of oil. Faced with such a complex situation, Yu Qiuli decided to transfer more manpower and continue the campaign in order to find the oilfield as soon as possible. After the campaign had gone on for nearly a year, only a few of the 37 exploratory wells that were drilled produced oil. One small oilfield with 100,000 tons annual production was found. Most wells either did not produce oil or produced oil sporadically. Although this campaign proved there was oil in Chuanzhong, the geological studies found that the oil reservoirs were compact, fractured, and had disadvantageous physical characteristics. They did not find any continuous, relatively high-yield, stable oilfields. Thus in March 1959, the Ministry of Petroleum decided to end the Chuanzhong Oil Campaign. In April, the oil teams left, leaving the employees of the Sichuan Petroleum Administration Bureau to continue work on their own.

7 Mao Zedong Encourages Yu Qiuli, "When the East Is Dark, the West Is Light"

From April 2 to April 5, 1959, the Seventh Plenum of the Eighth Central Committee of the Chinese Communist Party was held in Shanghai. During the meeting, Mao Zedong asked Yu Qiuli, "How is the situation in Sichuan?" With a heavy heart, Yu said, "I am sorry to report to the Chairman that the situation in Sichuan is not good. After exploration, we have learned that the oil reservoir is shallow, and that the output is low and decreases fast. No big oilfield has been found." Mao responded, "If no oilfield is found there, then look elsewhere. When the east is dark, the west is light. China is so big that oil will be found."

On March 27, 1958, Mao Zedong paid a surprise visit to the Longchang Gas Field. After visiting the carbon black shop, he commemorated it with calligraphy: "There is great hope in Sichuan." This hope has finally been realized. In 2014, Sichuan produced 13.7 billion m^3 of natural gas and 150,000 t of oil. Sichuan has become China's eighth largest oil and gas field.

CHAPTER 10

The Anti-Rightist Movement and Great Leap Forward Harm the Petroleum Industry

The Anti-Rightist Movement, which started in 1957, and the Great Leap Forward, which started in 1958, harmed the petroleum industry. The Anti-Rightist Movement attacked intellectuals and blocked the pathways of speech, silencing everyone. Afterward, in the oil industry, and especially among the oil geologists, there were no differing opinions or debates. Everybody followed the leader, which turned out to be extremely bad for the oil and gas industry. Even though the Great Leap Forward had a positive side in that it encouraged oil workers to construct socialism, its negative impact was huge. Exaggerating and setting unrealistically high standards became rampant in the oil industry. "The bolder people are, the more the wells produce." Oil companies were coming out with one after another unrealistic production plan. These were the pitfalls of a planned and centralized economy.

1 Deputy Chief Geologist Chen Ben Commits Suicide

I remember when, on June 1, 1966, the *People's Daily* printed the editorial, "Wipe Out Class Enemies of All Stripes (横扫一切牛鬼蛇神)," on its front page. On the heels of this bit of agit-prop, the Cultural Revolution came fast and furious. At the time, I was working in the office of the Political Department of the Ministry of Petroleum, in charge of gathering information on political work in all oilfields. To this day, I remember clearly that on June 16, 1966, I received an unexpected phone call from the Political Department of the Qinghai Petroleum Administration Bureau: the night before, Chen Ben had hung himself outside of his tent in Lenghu Oilfield.

Chen Ben, who had been an underground Communist during his student years, had graduated from the Geology Department of Qinghua University in 1939. After graduation, he worked for the Central Geological Survey of the Ministry of Agriculture and Commerce under the Nationalist government. The Survey Director, Dr. Weng Wenhao, sent him to work in the newly-discovered Yumen Oilfield. He then worked in the United States as a trainee for two years. With this solid technical background, he perfected his skills and had new and unique perspectives. As one of the first important group of Chinese geologists

 | DOI:10.1163/9789004402737_011

FIGURE 10.1
Geologist Chen Ben

who had propounded the "terrestrial facies of petroleum" hypothesis, he contributed significantly to our understanding of petroleum geology, especially the petroleum geology in northwestern China. When the Ministry of Petroleum was founded in 1955, he was appointed its first Deputy Chief Geologist. He was so talented that people were jealous of him, so in 1957, during the Anti-Rightist Movement, he was branded as a rightist. He was expelled from the Chinese Communist Party (CCP), demoted to a regular geologist, had his salary reduced, and exiled to the remote Lenghu Oilfield in Qaidam Basin to work and be "reformed" under supervision.

As for his later life in the *laogai* 劳改 (reform through labor) system:

> Of the more than 2,000 wells that had been drilled in Qaidam Basin in the 30 years beginning from the 1940s, none had been drilled to the Lower Jurassic stratum. Chen Ben, during his exile in Lenghu Oilfield, quietly studied the Jurassic system and announced that if a big oilfield were to be discovered in Qaidam, it would definitely be in the Triassic and Jurassic strata. The recent oil discoveries of Qaidam Basin verified his hypothesis, so that people now regard him as the "Father of the Jurassic System." Thus in the public graveyard in Lenghu Oilfield lies Chen Ben, who gave his life for China's oil industry.[1]

1 Bainian shiyou bianxie zu 百年石油编写组, ed., *Bainian shiyou* 《百年石油》 [*One Hundred Years of Petroleum*] (Beijing: Dangdai Zhongguo chubanshe, 2002), 417.

2 "Holding a Golden Bowl to Beg for Food"—Rightist Speech

Sichuan Petroleum Administration Bureau deputy chief Jiao Yiwen was branded as being anti-Party for criticizing Soviet experts as "holding a golden bowl to beg for food." This is a perfect example of how during the Anti-Rightist Movement people who held divergent points of views from those in power were treated as enemies.

On April 19, 2000, while I was visiting Sichuan Petroleum Administration Bureau in Chengdu, Dai Chongli 代崇礼, the manager of the Historical Records Division, Yi Huiting 易辉庭, manager of the Personnel Division, and senior engineer Xiao Zhisheng 肖之盛 told me the whole story.

After the founding of the new China, Soviet petroleum geologists greatly assisted and shaped oil exploration in Sichuan. According to these experts, in order to look for big oil and gas fields, one had to go onto "the platform," i.e., Chuanzhong or central Sichuan (most of Sichuan's natural gas is in Chuannan or southern Sichuan): the oil and gas are normally in porous reservoirs, and since Sichuan already had natural gas, it should also have big oilfields. Following the Soviets' advice, exploration shifted from its initial focus on natural gas and instead refocused on oil. This shift is illustrated in official records: between 1950 to 1957, total investment in oil and gas exploration was 210 million yuan, of which 60% was used for oil. Starting in 1956, the focus of oil exploration was turned to Chuanzhong.[2]

It was necessary to make the practical choice between focusing on gas or oil in Sichuan Basin. A debate on the matter would have been healthy under the circumstances. At the time, Jiao Yiwen held the view that the focus should be on gas—and it must be added that people in the Sichuan Basin have always focused on gas. History has proven Jiao Yiwen's view correct.

Jiao Yiwen, who was born in 1924, came to Sichuan with the PLA's Second Field Army in 1949. At the end of 1949, he got transferred to a job in the oil industry in Sichuan. He already had some education, but after his transfer, he studied hard to learn more about petroleum geology. He explicitly proposed that Sichuan should concentrate on finding natural gas and even wrote a rhyming essay in the style of *The Three-character Classic* about it, but he mainly criticized the Soviet experts' focus on finding oil, instead of gas, saying it was

2 Statistics from Sichuan youtian shiyou dizhi bianxie zu, ed. 四川油田石油地质志编写组, *Zhongguo shiyou dizhi zhi (juan shi): Sichuan diqu* 《中国石油地质志（卷十）：四川地区》 [*Petroleum Geology in China (Vol. 10): The Sichuan Region*] (Beijing: Shiyou gongye chubanshe, 1989).

like "begging for food with a golden bowl," by which he meant that since it was already known that Sichuan was rich in natural gas, they should concentrate on that. At the time, the foreign policy of the Central Committee of the Chinese Communist Party was overwhelmingly pro-Soviet: "Today's Soviet Union will be our nation's tomorrow." If you criticized the Soviet Union, then you were anti-Soviet, and if you were anti-Soviet, then you were anti-Party. Thus Jiao Yiwen was branded a rightist, expelled from the Party, and demoted to do manual labor. His wife died and his family broke up. A few years later, even though his rightist "cap" was taken off him, the record was still in "other books" (*lingce* 另册), meaning that he was still regarded as disreputable. In 1982, Jiao pleaded to Deng Xiaoping through his brother-in-law, the Deputy Foreign Minister Fu Hao 符浩. After Deng's intervention, Jiao was appointed as Vice-president of Shandong Petroleum Institute. After he retired, he returned to Chengdu. Despite this resuscitation, all the previous calamities had taken their toll, and he committed suicide in 1985.

3 Skewed Logic

In 1982, Zhao Yi 赵一, the Director of the Training Division of Huabei Oilfield, told me this story. Zhou Shicheng 周世城, the Secretary of the Communist Party Committee of the Machinery Plant of Yumen Oilfield, was branded a rightist. Zhou had originally been the political commissar of the 3rd Regiment of the PLA Petroleum Division. In 1952, he retired from the PLA and was transferred to Yumen Oilfield as the Party Secretary of the Machinery Plant, which was Yumen Oilfield's largest subordinate unit. During the later part of the Anti-Rightist Movement in 1957, Xinhua Bookstore in Lanzhou, Gansu Province held a conference at Yumen Oilfield, soliciting suggestions. Zhou attended the conference and proposed, "I would like to ask that the provincial Xinhua Bookstore come more often to our machinery plant, so that our employees could buy books they want to buy …" The Party Committee of the oilfield learned of his suggestions. After discussion, they decided to label him as a rightist in order to satisfy the rightist quota that the provincial authorities had given them. He was expelled from the Party and sentenced to "reform through labor." The rationale used was that Xinhua Bookstore was a Party-led institution, so that if you had an opinion about it, it meant that you were opposed to the Party and a rightist. After the fall of the Gang of Four in 1976, Zhou was reinstated and transferred to act as Director of the Supply Division in Huabei Oilfield.

4 "If the Party Has an Extra Bun, How Can You Object?"

In 1961, while I was working in the Political Department in Daqing Oilfield, I personally recorded the following story, which happened in an oil production crew in the First Oil Production Plant in Daqing Oilfield. At the time, these oil production crews were scattered across the wild prairie. China had just experienced three years of economic difficulty. Each crew had its own kitchen, and food was rationed. At one crew meeting, one young apprentice worker said, "The chef in our kitchen is no good. At every meal he always gives the Party Secretary an extra steamed bun ..." The master worker under whom he was apprenticed, who had come from Shaanxi, scolded him, "You insolent brat, how could you be so thoughtless? If the Party has an extra bun, how can you object?" The master worker had equated objecting to the extra bun to being anti-Party and was trying to shield his apprentice from getting into trouble.

5 Too Fast, Too Soon at the Ya'erxia Oil Deposit

At the height of the Great Leap Forward in 1958, the Ministry of Petroleum set unrealistically high production targets for both Yumen Oilfield and Karamay Oilfield. Thus, for instance, at Ya'erxia 鸭儿峡 oil deposit, which had just been discovered in Yumen in 1956 and had the highest single-well yield, the choke was opened up to let more oil gush out, in order to reach the exaggerated goals. Within half a year, the output from this oil deposit dropped off dramatically. Its director, Zhao Zongnai 赵宗鼐, opposed this practice, so he was labeled a rightist.

Because the newly-discovered Karamay Oilfield also extracted too much oil too quickly, in 1959 it had to adjust its production plans completely.

6 The Chuanzhong Oil Campaign: Emblematic of the Great Leap Forward

In the first half of 1958, as the Great Leap Forward was picking up steam, everyone, everywhere was reporting good news in the oil industry. Such boasting and exaggeration was emblematic of the Great Leap Forward. That year, there were roughly 250 items of good news: oil seepage found on a hill at such-and-such place, tar sand found in such-and-such place, and oil shale found at such-and-such cliff. In Chuanzhong, three of Sichuan Petroleum Exploration Bureau's exploratory wells were gushers, which greatly excited

Yu Qiuli and Kang Shi'en, so they hurriedly organized the Chuanzhong Oil Campaign. The Great Leap Forward was the campaign's ideological foundation, so it reflected all the idealism and flaws of that movement.

More than ten years later, in the early 1970s, Kang Shi'en proposed "building a dozen Daqings by the end of the twentieth century," each with an annual production of 50 million tons. The historical origin of this conception, however, was the Great Leap Forward of 1958. On the one hand, it showed that oil workers, like everyone else in the country, longed for the country to become prosperous, but on the other hand, it was the inevitable outcome of a planned economy and monopoly system. We shall discuss the origin of "a dozen Daqings" in more detail in Part 4, Chapter 18 below.

PART 3

The Daqing Era (1960–1969): China No Longer "Poor in Oil"

During this decade, the Chinese petroleum industry was led by the Daqing Oil Campaign, which started in 1960; the industry concentrated its forces and pressed on without letup. After achieving a victory with the Daqing Oilfield in the Songliao Basin, it went on to other victories with Shengli 胜利 Oilfield in Shandong and Dagang 大港 Oilfield in Tianjin. The nation's oil production grew to 11.31 million tons in 1965, disproving the assertion that "China is poor in oil." The country became self-sufficient in crude oil and all oil products. Although the Cultural Revolution started in 1966, oil workers continued to move ahead. In 1970, the nation produced 30.65 million tons of oil. During this decade, annual production increased on average by 2.8 million tons per annum.

China and the Soviet Union were at odds during this period. An Anti-Soviet Revisionist Movement swept through the country. Under its influence, enterprises in the oil industry abandoned management methods and regulations learned from Soviet Union and suffered as a result.

CHAPTER 11

The Search for Daqing Oilfield

1 Geologists Say the Songliao Basin Has Oil

Petroleum and natural gas are found mostly in sedimentary basins, and this is also the case with Daqing Oilfield. The Songliao Basin 松辽盆地 is a large sedimentary basin surrounded by the Greater Khingan Range 大兴安岭, Lesser Khingan Range 小兴安岭, and Changbai (Paektu) Mountains 长白山. The basin spans the provinces and regions of Heilongjiang, Jilin, Liaoning, and Inner Mongolia, with its main part in Heilongjiang and Jilin, covering a total area of 260,000 km^2. It is called Songliao Basin because it has Songhua River to the north and Liao River flowing in the middle of it in the south. Geologically speaking, 70 million years ago, during the Jurassic and Cretaceous Periods of the Mesozoic Era, this area used to be a gigantic inland lake. Inside and around the lake, primitive plankton, flora, and fauna were abundant. By the Cenozoic Era, a lot of sediment had accumulated, providing the rich material foundation that would become petroleum. Later, the lake gradually turned into grassland and marshes.

This bountiful basin is situated in our country's northeast. On its vast land, rich pastures and resources abound. However, the question as to whether oil and gas lay beneath this territory drew the interest of many petroleum workers and geologists.

In the early twentieth century, geologists from Japan and the American Mobil Corporation conducted geological surveys and exploration in this basin and the surrounding area but did not find any oil or natural gas field. Since the big oilfields of the major oil-producing countries sit on marine sediment and such areas are few in China, which sits mostly on continental sediment, some petroleum geologists concluded that "China is poor in oil" and "continental facies are poor in oil." Because Songliao Basin sits on the continental sediment of an inland lake, they therefore predicted that, judging from rock type and age, most of China's northeast area had no oil-bearing possibilities.

However, many geologists held a different point of view. In the spring of 1999, Yang Jiliang 楊继良, Daqing Oilfield's deputy chief geologist, told me a few years earlier he had come up with a rough estimate that since the beginning of the twentieth century, more than 30 eminent geologists had written books or articles rebuffing statements that "China is poor in oil" and "Songliao is poor in oil." For instance, Li Siguang, the famous geologist who became new China's

 | DOI:10.1163/9789004402737_012

first Geology Minister, in an article written in 1928, said that "Mobil's failure does not prove that there is no oil to extract in China."[1] Pan Zhongxiang, Huang Jiqing, Zhang Wenyou, Xie Jiarong, and Weng Wenbo also predicted that there was oil in China's continental sediment and in Songliao Basin.

The co-existence of two different theories brings into relief the pros and cons of both. There is a popular saying that goes, "If everyone stands on one side, it is not necessarily a good thing—this is the case on a ship as well as in a society." This is also the case in petroleum geology. The way I see it, the question of whether China had oil or not was a technical, practical question, and no matter who was right or wrong, no one had a corner on the absolute truth. The key with such issues is that decision-makers should comprehensively survey the whole situation, listen to different ideas and suggestions, endeavor to make realistic decisions and when needed, modify them in a timely fashion. They should not always lean towards one side or politicize the issues. This is the correct way of resolving theories in petroleum geology.

2 The Nan-17 Well: the Spring-Heralding Flower for Songliao Basin

While the debate raged as to whether or not Songliao Basin had oil, people started exploring, especially after 1949. In 1951, Liu Guochang 刘国昌 of the General Petroleum Administration Bureau of the Ministry of Fuel Industries conducted a tar sand survey on the banks of Zhalai Lake (Dalai Nur) 扎赉湖 in Manzhouli, Inner Mongolia. In 1953, the Ministry of Geology and the General Petroleum Administration Bureau invited the geologists Huang Jiqing, Xie Jiarong, Weng Wenbo, et al. to make a "Map of Prospective Hydrocarbon-bearing Areas of China" (中国含油气远景区划图); they listed Songliao Basin as the most promising area of hydrocarbon concentration. In 1953 and 1954, following leads from many local reports, the General Petroleum Administration Bureau's Zong Pisheng 宗丕声, Qiu Zhenxin 邱振馨, Zhang Chuangan 张传淦, Chen Lianghe 陈良鹤, Tang Zukui 唐祖奎, and other geologists conducted many geological surveys in Shangzhi and Yilan Counties, Heilongjiang Province, Fuxin County, Liaoning Province, and Antu County, Jilin Province. Besides confirming the presence of oil seepage, tar sand, and oil shale, they developed a preliminary understanding of the strata on the eastern border of Songliao Basin.

1 Li Siguang 李四光, "Ranliao de wenti 燃料的问题 [The Fuel Problem]," *Xiandai pinglun* 《现代评论》 7.173 (1928): 8–12.

In 1955, the Ministry of Petroleum was founded. At the same time, the State Council specifically divided up the work of oil exploration throughout the country: the Ministry of Geology and Chinese Academy of Sciences were responsible for general geological surveys and scientific research, and the Ministry of Petroleum was responsible for detailed surveys and drilling in key areas. Consequently, the Ministry of Geology further increased its general geological surveys of this basin. In August 1955, the Northeast Geological Bureau of the Ministry of Geology formed a six-person group. Headed by Han Jingxing 韩景行, the group did geological reconnaissance, measured profiles, and investigated oil seepage. In February 1956, the Ministry of Geology established the Songliao Petroleum Survey Team, which used various geophysical methods to conduct large-scale geological surveys in the basin. In 1957, the Geophysical Prospecting Bureau of the Ministry of Geology dispatched more technical personnel to form the Changchun Geophysical Prospecting Team. By the end of 1957, the Ministry of Geology had finished its preliminary overall survey of Songliao Basin. Combining geophysical prospecting and drilling information, they drafted a report on Songliao Basin's geotectonic outline and oil prospects. On April 17, 1958, the Ministry of Geology's Songliao Petroleum Survey Team, led by its Party secretary Zhang Liye 张立业, found oil sandstone at the Nan-17 well in Qian Gorlos County, Jilin Province (the Mongolian name, meaning "south of the [Songhua] River," is also the source of the well's name [*nan* 南 = south]) and oil-bearing strata at Huaide 怀德, Gongzhuling 公主岭, and other locations. In complete contrast to previous sightings of oil seepage, tar sand, and oil shale, this discovery proved that underneath this basin there existed a large oil reservoir. It was the first "spring blossom" announcing that Songliao Basin had oil.

Meanwhile, the Ministry of Petroleum also increased its efforts in the basin. In early 1957, it ordered its subsidiary, the Geological Investigation Division in Xi'an, to form a comprehensive geological research team—the 116 Geological Team—to be in charge of petroleum geological investigation and research in the Songliao Basin. Starting in March 1957, this seven-person geological team, under its leader Qiu Zhongjian 邱仲健, braved continuous rain and severe cold as it made ground reconnaissance of mountains, rivers, marshes, and lakes throughout the northeast. In Beijing and Changchun, they worked day and night, building upon rich and solid information gathered by colleagues from the Ministry of Geology; they pooled relevant materials in the Ministry of Petroleum, the Ministry of Coal, the Chinese Academy of Sciences, as well as materials gathered in the northeast area prior to the Liberation of 1949. By the end of 1957, after repeated comparisons and analysis, they produced a "Map of Songliao Basin Oil Prospects" (松辽盆地含油远景图), proposed specific deployments for geophysical prospecting, and offered suggestions on the location of key exploratory wells.

In 1987, Yang Jiliang, the geologist who rendered outstanding service in finding Daqing and the former Deputy Chief Geologist for Daqing wrote, "The work done by the Ministry of Geology's Songliao Petroleum Survey Team and the Ministry of Petroleum Xi'an Geological Investigation Division's 116 Geological Team played a pioneering role, which set the stage for larger scale petroleum prospecting in Songliao Basin."[2]

3 Songliao Basin Becomes a Focus

The achievements of the various geological survey teams described above prompted the Ministry of Petroleum to shift the focus of its oil-finding efforts to the Songliao Basin. At the time, the geologists in the Ministry of Geology and the Ministry of Petroleum often exchanged information on the Songliao Basin and discussed plans and ideas on how to conduct further petroleum prospecting. Based on those ideas, starting in 1956, experts and leaders at the Ministry of Petroleum would write articles, give speeches, and write formal reports urging that the Songliao Basin become a focus area for oil prospecting.

For example, in January 1956, at The First National Petroleum Prospecting Meeting held by the Ministry of Petroleum, Kang Shi'en, the Assistant Minister in charge of petroleum exploration, proposed in his report that in Songliao Basin, as one of the oil-bearing areas in the nation, "geological investigation should begin immediately."

In February 1956, when Petroleum Minister Li Jukui made a report to Mao Zedong and other top leaders and discussed the prospects of the nation's petroleum industry, he named fifteen areas that possibly had oil, including Songliao Basin.

In 1957, during the Ministry of Petroleum's annual meeting on oil exploration, the Deputy Chief Geologist Chen Ben presented a report entitled "Experiences from the Past Seven Years of Exploration, and Future Directions." He proposed that Songliao Basin be listed as one of the five key areas for increased exploration in the Second Five-year Plan.

2 Yang Jiliang 杨继良, "Daqing youtian de faxian 大庆油田的发现过程 [The Process of Discovering Daqing Oilfield]," in *Daqing youtian de faxian* 《大庆油田的发现》, edited by Daqing shi zhengxie ziliao yanjiu weiyuanhui 大庆市政协资料研究委员会 (Harbin: Heilongjiang renmin chubanshe, 1987), 24.

On February 27 and 28, 1958, Vice Premier of the State Council Deng Xiaoping presided over the Ministry of Petroleum's meeting on the coordination of oil exploration during the Second Five-year Plan; this was also a transitional meeting for the outgoing and incoming ministers. The newly-appointed Petroleum Minister Yu Qiuli attended the meeting. At the meeting, the outgoing minister Li Jukui first reported on the petroleum industry's basic status. Tang Ke, chief of the Geological Exploration Bureau of the Ministry of Petroleum reported on coordinating petroleum exploration during the Second Five-year Plan. Consensus was reached that that the Ministry should shift the focus for its oil prospecting to the northeast for the Second Five-year Plan. In the two weeks that followed, Li Jukui and Yu Qiuli began making the arrangements and arrived at two concrete decisions:

First, for 1958 and for the Second Five-year Plan, ten areas nationwide would be the foci for oil exploration efforts. Of these ten, five were the old areas of Junggar (Karamay) in Xinjiang, Qaidam in Qinghai, the Hexi Corridor (河西走廊 i.e., Yumen) in Gansu, Sichuan Province, and the Ordos Basin, which stretches across Shaanxi, Gansu, Shanxi, Ningxia, and Inner Mongolia. Five were new areas: Songliao Basin, northern Jiangsu and Shandong (both belonging to Bohai Bay Basin), Guizhou, and Turpan Basin. Among the five new areas, primary focus would be on the Songliao and Bohai Bay Basins.

Second, exploration infrastructures were created and exploration teams dispatched to the new areas. In 1958, the Xi'an Petroleum Geological Investigation Division was dissolved. In its place, four new petroleum exploration divisions were established: Songliao Petroleum Exploration Division, Huabei Petroleum Exploration Division, Ordos Petroleum Exploration Division, and Guizhou Petroleum Exploration Division. In June of the same year, Songliao Petroleum Exploration Division was promoted in status and became the Songliao Petroleum Exploration Bureau. The Ministry's Labor and Salary Bureau Chief Li Jinghe 李荆和 (senior underground Party member and prefectural Party secretary when new China was founded) was transferred to become the Director and Party Secretary of the new bureau, whose office was located in Changchun, Jilin Province. Since there were few geological personnel in Songliao Petroleum Exploration Bureau, the Ministry of Petroleum decided to transfer a group of outstanding geologists from Beijing Petroleum Institute, including its Chief Geologist Yu Boliang 余伯良. Thus within three months, its institutional norms were upgraded twice, and it had formed its own geological research team, which demonstrated the Ministry of Petroleum's eagerness to make advances in Songliao Basin. Meanwhile, construction teams from several oilfields in the northwest were transferred; the Ministry of Petroleum and the three northeastern provinces also transferred cadres and workers to support the exploration efforts. Songliao Petroleum Exploration Bureau formed 32 technical teams (e.g., geological survey teams, seismic teams, gravity and magnetic survey teams, electrical teams, geophysical data research team, drilling teams) that comprised nearly 2,000 people. They had two big rigs, one medium rig, and a wide array of other equipment. When the Ministry of Petroleum finished with its deployment of personnel, the Songliao Basin had become one of the key areas for oil prospecting.

4 Prelude to the Eastern March: Capturing Songliao's Central Uplift Belt

The Ministries of Geology and Petroleum worked hand-in-hand throughout 1958 and achieved remarkable success. The Ministry of Geology's Songliao Petroleum Survey Team and Changchun Geophysical Prospecting Team did much general surveying with various seismic, electrical, and gravitational geophysical prospecting methods, and they drilled 276 shallow wells, totaling 71,400 m. Crews from the Ministry of Petroleum conducted detailed surveys and drilled three deep wells, totaling 4,360 m. The large-scale effort of the Ministry of Geology and the detailed surveys of key areas by the Ministry of Petroleum in 1958 resulted in the discovery in the middle of Songliao Basin of a 60,000 km^2 central uplift belt, which starts in Lindian 林甸, Heilongjiang in the north and extends to Changling 长岭, Jilin in the south, and from Taikang 泰康, Heilongjiang in the west to Fuyu 扶余, Jilin in the east. This uplift belt had multiple structures, and what later came to be called the "Daqing Placanticline" (*Daqing changyuan* 大庆长垣) lies on this gigantic uplift belt. In November 1958, a group of eminent Soviet and Chinese petroleum geologists that included people like Onegin Ivanovich Brod and Weng Wenbo made a special trip to Songliao Basin. Foreign and domestic experts agreed that the Songliao Basin had many advantageous conditions for finding oil, and that its central uplift belt was the most promising area. They suggested the use of all kinds of prospecting means: creating a few big profiles across key areas of the basin, choosing a group of local structures, and starting drilling in hope of obtaining commercial oil flow. Kang Shi'en thought their suggestions were insightful and arranged for relevant experts to invite colleagues from the Ministry of Geology to study the plans carefully together. In early 1959, the Ministry of Petroleum convened a petroleum industry conference, in which Songliao Petroleum Bureau Chief Li Jinghe 李荆和, Deputy Song Shikuan 宋世宽, Chief Geologist Yu Boliang, and Directing Geologist Zhang Wenzhao 张文昭 put forward the bureau's exploration working plan for 1959. The main content of the annual plan was to make four large, comprehensive profiles across the central uplift belt, open up two exploration areas, conduct detailed surveys on ten structures, and drill eight key wells. The Ministry of Petroleum approved the plan. Thus played the prelude to the eastern march to Songliao Basin.

5 Determination of Songji-3: a Concerted Effort by the Ministries of Petroleum and Geology

On September 26, 1959, oil gushed out of Songji-3 松基三井. As the sound of thunder announces spring, this announced, "There may be an oilfield here." Daqing Oilfield was born!

Songji-3, which is an abbreviation for "Songliao Basin Key Well No. 3" 松辽盆地第三号基准井, is situated near Gaotaizi Village, Datong Town, Zhaozhou County (肇州县大同镇高台子村) in Heilongjiang. To determine its position, geologists from the Ministries of Geology and Petroleum walked across vast areas on both sides of Songhua River and traveled back and forth more than 10,000 km between Harbin, Changchun, Beijing, and even to Yumen in Gansu and Nanchong in Sichuan.

The actual process of determining the position of this single well took one year. It began in February 1958, when the Ministry of Petroleum's Xi'an Geological Investigation Division's 116 Team submitted its "Songliao Basin Geological Survey Report and Maps (松辽盆地地质调查研究报告及其图件)," in which they predicted that Songliao Basin had oil and suggested that a key well be drilled in the Songhua River area, at a position a little south of the later Songji-3. In April 1958, when the Ministry of Petroleum held an "on-site meeting" at Nanchong, Sichuan, it discussed enhancing Songliao Basin oil exploration and where to drill the key wells, based on the 116 Team's recommendations.

In June 1958, the Ministry of Petroleum set up the Songliao Petroleum Exploration Bureau and Geological Research Team, which was headed by Chief Geologist Yu Boliang. The Key Well Research Team, headed by Zhong Qiquan 钟其权 (1933–1982), raised a plan to drill a group of key wells in Songliao Basin. Songji-1 and Songji-2 were drilled that year, but with no result. In July 1958, at the Ministry's meeting in Yumen, the Songliao Petroleum Bureau proposed drilling eight more key wells in 1959; of these, Songji-3 was the first.

In August 1958, the Ministry of Geology's Changchun Geophysical Prospecting Team wrote to Songliao Petroleum Bureau, saying that after discussing the issue with relevant companies, they recommended changing the location of Songji-3 to Kaitong County, Jilin Province. Songliao Petroleum

FIGURE 11.1
Geologist Zhong Qiquan

Bureau immediately dispatched head geologist Zhang Wenzhao, geologist Yang Jiliang, and Key Well Team leader Zhong Qiquan, and others to discuss and review geological information with them. Even though they did not reach consensus on the location of the key well, all agreed to let the Ministry of Geology's Changchun Geophysical Prospecting Team immediately deploy survey lines, in order to accelerate the work and obtain as soon as possible the most up-to-date seismic information from the vicinity of the town of Datong.

On September 3, 1958, the Ministries of Geology and Petroleum held a joint forum in Changchun to issue of Songji-3's location. Han Jingxing of the Songliao Petroleum Survey Team and Zhu Dashou 朱大绶 of Changchun Geophysical Prospecting Team (both of the Ministry of Geology), as well as the Ministry of Petroleum Songliao Bureau's Zhang Wenzhao, Yang Jiliang and Zhong Qiquan, and others attended the meeting. At the meeting, geologists from the various units reached a consensus of placing the key well at an uplift detected by electronic prospecting near Gaotaizi Village, in Datong Town. The structure was later named Gaotaizi Structure. The Songliao Petroleum Bureau reported this suggestion to the Ministry of Petroleum, and Kang Shi'en gathered experts to discuss it. They believed that the evidence was still insufficient and asked Songliao Petroleum Bureau to submit more. From mid-September to mid-November, Zhong Qiquan and other colleagues from the Songliao Petroleum Bureau twice went to the Ministry of Geology in Beijing and the Geology Bureau of Jilin Province in Changchun and Datong Town, gathering information and conducting on-site geological reconnaissance. All three units confirmed that the new seismic and electronic prospecting data fully supported the presence of the uplift. On November 14, new data and supplementary opinions on the key well position were submitted to the Ministry of Petroleum.

On February 8, 1959, during the Chinese New Year, Kang Shi'en met with Petroleum Ministry geologists, bureau chief Li Jinghe, chief geologist Yu Boliang, head geologist Zhang Wenzhao, geologist Zhong Qiquan, and others from the Songliao Petroleum Bureau, who had come from Changchun to Beijing for the occasion. They devoted three days of their holiday to discuss the key well's position. On the third day of the New Year, the meeting lasted through the night until dawn. They were extremely cautious. On the fourth day, Yu Qiuli and Kang Shi'en, Shen Chen, deputy of Geological Exploration Bureau, and other geologists paid a special visit to the home of He Changgong 何长工, Deputy Minister of Geology and a Party founding member, to attend a meeting between both ministries, the "Songliao Basin Petroleum Exploration Cooperation Meeting." Deputy Minister of Geology Kuang Fuzhao 邝伏兆 and Petroleum Bureau Chief Meng Jisheng 孟继声 also attended. Besides discussing Songji-3's location, they discussed petroleum exploration and deployment throughout the Songliao Basin and the division of labor of the two ministries. After the two ministries reached a unanimity of views, Yu Qiuli pointed to the geological drawing of Songliao Basin and said: "We will overcome Songliao in three years!" He Changgong immediately responded, "Our goals are identical. You'll be responsible for drilling these key wells, including Songji-3, and we'll be responsible for completing these four seismic profiles. Let's overcome Songliao as soon as possible!" People burst out laughing. Mrs. He then took out fruit and desserts to celebrate the occasion.

The whole process of determining the position of Songji-3 was a process of concerted planning, repeated exploring, and cautious, rigorous scientific deliberation. In the ensuing oil campaign, Daqing Oilfield was found.

6 A Decision against the Teacher's Will

The 32118 Drilling Crew, responsible for drilling Songji-3, was an excellent team that had been transferred from Yumen Oilfield. After crew leader Bao Shizhong 鲍世忠 and crew political instructor (retired PLA major) Shen Guangyou 沈广友 took the assignment, they started moving their operation in the winter of 1958 from Songji-1, which was 126 km away, to Songji-3. Despite the lack of roads and large-scale transportation equipment as well as the necessity of operating in the open and enduring extreme cold and other hardships, they completed the move in early April 1959. Drilling of Songji-3 started on April 11. To expedite the process, cores were not extracted before 1,000 m were drilled. In July, after they reached 1,050 m, cores were extracted continuously. However, the design required the extraction of 411 m of cores, but because the coring tools were dated, they extracted only 202.51 m of cores, of which 3.15 m were saturated with hydrocarbons and smelled strongly of oil. Other logs and electronic prospecting data also showed hydrocarbons. When a depth of 1,400 m was reached, there was a 5.7° hole deviation, which severely impacted the drilling speed.

At this time, in mid-July, Kang Shi'en accompanied Mikhail Mirchink, chief geologist of the Soviet Ministry of Petroleum Industry, and a group of Soviet experts in our Ministry of Petroleum on a trip to Songliao Basin. They had first heard reports from the Songliao Petroleum Bureau in Changchun. Then in Harbin, they met with geologists who had come from Songji-3, and they reviewed information and oil sand samples. Not considering the information and samples complete, Kang instructed Songji-3 to immediately start sidewall coring and electrical testing. In two days, samples and information were sent to Harbin again and suggestions were made to finish drilling ahead of schedule and test for oil.

After seeing the information and samples, Mirchink and other Soviet experts were very excited, telling Kang that if such a situation presented itself in the Soviet Union, it would be an occasion for a toast. In terms of whether that well should finish its drilling, Mirchink had a disagreement with Kang. He believed that the design of the well should not change; 3,200 m must be finished, and then each stratum should be tested, so as to complete the key well's task of understanding the strata, whereas Kang agreed with the drilling crew, believing that they should stop drilling and start testing for oil. Mirchink was the most authoritative petroleum geologist in Soviet Union. When Kang visited and studied in the Soviet Union in 1955, the two of them were good friends and Kang regarded him as a teacher. He patiently explained to Mirchink that since this well had such good signs of hydrocarbon that they should find out as soon as possible whether commercial oil flow was present. As for the well's other functions, they could be delegated to other key wells to finish. In addition, since the well bore was heavily inclined, it would take more than a year to drill to 3,200 m. If the well bore was soaked by mud for too long, the oil reservoir would be clogged, delaying the time for finding the oilfield. Mirchink turned a deaf ear to Kang and became very angry, while Kang stood his ground. It was midnight, Kang respectfully asked Mirchink and other Soviet experts to retire to their rooms. As they reached the door, Mirchink still commented in parting: "This well has to be drilled to 3,200 m." After seeing off Mirchink, Kang said to people nearby, "He says what he says, we do what we do." Late at night,

Kang telephoned Yu Qiuli in Beijing, reporting his argument with Mirchink and their plan to stop drilling and start testing for oil. Yu replied with certainty, "I completely agree with you. Songji-3 should stop drilling right away and start testing for oil." In my view, the fact that Kang made the decision against the will of his teacher was a result of the responsibility he shouldered. This became an oft-repeated tale of the early days of Daqing Oilfield.

7 After Water Came Oil

During this time, Songji-3 was causing great excitement and concern among many people, as if it were a long-expected baby just before birth. To ensure the quality of the work of finishing the drilling and testing for oil and thereby ensure that the "baby" was born safe and healthy, a whole set of operational procedures were put in place under the auspices of the Ministry of Petroleum. It dispatched a group including Zhao Shengzhen 赵声振, Qiu Zhongjian 邱中健, and Jiang Xueming 蒋学明, while Songliao Petroleum Bureau dispatched a group including Zhang Wenzhao and Jiao Yabin 焦亚斌. They formed a group that worked at the site long-term and directed the well completion operations. The first crucial work was to fortify the well with cement. If the quality of fortification were not good, then all the previous work would be in vain. To guarantee fortification quality, the best drilling engineer in the country, Peng Zuoyou 彭佐猷 from Yumen Oilfield, was transferred to take on the task. The group had to report daily to the Ministry by cable and submit a written report weekly. They took utmost care to make sure that nothing would go wrong. On September 7, 1959, well cementing was completed at last. After an inspection, it was deemed good. Since they did not have advanced gas-lift equipment to force the water out of the well bore, they had to find whatever they could find on site. Thus they made their own very practical "long-handled bailing bucket" (*chang tilaotong* 长提捞筒), which attached to a drill. Starting on September 8, they bailed water continuously from the well 24 hours per day. At first, the water had some oil in it, but after a few days, the oil in the water increased. People at the site became extremely excited and wanted to stop bailing water and extend the bailer further down to scoop up the oil, in order to determine the well's production capacity. Kang Shi'en did not think that plan was appropriate. He immediately sent a telegram, stating his opinion that the oil reservoir of this well had been soaked in mud for too long, so that the water-bailing had to continue. The mud in the well bore had be drained to unblock the oil reservoir. He gave strict orders to only bail out the water, not the oil, "so that the oil will come after the water is drained." The workers on-site resolutely followed these orders. The bailing went on for seven days and seven nights, one bailer after another, tens of thousands of them. Even screws that had fallen on the well bottom were scooped up. After the ladling stopped, the well was closed up to accumulate pressure. On September 26, the well was opened, and oil gushed out of Songji-3—now a high-yield well.

On November 7, 1959, Ouyang Qin 欧阳钦, one of the founding members of the Chinese Communist Party and the Party Secretary of Heilongjiang Province at the time, travelled from Harbin to the town of Datong on more than 200 km of very bad roads to inspect the oil-producing Songji-3 and visit the oil workers. The next day, at a celebratory meeting in Datong, he gave a passionate speech. Since Songji-3 produced oil right before the big celebration (*daqing*) of the tenth year of the founding of the new China and the town of Datong had the same name of the City of Datong in Shanxi Province, he proposed changing the name of Datong to Daqing. This is how the name of Daqing Oilfield came to be.

8 Using Past Lessons and Empirical Evidence to Decide Whether to Organize a Campaign

The fact that Songji-3 produced oil created an extremely challenging situation for the Ministry of Petroleum. Kang Shi'en, writing in his memoirs, said:

FIGURE 11.2 Songji-3 gushes oil, September 1959

> Reflecting on past lessons, we could not regard a single gusher as [the equivalent of] having found an oilfield. To avoid firing empty cannons again, as had happened with the Chuanzhong Oil Campaign, comrades from the Ministry's Party Group and I decided that we would not be in a hurry to promote the discovery of oil at Songji-3; on the contrary, we would first immerse ourselves in genuine hard work.[3]

After repeated discussions, they realized that they needed to expand the success of the Songliao Basin exploration to find out whether it met the conditions for organizing an oil campaign. As a first step, on September 30, October 2 and 5, the Party Group of the Ministry of Petroleum held three meetings to discuss expansion of the Songliao Basin exploration. Based on the Ministry of Geology's most recent seismic information, Kang Shi'en proposed a plan to expand exploration. Its central idea was to establish the operation across the whole Daqing Placanticline and drill 63 exploratory wells, with the first group fanning out around Songji-3; the second group would proceed in 1960, depending on the results of the first group. Accordingly, another batch of oil workers were transferred from Yumen, Xinjiang, Qinghai, and Sichuan to Songliao Basin. By the end of 1959, seven exploratory wells drilled to the south of Songji-3, in what came to be known as the Putaohua Oilfield 葡萄花油田, produced oil. In early 1960, six more exploratory wells produced oil. These 13 wells covered a 200 km^2 area of the oilfield and showed that the oil reserves would be more than 100 million tons.

Secondly, the Ministry sent a strong lineup of experts to the site, headed by the chief of the Geology Bureau, Zhang Jun. The group included, among others, the geophysicist Weng Wenbo, petroleum exploration expert Li Desheng 李德生, oilfield development expert Tong Xianzhang, petroleum geologists Wang Gangdao 王纲道 and Dong Enhuan 董恩环, drilling expert Jiang Fuzhi, and Drilling Division head Deng Lirang 邓力让. For more than 20 days, they met with dozens of experts from the Ministry of Geology and Songliao Petroleum Bureau, carefully studying the geological situation of Songliao Basin. Based on the gathered information, new assessments and calculations were made in terms of exploration and future development. The opinions and suggestions of the expert group played a crucial role in organizing the Daqing Oil Campaign.

From February 1 to 8, 1960, the Party Group of the Ministry of Petroleum held a conference to study and analyze oil exploration in Songliao Basin. People

3 Wang Tie 王铁, ed., *Kang Shi'en huiyilu: haiyang shiyou bufen* 《康世恩回忆录：海洋石油部分》 [*Memoirs of Kang Shi'en: Offshore Oil Part*] (Beijing: Dangdai Zhongguo chubanshe, 1998), 110.

from the expert group spoke about having secured the oilfield with 100 million tons of reserves. They noted the good conditions, such as a relatively thick oil reservoir, relatively high-yield wells, good oil-bearing conditions, stable strata, oil in relatively shallow strata, and loose strata. They also noted that the crude oil had little gas, high paraffin content, and that storage and transportation were difficult in the frigid weather of the region. They emphasized that to the north of this oilfield, within a several thousand square kilometer area on the Daqing Placenticline structure, there might be even better prospects and that bigger and better oilfields might be found. Experts made specific proposals on exploration deployment and oilfield development for 1960. After full discussion by participants, a consensus was reached to put together the best forces to wage an oil campaign in Daqing. At the closing of the conference, Yu Qiuli said, "Whether we can change the petroleum industry's backward situation hinges on this act. We must be resolute, fight with our back to the wall, go all out, and capture this big oilfield as soon as possible."[4]

9 Three Exploratory Wells Confirm an Oil-Rich Area

After Songji-3 started producing oil early in the winter of 1959, dozens of top Chinese petroleum geologists assembled in Datong Town to figure out the location of the main portion of the oilfield, i.e., the oil-rich area. They lived in rough living conditions—crude adobe houses and cellars. At the end of December 1959, Yu Qiuli went to Datong and held a two-day geological meeting to discuss this issue. Many geologists at the meeting said that after having made comparisons they noticed that on the structure south of Songji-3, relevant oil reservoirs had become thinner. Using "reverse thinking," they predicted that the oil reservoirs north of Songji-3 (in the three structures of Xingshugang 杏树岗, Sartu 萨尔图, Lamadian 喇嘛甸) would be thicker, that it might be the oil-rich area. The young geologist Wang Yujun 王毓俊 added that judging from the information obtained from the shallow wells that the Ministry of Geology had drilled, the oil reservoirs might become thicker at the Sartu structure to the north.

Li Ye, Yu Qiuli's secretary at the time, recalled in 2004 that a young man attending the meeting shouted, "I think there is a big oilfield in the north with geological reserves possibly amounting to more than 2 billion tons." Goodness!

4 Yu Qiuli 余秋里, *Yu Qiuli huiyilu* 《余秋里回忆录》 [*Memoirs of Yu Qiuli*] (Beijing: Renmin chubanshe, 2011), 599.

Two billion tons at the time was an astronomical figure for our country. It was an unthinkable number, because the nation only had reserves of 56 million tons at the time. That remark shocked everyone present, causing quite a stir. On the spot, Yu Qiuli praised the young man for vigorously daring to think and speak, and encouraged him: "If we do find a 2 billion-ton oilfield, I'll let you be the chief geologist!" At the time, it was only a rash remark, but history has proven that the reserves for this oilfield were not merely 2 billion tons, but approximately 6 billion tons. Where is that young man now? There is no way to find out.

The discussions went on among experts from the Ministry of Petroleum such as Weng Wenbo, Li Desheng, and Tong Xianzhang; Chen Bingyin, petroleum expert and head of the Petroleum Division of the State's Planning Commission; and Yu Boliang, Zhang Wenzhao, Yang Jiliang, Zhong Qiquan, Zhang Tiezheng, An Qiyuan 安启元, and Qiu Zhongjian from the Songliao Petroleum Bureau. They gradually developed an exploration scheme whereby a few exploratory wells would be drilled in the Xingshugang, Sartu, and Lamadian structures further north of Songji-3. This at the time was a revolutionary move, breaking with convention. Conventionally, after a key well produced oil, new wells would be spread out approximately every two km in a cross-like layout, gradually expanding the area of the oilfield, but this time, three lone exploratory wells would be scattered 70 to 80 km to the north. Coincidentally, at this critical moment, the Ministry of Geology's Changchun Geophysical Prospecting Team sent over a new seismic structure map that they had made, which clearly showed that in addition to the structure of Songji-3 and the structures south of it, there were three big structures at Xingshugang, Sartu, and Lamadian, each measuring hundreds of square kilometers, that fit perfectly with the uplifts shown on that gravitational and electronic surveys. Of the three, Sartu was the biggest. Elated, the experts immediately made another important decision on the Daqing Oil Campaign: an exploratory well would be drilled at each of the high points of Xingshugang, Sartu, and Lamadian. Afterward, petroleum geologist Li Desheng and his group determined the precise positions for the three exploratory wells. This information was turned over to Deng Lirang, head of the Drilling Division of the Exploration Bureau of the Ministry of Petroleum, who personally dispatched drilling crews. These three exploratory wells confirmed that this was an oil-rich area.

But this maneuver was not without risks and mistakes. Geologically speaking, Songji-3 and the area south of it, had already been developed and was producing oil. The underground situation there was relatively clear; roughly 100 million tons of reserves lay there. On the other hand, the area to the north appeared to have three structures, but the underground situation there was not very well known. It was still a castle in the air. As for ground transportation, although there was a railroad, there were no villages, such as in Datong Town—just blue sky overhead and a vast grassland. There was no support infrastructure or amenities whatsoever. If a large number of oil workers and their equipment were rushed in all at once, one could imagine how difficult production and life would be.

Nonetheless, two months later, these three wells, one after the other, proved to be high-yield, which made evident to all that Daqing Oilfield's richest zone might be in Xingshugang, Sartu, and Lamadian. At a single order, ten thousand oil campaign workers descended on the region, bringing with them ten thousand tons of equipment. There was a severe shortage of trucks and tractors. They carried supplies and equipment on shoulder poles, in carts that they pushed themselves, and in carts pulled by horses and donkeys. Anyone with a bicycle was envied. They arrived three or five in a group, a dozen or so in a team, braving the chill of early spring, treading across the grasslands and marshlands. Traveling day and night, they went north some 70 or 80 kilometers to the area of Xingshugang, Sartu, and Lamadian. This became the key battleground of the campaign.

10 First Contributions and Wise Acts

First, the Ministry of Geology made the initial contributions toward the discovery of Daqing Oilfield. A debate had long been underway as to whether the 260,000 km^2 Songliao Basin had oil, so the Ministry of Geology first conducted quality investigations, surveys, and scientific explorations over an extended period of time. These efforts, which included the discovery of Nan-17 and the area of the Central Uplift Belt, suggested that there was oil in this basin. It provided precise geological information for subsequent detailed surveys and drilling by the Ministry of Petroleum. This resulted in the discovery of the Daqing Oilfield.

Second, the original Songliao Petroleum Exploration Bureau of the Ministry of Petroleum also made important contributions. The Bureau and its Geological Research Crew were formed in the beginning of 1958. Alongside their colleagues from the Ministry of Geology, they quietly immersed themselves in the work in this basin. After Songji-3, they found the Gaotaizi and Putaohua Oilfields. In the beginning of 1960, they found the crucial oilfields of Xingshugang, Sartu, and Lamadian. Their contributions were undeniable.

Looking back at this period, we can also see that the first Minister of Petroleum Li Jukui had also made outstanding contributions.

Third, reflecting on this history, it is apparent that that the geologists of the Ministry of Geology and Ministry of Petroleum thought along the same lines and worked towards common goals in their effort to discover oil. Treating one another with sincerity, acting in unison, and sharing information, they accomplished many impressive feats, such as the deployment of exploratory wells at Songji-3, as well as the exploratory wells at Xingshugang, Sartu, and Lamadian structures. Their sincerity, camaraderie, and enthusiasm for finding oil to benefit our country were commendable.

Fourth, Daqing Oilfield was the first oilfield in China to be discovered with large-scale geophysical technology and equipment. It initiated a new era in the use of more advanced technology and equipment (mainly artificial seismic surveying) for finding oil and gas fields in the middle of large sedimentary rock basins. To us, this was a milestone, though in comparison with Europe or the United States we were still thirty to fifty years behind. In addition, even though marshes and lakes were densely distributed on its surface, the Central Uplift Belt of Songliao Basin was as flat as a pancake. Its underground geological structure was regular, simple, and gigantic, with loose strata. Oil could be found at approximately 1,200 to 1,400 m in thick layers of sandstone. This was conducive to easy discovery and development. However, it would not be truthful to blindly announce to the world that the technology of Daqing was world-class, though there were a few individual exceptions.

Fifth, after Songji-3 gushed oil on September 26, 1959, Yu Qiuli and Kang Shi'en, learning from the failure of the Chuanzhong Oil Campaign in early 1958, were in no hurry to report the event to the CCP Central Committee and publicize it. Instead they got down to hard work. Once they had proven to their own satisfaction that the area, to be known as the Putaohua Oilfield, had oil reserves of 100 million tons and extended over 200 km^2, they formally made the request to organize the campaign to the CCP Central Committee in February 1960. This hard work, caution, and discretion resulted in a victorious outcome. Regretfully, in the oil campaigns that followed "In Industry, Learn from Daqing," decision-makers rarely acted this wisely.

Sixth, the starting time of this campaign is worth discussion. All previous historical records have put its date at February 1960, when the CCP Central Committee formally ratified it. I question the validity of that. That date does not take into account the full history, which should include the struggle to discover the Daqing Oilfield. The period of exploration and discovery, from the beginning of 1958 to February of 1960, should be considered not only an essential part of the Daqing Oil Campaign, but also perhaps its most important part. If there had been no discovery, there would have been no oilfield! Fixing the start time as February 1960 was a natural choice, due to the circumstances of that time, so it would be unreasonable to blame our predecessors for this. However, to continue to record February 1960 as the beginning of Daqing flies in the face of historical fact. To correct this error, I therefore propose to set the start time of the campaign in early 1958.

CHAPTER 12

Constructing Daqing Oilfield

1 The Party Central Committee Approves Full-scale Development of Daqing Oilfield

By February 1960, after five months of work, Songji-3 produced oil. Enthusiasm mounted for a big oil campaign in Daqing's vast area. The top priority was to formally submit a report to the Central Committee of the CCP and win support for the project.

On February 6, 1960, Yu Qiuli wrote to Li Fuchun, Vice Premier in charge of the State Planning Commission, and Bo Yibo, Vice Premier in charge of the State Economic Commission, reporting the Ministry of Petroleum's plan to organize an oil campaign in Daqing. He requested an increase of funds, equipment, supplies, and labor. Upon receipt of the letter, Li Fuchun, a CCP elder, had a talk with Yu Qiuli and expressed great concern: "All government agencies are now abandoning existing projects, but your Ministry of Petroleum wants to start a new project. This is a big move. You need to write a report to the Central Committee." On February 13, the Ministry's Party Group formally submitted a "Report on Petroleum Exploration in Songliao Area and the Issue of Future Work Arrangements (关于松辽地区石油勘探情况和今后工作部署问题的报告)" to Premier Zhou Enlai, Vice Premiers Li Fuchun and Bo Yibo and the Central Committee. Seven days later, on February 20, the Central Committee formally ratified the report, commenting, "A big campaign approach is good. It will play an important role in rapidly changing the backward situation of our country's oil industry." This memorandum from the Central Committee gave the green light to the Ministry of Petroleum. At a time when China was having a great shortage of capital and material resources, the relevant government agencies allocated an extra 200 million yuan in addition to the Ministry of Petroleum's regular annual funds. Other agencies supplied thousands of tons of steel and equipment, satisfying the pressing need for the campaign that was about to begin.

In early 1960, when the Ministry of Petroleum was still contemplating the campaign's organization, it realized the existence of a labor shortage. All hoped that the Military Commission of the Central Committee would assign 30,000 army veterans to the Daqing Oil Campaign, similar to what it had done at Yumen Oilfield. Zhou Wenlong, the Vice-minister in charge of human resources, made the request to Luo Ruiqing 罗瑞卿, Chief of Staff of

 | DOI:10.1163/9789004402737_013

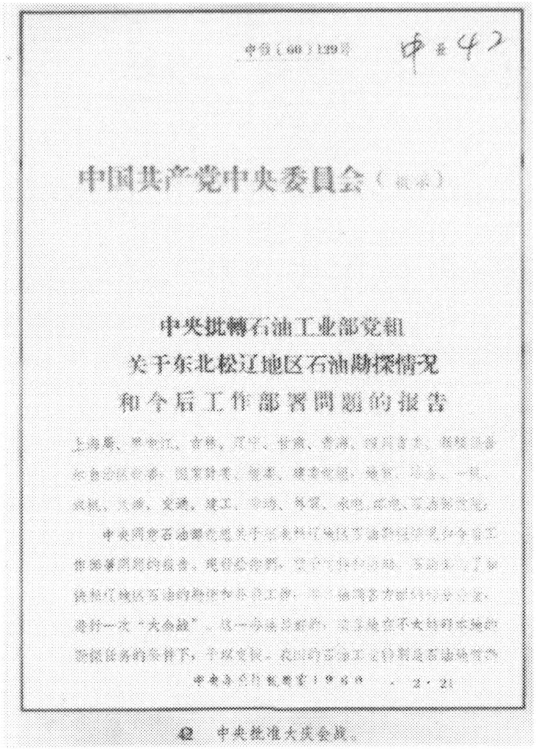

FIGURE 12.1 The Party's Central Committee ratifies Daqing Oil Campaign in February 1960.

the Military Commission, in writing. In February 1960, Yu Qiuli made the same request to Premier Zhou Enlai. Zhou responded, "Good idea! The Chairman is in Guangzhou for the enlarged conference of the Military Commission. You go to Guangzhou right away." Thus in mid-February, Yu Qiuli flew to Guangzhou on a plane especially dispatched from Premier Zhou's office. At first, Yu Qiuli made the request to the chief of staff of the Military Commission Luo Ruiqing as well as Marshals Liu Bocheng 刘伯承 and He Long 贺龙.[1] They all agreed, with Liu Bocheng wittily saying, "That's more like it. To capture a tiger, you need to rely on your brothers; to go on a military expedition, you need father

1 He Long (1896–1969), one of the Ten Marshals, was a Vice Premier and Director of the State Athletic Commission. He was overthrown during the Cultural Revolution and died under the harsh conditions of his indefinite house arrest. In 1975, Mao Zedong proposed rehabilitating his name, but that did not occur until later.

FIGURE 12.2 Daqing Oil Campaign mass rally on April 1, 1960

and son soldiers!" By this he meant that Yu Quili had come to the right place to find manpower. Afterward, Yu reported his intention of initiating a Daqing Oil Campaign to Mao Zedong and other senior generals. Mao listened attentively and commended him repeatedly: "Good, good, it is good to have a campaign!" On February 22, the Military Commission of the Central Committee issued an order to "mobilize 30,000 veterans for the Ministry of Petroleum." Beginning in March, these 30,000 veterans, together with 3,000 veteran officers, arrived at the campaign area one after another.

After the Party Central Committee ratified the campaign, the Secretariat of the Central Committee instructed Bo Yibo, Director of the State Economic Commission and Vice Premier, to be in charge of the supervision, inspection, organization, and support of the campaign. It further directed the State Planning Commission, the State Economic Commission, and the State Construction Commission to be responsible for daily organization and leadership. In February and March, Bo Yibo held many meetings and sent out many telegrams, arranging for all manner of support. Under the unified arrangement of the State Council, relevant municipal, provincial, and state agencies became responsible for assisting the Daqing Oil Campaign. Within a month, funds, goods, and materials arrived. Under the care and management of Ouyang Qin, the provincial Party Secretary, Heilongjiang Province set up a very strong team and apparatus that provided support to the campaign.

The PLA's support for Daqing came in many ways, ranging from welding rods and electrical wires to retired Soviet T-34 tanks, which were modified as tractors. According to incomplete statistics, during the Daqing Oil Campaign, Shenyang Military District cumulatively dispatched soldiers and officers more than 28,700 times, built 363 km of road, dug 778 km of pipe trenches, built 412 oil and water pools, welded 619 km of pipe lines, constructed 14 bridges, built 140,000 m^2 of housing, and loaded and unloaded 20,000 tons of goods and materials. In a nutshell, the PLA was at the forefront of this most exhausting and difficult campaign.

Within the petroleum industry, there was great excitement in the workforce. People longed to go to Daqing and eagerly signed up to take part in the campaign. The Ministry of Petroleum selected its best personnel for the transfer from among 37 petroleum units, colleges, and institutions. It transferred half of its personnel from its office in Beijing. Around 2,000 professors, engineers, technicians and new graduates from petroleum academies, schools, and research institutions were mobilized for the campaign.

On February 21, 1960, the first preparatory meeting for the Daqing Oil Campaign was held in Harbin. It was announced at the meeting that the campaign leadership team had been formed, with Kang Shi'en as the team leader and Tang Ke, Chief of the Ministry's Geological Exploration Bureau, and Wu Xingfeng 吴星峰, Secretary of the Ministry's Party Group, as deputies. Soon, the campaign leadership team was called the Campaign Working Committee, with Yu Qiuli as Secretary, Kang Shi'en as Commander, and Tang Ke, Wu Xingfeng, Songliao Petroleum Bureau Chief Li Jinghe, Xinjiang Petroleum Bureau Chief Zhang Wenbin, Yumen Petroleum Bureau Chief Jiao Liren, and First Petroleum Engineering Bureau Chief Chen Lizhong 陈李中, as deputies.

Now the campaign was in full swing, but it started so vigorously that chaos set in. For instance, some employees were not paid, even after having worked there for a few months. A few people could not endure the hardship and deserted. Along the approximately 70 km of railway from Anda 安达 County to Sartu Town, campaign goods, materials, and equipment piled up. However, in those days, local people were honest, and almost no theft occurred. Accidents were also frequent. For instance, in October 1960, in Sartu Oilfield a heater set off a fire at an oil well, and it burned for seven to eight hours; it almost went out of control. At the East Oil Storage Depot, the earthen oil pool—so constructed because of the steel shortage—exploded and almost turned into a disaster.

2 The Ironman Spirit: "Even If I Live 20 Years Less, I'll Risk It for this Big Oilfield"

In March 1960, Wang Jinxi led 1205 Drilling Crew to Daqing from Yumen Oilfield. They had left in a hurry, so they forgot to bring the rig's spare parts. While assembling the rig, they realized that they were short of this and that. Wang Jinxi shouted, "With or without, we'll do it!" Kang Shi'en was present and highly commended him. From then on, this slogan spread, becoming the campaign employees' pet phrase. Whether running into new problems or accepting new assignments, they too would say, "With or without, we'll do it!" and set to work without saying anything more. I came across this numerous times in 1960, since it encapsulated the employees' general spirit. A few years later, in October 1963, while publicizing Daqing, the Campaign Working Committee's Deputy Secretary Wu Xingfeng figured that this slogan was not refined enough. So he changed it to "If the conditions are right, do it; if the conditions are not right, then make them right and do it." But this overprecise version lacked the relentless energy of the original.

Wang Jinxi was born to a very poor family in 1923 in the village of Chijin 赤金, Yumen County, Gansu Province. Later, he went to Yumen Oilfield to be a laborer. After the Liberation of 1949, he became a driller. Later, he was promoted to crew leader. Due to his hard work, the drilling crew he led had repeated outstanding successes. He attended the National Heroes Conference held in Beijing in September 1959. When he learned that Songji-3 produced oil, he applied to go to Daqing again and again. In March 1960, the Ministry especially requested his drilling crew to join the Daqing Oil Campaign. As soon as he led his whole crew to Sartu railway station, he ran to the dispatcher's office to register. He did not even ask for food or lodging. He first asked three questions: "Has our rig arrived? Where's our well? What's the record time for drilling a well here?"[2] After he learned the position of their first well, he immediately walked 20 km to the site to see it. A few days later, the rig arrived. There was a shortage of cranes. Since trucks were scarce, they used a few automobiles—or his crew manually pulled and carried the equipment to the well site and installed it. When they started drilling, there was no water on the site so they broke through the ice at a nearby frozen pond and retrieved water with washbasins. This enabled drilling to start at the first well. He ate and slept at the drilling site, not leaving it for five days and five nights. At last they

2 Wang Jinxi's three questions were first mentioned in a speech that I wrote for Campaign Deputy Commander Zhang Wenbin, who delivered it at the CCP Central Northeast Bureau's Conference for Industries and Transportation held in Daqing in October of 1963.

FIGURE 12.3 Iron Man Wang Jinxi

finished the first well, setting the best drilling record at the time. His actions moved nearby villagers. An old woman carried a basket of eggs to the drilling site to show her appreciation. She would say this when she ran into workers: "Your Crew Leader Wang is a real Ironman!"

Wang Jinxi's deeds were noticed by Song Zhenming 宋振明, the manager of campaign teams from the Yumen Petroleum Bureau (he later became Petroleum Minister). Song, who was working at the site of 1205 Drilling Crew, reported all of this to Yu Qiuli. Very pleased, Yu promptly responded, "Good, this is a good model. We'll borrow the old woman's saying and call him 'Ironman Wang.'" The decision was made to coin the slogan, "Everyone Learns from the Ironman, Everyone Becomes an Ironman." At the "Petroleum Campaign Mass Pledge Meeting," Wang Jinxi and other model worker representatives, each wearing a big silk red flower and sash, entered and circled the meeting grounds while mounted on big horses. Wang Jinxi was pushed to the podium. Facing the excited oil workers, he shouted his oath: "Even if I live 20 years less, I'll risk it for this big oilfield!"

Afterward, the whole oilfield was bustling with activities related to the slogan, "Everyone Learns from the Ironman, Everyone Becomes an Ironman," which motivated the campaign employees as they pushed forward the production and construction work in Daqing Oilfield. As for Wang Jinxi himself, in the end he died in 1970 of sickness, thus realizing his oath, "Even if I live 20 years less, I'll risk it for this big oilfield."

3 Overcoming Rain: "Even If It Rains Knives, the Campaign Can't Stop"

In 1960, the rainy season came especially early to the Daqing area and lasted an especially long time. Starting in May, every few days it rained, sometimes for days at a time. However, in the tents, plank houses, cowsheds, horse stables, cellars, and other simple sheds where the workers lived, the rain fell all day long. If it rained heavily outside, it leaked in the house; even if the rain had stopped outside, it continued to trickle inside, making it difficult for people to rest after strenuous work. Even Kang Shi'en's cowshed leaked. Sometimes he had to move his bed and cover the shed with tar paper once or twice a night, which made it hard to sleep all night. In August 1960, I had to move my bed twice in the middle of night because my cowshed leaked.

The rain decreased our productivity. Daqing is in low-lying land. If it rained, it would be muddy at the construction and drilling sites, and every step became difficult on the few dirt roads. The name "Sartu," one of the main "battlegrounds" of the Daqing Oil Campaign, means "big bean paste jar" in Mongolian, meaning a muddy swamp. In Sartu, the drilling and construction sites, each with small crews numbering three to five people, were scattered across a hundred kilometers of wilderness. Daily necessities could not be transported to such locations, and support teams could not service them. Even communications were cut off. One five-person crew at the time was working deep in the wilderness and ran out of food. Instead of retreating, they kept on working and survived seven rainy days on wild herbs.

When confronting such situations, Commander Kang Shi'en said, "No matter how much or how long it rains, or even if it rains knives, the campaign can't stop!" Campaign employees proposed the slogan, "Work Extra on Clear Days, Do Battle on Rainy Days, Work Less in Heavy Rain, Work Hard in Light Rain." They devised many ways to deal with the difficulties brought on by the rainy season. To ensure that the urgent production needs of the front line were met, Deputy Commander Zhang Wenbin mobilized office workers on all levels to go down to the sites to figure out and summarize the lessons learned from battling mud, and they drafted more than 30 sets of plans to prevent trucks from getting bogged down in it. "Mud-chains" were made. Afterward, truck drivers were mobilized to keep trucks running continuously, with drivers taking turns, working night and day, and within seven days they transported thousands of tons of equipment and materials to scattered drilling and construction sites in the rain.

4 Overcoming Extreme Cold: Building Adobe Houses

While the campaign was encountering torrential rain, campaign leaders were already considering how to survive the winter. At latitude 46°N, the Daqing area has short summers and long winters. It starts freezing and snowing around the time of National Day on October 1, and in winter the temperature can dip as far as −40°C. But the campaign workers still had no actual houses to live in—they were still in temporary housing like tents, cowsheds, cellars, and plank houses. There was not even a garage where the thousands of tons of equipment and the hundreds of vehicles (in continuous operation) could be repaired. Without heating and proper housing to guard off the cold, the severe cold would result in frostbite and equipment breakdowns. At the beginning of the campaign, someone kindly suggested that the campaign workers, except for a few staying behind to take care of things, should retreat to nearby big cities such as Harbin, Changchun, and Qiqihar before the advent of winter and return the next spring. After much consideration, the Campaign Working Committee concluded that if they were to do this, the effective campaign time per year would be less than six months, delaying oilfield development and construction and causing more difficulties for the country. They decided to have the workers spend the severe winter in Daqing. Party Secretary of Heilongjiang Province Ouyang Qin suggested that they should learn from local villagers how to build adobe (*gandalei* 干打垒) houses. To build these houses, one could get the dirt on location to make the walls and gather straw and reeds for roofs. This required little lumber. What's more, the masses could be mobilized, and everyone could take part in construction.

Making adobe houses was a good solution. As a result, an adobe construction headquarters was set up, headed by Deputy Petroleum Minister Sun Jingwen 孙敬文. Special crews were dispatched to the forest to get lumber. All base production units assigned labor to form house-building crews. At scientific, research, and design units and offices of all levels, people would work during the day and, in addition to that, work extra hours in the evenings to build houses. By the end of the summer, most of the houses were built, but some people worried about whether they would withstand harsh winter conditions.

In September 1960, as I was researching the topic of winter preparation and insulation, I visited two nearby farmers living in adobe houses. They received me warmly, showing me the proper way of making the mud and straw bricks and walls. They demonstrated how to spread straw and reeds as roofs. They assured me that these houses were warm in winter and cool in summer, like the cave dwellings in the northwest. I took careful notes and summed up my findings in a thousand-word how-to campaign memo, which was well-received by the oil "troops," most of whom had come from the northwest. In early October, at the oilfield-wide "Winter Preparation and Insulation" meeting, which was attended by ten thousand people, Deputy Commander Zhang Wenbin said, "Lately a secretary in the political department visited local farmers and wrote a memo on how to make adobe houses warm in winter and cool in summer. Production units should print out the memo for everyone to read."

In the evenings from July to late October that year, lights could be seen throughout the grassland, as men and women, old and young, workers, office workers, experts, and even professors were busy building. With their sleeves and pants rolled up, people dug dirt, carried water, and mixed and packed dirt; they used wooden molds to make adobe bricks (mud mixed with dry grass) for partition walls, and for the external walls, which were thick at the bottom and narrower at the top, they put earth, dry grass, and a little water between wooden boards, and then they pounded the mix, layer by layer. People sweated together and were bitten by mosquitoes together. Construction sites bustled with activity. The "Adobe Construction Headquarters" monitored daily progress. After a few months, more than 300,000 m^2 of adobe houses had been built. Many new villages appeared on the grassland, thus realizing the goal of the slogan, "People in Houses, Machines in Sheds, Vegetables in Cellars, Trucks in Garages." As a result, everyone made it safely through their first winter in Daqing.

5 Overcoming Hunger: Clearing the Wasteland and Planting Crops

The biggest difficulty Daqing Oil Campaign encountered was the food shortage. In the first few months, grain for the "troops" was available, with non-staple foods being supplied by the southern provinces and Heilongjiang Province. In September and October in 1960, the food supply nation-wide became extremely tight. The grain supply at Heilongjiang, which was considered the nation's granary, fell beneath the danger line. Grain rations for oil campaign workers were slashed by the government. Rations for office workers were reduced to around 20 kg from around 30 kg; and for second-line workers they were reduced to around 30 kg; for drillers with the most labor-intensive work, they were reduced from the original 50 kg to around 40 kg. Non-staple foods were drastically reduced, and almost no meat was supplied. Working in the open in extreme cold sapped one's strength. People's health began to deteriorate. Under these circumstances, many people used their meager salaries to buy high-priced food in the market. Some even exchanged their treasured clothing for a pittance of food. Hunger drove some to drink salt and soy sauce water. Still, people kept on working.

In early 1961, the hardship became extreme. While the campaign continued, the famine dragged on and, as a result, many employees suffered from dropsy. At first, the symptoms of dropsy affected only a few dozen, but then, ten days or so later, more than a thousand people were affected. Kang Shi'en discussed the matter with his colleagues and adopted the following measures: unit managers would take care of both production and living standards, deputy managers would take care of living standards exclusively. Ten percent of all unit employees would be responsible for improving living standards; some trucks were allocated for foodstuff transportation. Twenty percent of the employees would, at any given time, be allowed to rest. Since all campaign employees dined at cafeterias, cafeterias had to be managed well. The campaign headquarters' big cafeteria was replaced by several small cafeterias. Furthermore, special "nutritional cafeterias," set up for dropsy patients, gave each patient two to three *liang* (= 50 g) of meat per day, and they were allowed to rest all day. I was diagnosed with third-degree dropsy (relatively mild), so I enjoyed a month of the nutritional meals, but I still did not get a day of rest. After two or three months, the number of dropsy cases decreased. Throughout the whole campaign, no one died of hunger. People endured the crises, and the campaign persisted against great odds.

However, the overall health of the campaign troops was still poor. Food shortage remained the biggest hurdle for the Daqing Oil Campaign. Confronted with the dire situation of directing a hard campaign when people's stomachs were empty, the Campaign Working Committee suggested that people try to do something about it—that hard work is better than suffering. The Heilongjiang provincial government was very supportive, allotting tens of thousands of *mu* (1 *mu* = 666.667 m^2) of grassland to Daqing. Employees cleared the wilderness, planted crops, and engaged in large-scale agricultural and foodstuff production. Many people were transferred from industrial posts to farm work. Meanwhile, ten thousand spouses of employees took part in farming. After the hard work of that spring and summer, more than two thousand *mu* of soybeans and large quantity of vegetables and other marketable crops had been cultivated. In the fall of 1961, 3.04 million *jin* (1 *jin* = 0.5 kg) of beans and 13 million *jin* of vegetables were harvested.

I remember that in the fall of 1961 I accompanied Deputy Commander Zhang Wenbin to inspect the Drilling Headquarters' large farm, Babaishang 八百垧. We were offered newly harvested potatoes, corn, cucumbers, tomatoes, and soybeans, and all were incredibly delicious. That winter, the Campaign Working Committee issued an order: 40 *jin* of soybeans were to be distributed to every employee in person by each unit's Party secretary. Since my wife and I both worked at the campaign headquarters, we received 80 *jin* of soybeans. We felt reassured as we put it under our bed. The campaign workers were delighted that they themselves had survived the famine.

6 Tracking Underground Data to Find Oil

From April 9 to 11, 1960, at Anda Railroad Club, which was the best building in Anda 安达 County at the time, the campaign headquarters held the first Combined Technical Seminar, inviting representatives from all levels (Ministry, Bureau, Division, Section, and Crew) and kinds (workers, scientists, and managers). One hundred eighty people attended. The purpose of the discussion was to figure out what kind of oilfield Daqing was. Presenters made use of lots of colorful geological maps, which they had hung up around the meeting hall. Geologists first reported results of some exploratory wells that had been tested for oil. Results were marked with differently colored dots on the graphs: red for oil-producing wells, yellow for gas-producing wells, and blue for water-producing wells. Even after they had scrutinized the graphs for a long time, the attendees could not figure out the significance of these patterns and dots. When asked, the geologists were also hard put to explain.

Kang Shi'en believed in order to solve the problem, the experts needed to discuss how to clearly explain these red, yellow, and blue dots. After much discussion, it came to light that they needed to concentrate on parameters having to do with conditions underground. Attendees aired their opinions on what kind of data to gather in the course of drilling and testing for oil. In the end, when summing up people's input, 13 kinds of information were deemed important for oilfield exploration, mainly oil-bearing area, thickness of oil reservoir, porosity, permeability, oil saturation, and original formation pressure. Later on, Kang Shi'en organized experts to discuss them, and the number of kinds of data was expanded to 20. Based on these 20 kinds of data, 72 key indicators were determined. Then the geologist Li Desheng, with others, wrote a guideline for geological work entitled "Geological Working Rules on the

Complete and Accurate Gathering of 20 Items of Data and 72 Indicators in the Exploration and Development of Daqing Oilfield (大庆油田勘探开发中取全取准 20 项资料、72 个数据的地质工作规范)," which was formally distributed by the Ministry of Petroleum so that oilfield geological workers could adhere to it.

In the chilly spring of early 1962, I investigated the Daqing Geological Research Institute, which managed all of Daqing's geological work, making its technical policies and development plans. I first visited a huge core storage facility. All these core samples retrieved from underground, which if placed end-to-end would measure hundreds of thousands of meters in length, provided direct evidence for geologists to study oil reservoirs. Then I visited the institute's Small Oil Reservoir Comparison Team. At the time, there were no computers in China. To understand oil reservoirs, the institute formed this 20-person comparison team, comprised of mostly female graduates from geological and petroleum colleges or trade schools. Their cramped low adobe dwellings doubled as their workspaces. Using abacuses and slide rulers, they marked numerous dots and lines on the seismic profile graphs and recorded each number. In this manner they provided Daqing Oilfield with oil reservoir information for its initial development plan. It was tedious work, but this team was one of Daqing's exemplary units. Thus I ended my report with an account of the deeds of this team, which worked behind-the-scenes, "tracking underground data to find oil," and it generated a positive response. They finally earned praised and admiration.

7 Comparison of the Soviet and Sartu Processes

Advanced oilfield development requires advanced technology. Now that crude oil was being extracted from underground in a steady stream, Daqing faced the issue of how to gather and transport it. Daqing oil had a high paraffin content, high solidification point, and high viscosity, but Daqing's climate is cold, with average annual temperatures 20°C lower than the crude oil solidification point. Even in July, when the temperatures are at their highest, the average temperature is still five degrees lower than the crude oil solidification point. This meant that the usual method of transporting it by pipeline was useless. What could be done?

In early summer of 1960, a Soviet expert known in China as "Baluoning" 巴洛宁 who had assisted in the development of the Xinjiang Oilfield was returning to the USSR via the northeast. He specialized in oil and gas collection and transport. He had designed many processes in the USSR as well as the one in Xinjiang's Karamay Oilfield. Kang Shi'en invited him for a short stay in Daqing, so that he might offer his advice. At the meeting, he admitted, "I have not come across this kind of oilfield. In this case, I suggest you use a Soviet collection and transport process, using a steam boiler to maintain the temperature of the well heads and using a steam pipeline to accompany the oil pipeline." When pressed if there were other ways, he shrugged his shoulders and replied jokingly, "There might be another way—by moving the oilfield to the tropics."

On hearing this, Kang Shi'en figured that we could not afford the Soviet process, using steam boilers and pipeline, as the Soviet expert had suggested. The 30 km^2 trial area alone would have needed close to a thousand boilers and a thousand tons of piping. Our country was poor and could not afford that, and even if it had the money, no one was making these things domestically. The boilers and piping were all foreign products.

At this key moment, the Campaign Working Committee decided firmly that China would have to design its own oil and gas collection and transport process suited for Daqing. It mobilized all of the design personnel, which then devised dozens of plans. A consensus was reached that a so-called "lantern-hanging" plan proposed by a young technician, Feng Jiachao 冯家潮, and his team was novel and practical. He proposed laying an oil pipeline linking well arrays directly to the oil gathering station, and attaching several heaters along the pipeline, like hanging lanterns. This bold, unprecedented design pleased Kang Shi'en immensely. He asked his secretary to drive Feng Jiachao to his office at 2:30 in the morning. The two spoke the entire night. Right then and there Kang ordered Deputy Commanders Zhang Wenbin and Jiao Liren to organize the labor and materials for the young man to work on the project.

Many questions remained to be answered concerning this project. How many heaters would be needed? To what temperature should the crude oil be heated? What should be the diameter of the pipeline? How deep should the pipeline be buried? All of these questions revolved around a certain key datum—the K value, i.e., the overall heat transfer coefficient. The K value appearing in Soviet textbooks definitely did not suit Daqing Oilfield. If the number was set too high, it would incur a lot of waste; if too low, the crude oil would freeze in the pipeline, and the resulting "sausage-stuffing" effect would cause accidents. To test the K value for Daqing, the Design Institute assigned a crew led by the young engineer, Tan Xueling 谭学陵. Starting in the winter of 1960, he and his crew crawled on ice and on snow and endured extreme hardship to accomplish the task. They dug holes in the ground every 50 to 100 m, took temperatures every two minutes, and worked in the wilderness more than ten hours every day. To get accurate temperature readings, they would even squat in those holes on winter nights when the temperature was −30°C to −40°C. In summer, they endured rain and mosquitoes. What they endured defies description. After ten months, during which time they traveled more than 6,000 km, took measurements at more than 1,600 holes, obtained more than 25,000 items of data, and conducted more than 11,000 comparisons, they at last determined the soil heat transfer coefficient for the Daqing area during the most frigid conditions. This provided the scientific basis upon which to design the oilfield's oil collection and transport process. (As for Tan Xueling, this fine representative of Daqing's outstanding scientific and technological personnel finally succumbed in Daqing to work-related illness in 1983.)

This "lantern-hanging" process came to be called either the Sartu Process or Airtight Single-line Mixed Oil and Gas Transport Process. It was first successfully carried out in more than 110 well arrays and 2,300 wells in Sartu and Xingshugang Oilfields. Not only did this invention satisfy the need for oil and gas collection and transport, it also greatly reduced capital costs and steel. It received the National Invention Certificate in 1965 and the National Science Commission's First Class Invention Award in 1985. Such examples abound in the history of Daqing. For instance, the hydraulic separation technique, perfected after 1,080 trials, which has contributed to Daqing Oilfield's long-term high yield, also won the National Science Commission's First Class Invention Award in 1985.

8 Political Work at Daqing: Extolling the Spirit of Hard Work

Given the context of this time, ideological and political work in Daqing was always paramount, in line with Mao Zedong's call to "put politics in command" and Lin Biao's 林彪 "stress politics." The ideological and political work in the first four years in Daqing was lively and varied, but it was also sometimes full of empty talk, formalities, and showboating. If we set aside its multifarious guises, its only valuable content can be summed up in single sentence: by all possible means encourage the oil campaign workers to work hard and contribute to the development of Daqing Oilfield.

As one of Daqing's ideological and political workers, I experienced this firsthand. I clearly remember that starting in the spring of 1960, we launched a steady stream of activities among the oil campaign workers, such as "Study Mao's Works, Study Mao's Two Theories," "Learn from the Ironman, Become an Ironman," "Five Red-Banner Pacesetters," "On-the-Job Training," "Job Responsibility System Inspection," and so on. These followed one after the other non-stop for years. One of the primary responsibilities of political cadres was to discover exemplary people and then publicize and popularize their deeds and experiences. Starting in May 1960, the most important political activity revolved around the "Five Red-Banner Pacesetters." These five red flag pacesetters were five grassroots unit managers who had originally been workers. They were Wang Jinxi, the drilling crew manager from Yumen, Ma Deren 马德仁, a drilling crew manager from Xinjiang, Duan Xingzhi 段兴枝, a drilling crew manager from Qinghai, Xue Guobang 薛国邦, a production crew manager from Yumen, and Zhu Hongchang 朱洪昌, a construction crew manager who was originally a welder. Following this example, units on all levels had their own pacesetters and model workers. At the time, I wrote a rhyme to describe this activity's influence and function: "Five red banners flutter in the wind, / The red spreads over a whole oilfield."

Frigid weather would halt most outside activities in the winter, so the oilfield held most of its political training during the winter season. In the winter of 1961, since there was a "do it alone" trend (*dang gan feng* 单干风) in places such as Anhui Province in which collective land was divided among individual families, and there were a number of army veterans from Anhui among the oil campaign workers, there emerged sayings like, "Workers are inferior to farmers, farmers are inferior to merchants, and a level-eight worker [i.e., a worker of the highest salary grade] is inferior to a bunch of green onions." A few Daqing employees even deserted, going back to their homes either to farm or to do small business. To curb such behavior, Daqing studied the PLA's approach and launched a grassroots self-education campaign to "contrast past misery with present happiness," which helped people to put their current difficulties into proper perspective and encourage them to view themselves as masters of their own destiny. This campaign effectively stabled the situation in Daqing.

Winter trainings held from 1962 on focused on appraising merits and summing up the activities of the year. In December 1962, I took part in the summary and merit appraisal of a work unit in the Fourth Construction Team. For one week, during the day, I participated in group discussions; in the evenings, I arranged information and notes until one or two o'clock in the morning. Afterward, I wrote a brief report, systematically introducing the steps, content, and course of action. The Deputy Party Secretary Wu Xingfeng commended my work and instructed each unit throughout the oilfield to copy this procedure. The purpose in making summaries and recognizing special achievements was that we needed to mobilize the masses to make suggestions, identify issues, problem-solve, and make plans for the next year. After the Daqing Experience was publicized nationwide in 1964, some units in Beijing also engaged in awarding merits, claiming that they had learned the practice from Daqing. The problem was that the activity in Beijing only superficially resembled what was done at Daqing. In Daqing, issues as well as breakthroughs were identified and examined, and suggestions for improvement were made on that basis, but in Beijing this was not done.

CHAPTER 13

"In Industry, Learn from Daqing" Reconsidered (Part One)

1 The Origins of the Slogan "In Industry, Learn from Daqing"

By the latter part of 1963, the mass campaign to find oil at Daqing had won a decisive victory. Even though the oil industry had long been abuzz about it, it was still little known among the general public. At the time, most Chinese industries operated under a cloak of secrecy so as to not attract foreign attention, and the petroleum industry was no exception. Thus, to the eyes of the outside world, the oil campaign headquarters went by the name of "Sartu Agricultural Land Reclamation General Farm (萨尔图农垦总场)," and its subordinate units were "Agricultural Land Reclamation Branch Farms (农垦分场)." Since the workers there wore padded winter work clothes similar to those used for prisoners, it sometimes occurred that when they stepped out of a train station, say, on their way to visit family, they were mistaken for prison escapees and taken into police custody. This vividly reflects how their real work was kept secret.

At the end of October 1963, after a feasibility study, the State Economic Commission and the Economics Committee of the CCP Central Northeast Bureau decided to hold a forum at Daqing for the purpose of exchanging information among various industrial enterprises. The name of the forum was simply "Northeast Area Base Industrial Enterprises Information Exchange." More than one hundred people attended the meeting, including executives of large state-owned enterprises in the northeast, experts and managers from the State Economic Commission, and the Economics Committee of the Northeast Bureau. The Deputy Director of State Economic Commission Gu Mu 谷牧 and the Director of the Economics Committee of the Northeast Bureau Gu Zhuoxin 顾卓新 presided over the meeting.

According to the agenda, attendees started out by visiting 20 different base-level units of the oil industry, including drilling teams, production teams, construction sites, oil gathering stations, sewing shops that employed workers' wives, the petroleum geological exhibit, oil storage units, and a newly built oil refinery. Then, the Deputy Commander of the Daqing Oil Campaign and concurrent Production Office Manager Zhang Wenbin delivered the speech, "The Development of the Three-Year Oil Campaign," Deputy Commander

 | DOI:10.1163/9789004402737_014

Chen Lizhong gave the report, "Ground Construction of the Oilfield and Gathering the Best Forces for Victory," Assistant Manager of the Production Office Song Zhenming delivered the report, "Maintaining the Base-level Job Responsibility System," and Assistant Communist Party Secretary Wu Xingfeng gave the report, "Strengthening Ideological and Political Work." After visiting Daqing and hearing the reports, participants of the forum grew enthusiastic. All present praised the Daqing Oil Campaign as heralding a great advance for China's industrialization. They commended the fortitude and perseverance of Daqing workers, as well as the valuable lessons and achievements gained in scientific research and production management. The day prior to the conclusion of the forum, Yu Qiuli came from Beijing to give a short speech. He repeatedly stressed that the lessons of Daqing could be summed up in a nutshell: "The socialist modernization of enterprises must include revolutionization."

News of the forum soon reached Beijing. The State Economic Commission decided to promote the achievements of Daqing and directed Kang Shi'en to give a report. Kang brought me with him to Beijing to help write the report. At the National Meeting for Industries and Transportation on November 6, 1963, held by the State Economic Commission, Kang delivered a speech on Daqing that met with an enthusiastic response. Then on November 19, basing his speech on Kang's report with some minor revisions, Yu Qiuli addressed the Fourth Plenary Session of the Second National People's Congress. In December, the Communist Party Group of the Ministry of Petroleum formally submitted its "Summary Report on the Daqing Oil Campaign" to the CCP Central Committee. On December 24 and 28, at meetings held at the Great Hall of the People, presided over by Politburo member Peng Zhen 彭真, Kang Shi'en and Yu Qiuli again gave reports to tens of thousands leaders of the Central Committee, national government institutions, and Beijing municipality.

On December 25, 1963, Chairman Mao Zedong put forward the slogan, "In Industry, Learn from Daqing." On February 5, 1964, the Party Central Committee issued a special notice affirming the success of Daqing. From that point on, the fervor to learn from Daqing spread throughout the nation. It is said that Mao Zedong, upon hearing the report on Daqing, urged: "Comrades in the Politburo should all go to Daqing and take a look." In fact, except for Mao Zedong himself and Lin Biao, all the members of the Politburo, as well as many other political and military VIPs, visited Daqing, some more than once. For example, Premier Zhou Enlai visited three times. These visits were interrupted by the turmoil of the Cultural Revolution, but after it had run its course, successive leaders of the Party and state began visiting Daqing again. At the beginning of the new millennium, Deputy Petroleum Minister Chen Liemin 陈烈民, who had once

FIGURE 13.1
In Industry, Learn from Daqing (1964)

been the Party Secretary in Daqing Oilfield, recalled, "After Chairman Mao came out with the slogan, 'In Industry, Learn from Daqing,' more than 1.25 million people from around the country made the trek to Daqing."

I remember that from 1964 to 1966, before the onset of the Cultural Revolution, hundreds or even more than a thousand people visited Daqing every day. Daqing Oilfield built two gigantic temporary reception centers, manned by hundreds of people, which offered good food and free lodging for the continuous stream of visitors.

2 The "Charter of the Anshan Iron and Steel Company" and the Daqing Experience

In December 1964, at the First Plenary Session of the Third National People's Congress, Zhou Enlai assessed the basic lessons of the Daqing Oil Campaign in his "Government Work Report." He pointed out:

> The construction of this oilfield was a model application of Mao Zedong Thought ... This oilfield is a good example of assiduously learning from the PLA and putting into practice the PLA's lessons in political work. This oilfield has from beginning to end adhered to the principle of combining leadership with mass movements, adhered to the principle of

FIGURE 13.2 Former No. 2 Yard—Daqing Oil Campaign Headquarters

> combining high revolutionary spirit with strict scientific attitude, and adhered to the principle of combining revolutionary spirit with exploration and construction. It has fully exemplified the general line of socialist construction for achieving greater, faster, better, and more economical results.[1]

Zhou Enlai's assessment of Daqing reflected the reality at the time. In April 1960, when the Daqing Oil Campaign had just started, Yu Qiuli indeed instructed Wu Xingfeng, who was in charge of political work, to draft on behalf of the Ministry of Petroleum Party Group, "A Resolution to Study Mao Zedong's 'On Practice' and 'On Contradiction,'" which was published in the inaugural issue of the oilfield's newspaper *Campaign Report*. Yu Qiuli also instructed the Ministry of Petroleum to purchase hundreds of copies of Mao's essays, "On Practice" and "On Contradiction," fly them to Harbin, and have them distributed throughout the oilfield for employees to study. These were read and discussed. As for how effective this study was in improving the Daqing Oil Campaign, that would depend on whom you asked.

As mentioned above, the fervor that lead to the "In Industry, Learn from Daqing" movement began at the Northeast Area Industrial Enterprises Information Exchange held in October 1963 in Daqing. Kang Shi'en and the

1 As quoted in Bainian shiyou bianxie zu, ed., *Bainian shiyou*, 160.

Daqing Oil Campaign Working Committee took the forum seriously. At the time, Daqing did not even have a guesthouse or hostel, so they moved out of their adobe offices, which doubled as their living quarters in Courtyard No. 2 (these were only slightly larger than the usual), and moved into even lower, smaller, and damper offices. I was dispatched as the only liaison from the oilfield to take part in the forum's Secretarial Group's activities. My job was to go to meetings of the Secretarial Group held between 8:00 and 10:00 every evening. The Secretaries would have gathered participants' reactions throughout the day, and I in turn would compile the information from these meetings to put into a report. Then, after 11:00, I would go to Kang Shi'en's adobe office. He and his colleagues would eagerly listen to my report. On such occasions, Kang Shi'en would respond, in effect, that "participants of the forum are all from 'dragonhead elder brother' industries, i.e., Anshan Iron and Steel Company, First Automobile Company, and Harbin's three major power companies,[2] which were built in the early 1950s with the help of the Soviet Union, and their management style was also copied from the Soviet Union. Even though our oil production has increased, our methods are different from those of the Soviets. I have no idea whether our elder brothers will approve of our ways." When Kang heard praise, he would be very excited, but would still say modestly, "Actually, we just did things strictly according to the 'Charter of the Anshan Iron and Steel Company.'"

What was the "Charter of Anshan Iron and Steel Company?" It was Mao Zedong's March 1960 memorandum on a study of Anshan Iron and Steel Company. The study was written by Li Lisan 李立三, who was one of the founding members of the Party and a leader of the labor movement, after he stayed at and researched the company. Mao's memorandum stated five guiding principles for running a good enterprise: "Keep politics firmly in command and strengthen Party leadership; launch vigorous mass movements; have cadre participation in productive labor and worker participation in management, with reform of irrational and outdated rules and regulations and maintenance of close cooperation among cadres, workers, and technicians; and go all out with technological innovation and technological revolution." These five principles, i.e., the "Charter of Anshan Iron and Steel Company," became the basic rules for running Chinese industrial enterprises at the time and for decades thereafter.

2 "Harbin's three major power companies (哈尔滨三大动力厂)" were Harbin Electrical Machinery Works (哈尔滨电机厂), Harbin Boiler Works (哈尔滨锅炉厂), and Harbin Turbine Works (哈尔滨汽轮机厂).

What were the main lessons from the Daqing experience? These were all written in Kang Shi'en's report, which I helped write, and which he delivered at the National Meeting for Industries and Transportation on November 6, 1963. Even now I remember the report included nine lessons: (1) revolutionary spirit; (2) scientific attitude; (3) mass movement; (4) three fundamental areas of work ("base work" [quality and equipment], basic skills, the base-level job responsibility system); (5) leaders' personal work on the frontline; (6) mentoring and advancing young cadres; (7) a strict and detail-oriented style of work; (8) full attention to workers' livelihoods; (9) ideological and political work. Of the nine lessons, six were about politics and revolution, and only three related to technology, management, and livelihood—and even these three were discussed from a political standpoint.

If one compares the nine items of the "Daqing experience" with the five of the "Charter of Anshan Iron and Steel Company," one would find how similar they are! It goes without saying that the "Daqing experience" was the direct descendent and concrete realization of the Charter.

3 Regressive Management Methods: a Critique of the "Charter of Anshan Iron and Steel Company"

The management of industrial enterprises worldwide has gone through three stages: traditional management, scientific management, and modern management. The first stage of traditional management started with handicraft workshops in the latter part of the eighteenth century and lasted until the end of the nineteenth century. This traditional management is the tradition of small-scale production. Its model is that of the master passing on his skills and experiences to the apprentice. The second stage of scientific management started in the latter part of the nineteenth century and ended in the 1970s. It was due to accelerated industrialization, the replacement of manual labor with machines, and great advances in production technology. The American engineer Frederick Winslow Taylor proposed "scientific management," i.e., the Taylor system. He used "time study" to devise work quotas and work time quotas; through the study of work, he proposed the punch card system; by studying simple piece-rate wages, he proposed a bonus wage system; from his study of the organizational structures of enterprises, he proposed the establishment of functional organization, i.e., workshops and departments. Lenin commented that the "Taylor system" embodied "the richest scientific achievements." As for the third, modern stage of management, that began in the 1970s when computers became widely used.

The Jin dynasty poet Tao Yuanming 陶渊明 (ca. 372–427) said, "Sensing that today is right, we renounce yesterday." When comparing the "Charter of Anshan Iron and Steel Company" and the Daqing experience with Soviet-style management and that of the three Harbin enterprises, we may ask which conformed to the principles of scientific management, i.e., the Taylor system. Which went backwards in terms of enterprise management?

To demonstrate Party leadership, Daqing implemented collective leadership of the Party Committee, which contrasted with the Soviet one-leader system, i.e., the system wherein one plant manager is in charge.

To demonstrate that it was "keeping politics in command," Daqing criticized the use of material incentives. To this effect, they eliminated piece-rate wages and rewards based on work accomplished, which was in stark contrast to the Soviets who emphasized economic results, auditing, and paying wages based on work accomplished. People at Daqing paid little attention to auditing, and concerns about economic results were secondary.

Everything in Daqing was a noisy mass movement. Systems of rules and regulations were seen as expressions of Soviet revisionism's interference, obstruction, and oppression.[3] During the early part of the oil campaign, even the quest for technological innovation and technological revolution was turned into a mass "dual revolution" movement. The Party Committee even set up a "Dual Revolution Office" to lead this movement in opposition to the normal Soviet production procedures, rules, and regulations.

Daqing, following the PLA organizational model, formed "campaign troops." Bureaus and divisions, under the Party's collective leadership, had production offices (similar to headquarters in the PLA) and political departments. Even the most basic work units of the organization had their own political teachers and advisors. This model, which contrasted with the system of workshops and departments in the big northeastern enterprises built with Soviet assistance, was affirmed and praised by Mao Zedong. In 1976, after the Gang of Four's downfall, Deng Yingchao 邓颖超, the wife of Zhou Enlai and Chairman of the All-China Federation of Trade Unions, was surprised to learn that plants had political departments and that unions were subject to their leadership; she thought this was very inappropriate.

3 In 1960, the Soviet Union and China had a serious ideological disagreement. The CCP considered itself Marxist and argued that the Soviet Union had betrayed Marxism and had become revisionist. In China, there was a movement criticizing Soviet revisionism, denouncing all regulations and lessons learned from the Soviet Union as forms of interfering with, obstructing, and oppressing the masses.

As Francis Bacon (1561–1626) once said, "Truth is the daughter of time, not of authority." If we set aside the background to the trends of that time and get beyond the ideological disputes, however, we can see that Daqing Oilfield went backwards by following the management approach of the "Charter of Anshan Iron and Steel Company."

4 The Construction of Little Oil Towns

A 1998 biography of Kang Shi'en quotes him as saying, "Once Chairman Mao heard the report on Daqing, he said that Daqing was breaking several of the frameworks of the past."[4] In the previous section, I have mentioned that Daqing, in terms of management, used the "Charter of Anshan Iron and Steel Company" as its blueprint for eliminating the regulations and restrictions that had been promoted by the Soviets.

For example, Daqing dispensed with Soviet models in terms of construction in oil zones. One such model concerned workers' living quarters. For example, the Dushanzi Oilfield in Xinjiang and the Lanzhou Oil Refinery in Gansu had been both designed by the Soviets in the 1950s. The production areas of these two plants were situated relatively far from the offices and living areas. Basically, the living areas and offices were in cities, whereas the production areas and factories were in the outskirts. At the time, Yu Qiuli and Kang Shi'en thought that this kind of layout separated production from the masses and was therefore an expression of Soviet revisionism in oil zone construction.

The Daqing Oil Campaign occurred during the three years of national economic hardship caused by the Great Leap Forward, and tens of thousands of employees were suffering from hunger. This was the most pressing issue faced by Yu Qiuli and Kang Shi'en. With the support of Heilongjiang Province, tens of thousands of *mu* of land were allocated to farming. Some employees and employees' wives reclaimed wasteland and planted crops. This effort was very effective. Zhou Enlai praised it as a new model for "combining industry with agriculture, combining city and country, beneficial to production and making life more convenient." Now when we look back on it, this mode of oil zone construction was a rather good approach, given the conditions of its time and location. However, as a model for modern industrial enterprise, it was of dubious benefit.

4 Kang Shi'en zhuan bianxie zu 康世恩传编写组, ed., *Kang Shi'en zhuan* 康世恩传 [*Biography of Kang Shi'en*] (Beijing: Dangdai Zhongguo chubanshe, 1998), 143.

But given the above circumstances, Yu Qiuli and Kang Shi'en used the slogans, "Revolutionize the Offices of Enterprises" and "Move Leadership Closer to the Frontline," and insisted on setting up the campaign headquarters not in a large- or medium-sized city like Harbin and Qiqihar or even a town like Anda and Sartu, but rather right by the oil wells. All the second-tier units were situated anywhere from three to more than 100 km away from the headquarters. Subsidiary teams and production crews of these second-tier units were also scattered around the oil gathering stations or oil wells for which they were responsible. Thus scattered across the several thousand square kilometer area of the oilfield were these small but complete facilities for production, living, education, entertainment, and medical services; the "Bureau" became Daqing City, "offices" became towns, and "oil teams" became villages. It was the "rural peasant economy" translated into oilfield production!

If they had had the foresight, if they had had the slightest idea of what a modern enterprise really is, they would have put the campaign headquarters in nearby Harbin to the east, or Qiqihar to the west. That would have benefitted not only the construction and development of those two cities and the oilfield, but also hundreds of thousands of oilfield employees and their families. The money saved would have amounted to hundreds of millions of yuan, and the benefits would be too enormous to assess.

What is exasperating is since the advent of the present century, continuous clusters of buildings and crisscrossing highways with unending streams of traffic burgeoned across the entire Daqing area, creating sprawl that irreversibly swallowed up the green, grassy plain. This outcome had never been anticipated in the early days of the campaign. These comments are not meant to scold the older generation, but rather to remind the younger to pay special attention to the environment of this grassland.

5 Children Speak the Truth

In the middle of the winter in 1961, a few colleagues of mine were chatting in the office. Also present was an eight- or nine-year-old boy, the son of a division manager. Someone asked him: "When you grow up, what would you like to be?" Without hesitation, he answered: "I want to be the head chef in the little cafeteria of No. 2 Courtyard." That courtyard held the offices that doubled as living quarters for the various heads of the campaign headquarters. The cafeteria was only a dozen meters away from the window of our office. Its sewer drained into an outdoor cesspool. Leftover fish and meat often flowed out into this cesspool, and because it was winter, the inflow would quickly turn into layers of white fat—at a time when there was an extreme food shortage. No wonder even a child would want to be a chef there!

In Daqing, starting in 1960, little cafeterias were set up in the office buildings of the bureau and second-tier divisions. The little cafeteria in No. 2 Courtyard cooked meals for a dozen bureau directors and 30 or 40 division managers. Meanwhile, the big cafeterias used by everyone else offered only corn meal, sorghum, millet, and cabbage for months. Each person's quota was some 20 *jin* (10 kg) per month or 8 *liang* (400 g) per day. A typical meal might be one steamed cornbread bun weighing 2 *liang* (100 g), a lot of cabbage, and a 1 *liang* (50 g) bowl of millet porridge so thin that the diner could see his or her own reflection in it. It very seldom served meat. In the little cafeteria, on the other hand, the rice and flour were refined, and fish and meat abounded; there were five or six dishes per meal—often northwestern dishes, since many of the diners had come from the northwest. As long as one paid enough grain coupons, one could eat until he was full. The cost of meals was only one-fourth or one-fifth of the actual cost.

During the four years I was in Daqing, I was lucky enough to set foot in there twice. Once was in the winter of 1961. I was helping a deputy commander write a speech, and we finished up around one o'clock in the afternoon. The big cafeteria was closed. He took me to the little cafeteria and asked the chef to make me a bowl of lamb soup with a pancake. What a hearty and memorable meal! Another time, on some holiday in the first half of 1962, Kang Shi'en entertained some executives and experts. Since I was the rapporteur for the meeting, I was included. We drank liquor and ate lamb kebabs to to our hearts' content. That was the first time that I, as a "Hunan barbarian," first tasted these two fine northwestern dishes. That is why, to this day, I still remember it. It is said that the practice of having little cafeterias for bureau- and division-level managers was learned from the PLA. During wartime and when the People's Republic of China was first founded, there were indeed little cafeterias in the army, but they were under strict management. The little cafeteria in Daqing was badly managed. It became a hospitality center where people ate better and took much more than their share, creating a sharp contrast to the extreme hardships endured by the masses.

During the Daqing Oil Campaign, it became routine for bureau and division chiefs to have their own personal secretaries. This "tradition" lasted to the end of the twentieth century throughout the oil industry. A few bureau chiefs who had come from the Ministry of Petroleum in Beijing and worked and lived in the No. 2 Courtyard not only had personal secretaries but even messengers and female attendants.

All kinds of business meetings and report meetings were held day and night at the No. 2 Courtyard. For nearly twenty days each month, there were meetings, and they rarely ended before midnight. Employees like us could only to stay up with them through the meetings. However, after the meetings ended, the leaders could first go to the little cafeteria for a "midnight meal" and then sleep until nine or ten o'clock the following morning, whereas employees like us could only go to sleep hungry and had to go to work the following morning at 8:00. Even worse was that Yu Qiuli and Kang Shi'en stipulated without authorization that the whole oilfield had to observe long weeks, meaning that employees could have one day off every ten days. This regulation remained in effect until the campaign at Huabei Oilfield in 1976 and was widely promoted as a sign of the oil workers' endurance and perseverance. They did not see the negative effect it had on workers and their families. They deprived the workers of the basic right of one day of rest every seven days. It was really inhumane.

During the first two years of the campaign, a large ball would be held once every ten days in the big conference room of the No. 2 Courtyard. It was said that they borrowed this tradition from Zhongnanhai, where leaders of the Chinese central government reside in Beijing. Orchestras and beautiful young women from performing troupes would provide the entertainment. Division-level managers and higher could attend the ball with their dance partners, but other people were not entitled to go. The event was heavily guarded to turn away young people who sought entry. Meanwhile, there was hardly any entertainment to speak of in those days other than a free film in the big cafeteria or out in the open in the summer. The balls were held most often when Tang Ke and Wu Xingfeng were at the oilfield, because the two were very fond of them. Once the campaign began, Yu Qiuli never stayed on site for more than a few days, but Kang Shi'en stayed on site year-round during the four years he resided there. Only on the rare occasions when he left would there be balls, because he did not like them. I did not go to these balls even once, mostly because I lacked the status and would not be allowed in, but in any case, I did not know how to dance and loathed the noise. I lived close to where the balls were being held, and they kept me up late into the night.

All of these things show that from 1960 to 1963, the cadres in leadership positions at the department level and higher led privileged lives. Those who were excluded really resented that.

6 Exaggerated Achievements and Other Problematic Aspects of the Daqing Experience

The first exaggeration related to the calculation of the size and oil reserves of the Daqing Oilfield. Since oil reserves are hidden many meters underground, one cannot see or touch them. One can only get information based on indirect methods, calculating according to certain formulas and analytical methods. Furthermore, the underground data are dynamic, and even the set formulas one uses are dynamic, because of technological advances. The data on Daqing Oilfield were gathered within a relatively short time, but then its size and its world ranking were announced to the world, though it also was boldly declared that the calculations were based on only one-and-a-half years' worth of data, while an American oilfield used nine years' worth of data, and a Soviet oilfield used three years of data. Although laymen were pleasantly surprised, oil insiders said that the size and reserves for a large oil and gas field like Daqing could not be calculated in the first year or two. Furthermore, the information on the oil reserves for those two American and Soviet oilfields had been collected many years earlier—it was dated. Using those as reference points was very inappropriate.

A second set of exaggerations revolved around the development in 1960 of a 30 km^2 trial area in the center of Sartu Oilfield, which measured a few hundred square km. This trial area was marked for immediate drilling—which was opposed by many scholars and experts. They said it was like making underwear using a high-quality piece of material, good enough for a suit. This remark was caustic, but it really meant that the oilfield's development should begin only after the oilfield exploration had been basically completed, and then the development should proceed in an orderly manner, according to a comprehensive plan. Only then would one get the best results. Drilling should not be done haphazardly, without a good plan. Not only did Yu and Kang not abide by that train of thought, they waved Mao Zedong's banner of "All experience needs to be tested," and expended many words to argue that drilling was necessary, scientific, and advanced. They went so far as to call it an "innovation" of Daqing. The truth is that at the time, production in several old oilfields in the northwest had not gone up, which meant that the targets in the annual national crude oil production plan could not be reached. This mattered a great deal for the political aspirations of Yu Qiuli. Once this "big chunk of juicy meat" was discovered in Daqing, even if the right conditions for drilling did not exist, Yu was like a hungry person who could not be choosy about his food. He and the other leaders needed to fulfill that production quota. That is why they set aside an area for drilling and called it by the fine-sounding name, "Experimental Zone for Launching Production." They did not have enough time to construct oil collection, transport, and storage facilities, but they drilled anyway, so the zone was soon full of oil ponds big and small, all filled with black crude oil. Not only did it pollute the grassland but it also caused tremendous waste. Thus in 1960, Daqing produced 970,000 tons of crude oil, which guaranteed that the Ministry of Petroleum met the targets of its annual plan. Victory was reported to the State Council. Looking back on it now, I would concede that one could do things this way, but it is completely unwarranted to describe this as an "innovation."

Thirdly, the "job responsibility system" was publicized as one of the main lessons of the Daqing experience. Actually, it was merely a means to rectify the chaotic situation after a fire broke out at a water-injection station in 1961. Posts were defined, responsibilities clarified, regulations made. It should not be considered as some "innovation." It is said that Zhang Chunqiao 张春桥, one of the members of the Gang of Four,[5] once said privately, "Daqing's job responsibility system was nothing new. Even the maids in the Grand Prospect Garden in *Dream of the Red Chamber* had a system of personal job responsibilities." Well said! We should not trash the remark because of the man.

Fourth, a key component of the Daqing experience was fostering the oil teams' "strict and detail-oriented style of work." The best example of the emphasis on this would be the following. At the Petroleum Executives Meeting held in February, 1964, in Beijing, Yu Qiuli delivered a report in which he discussed how to cultivate the teams' style. He said that being "strict" or "strictness" (*yan* 严) should be paramount. In fact, he used the word 15 times:

> Strictness can arouse enthusiasm; strictness can bring out a responsible mindset; strictness can strengthen fighting power; strictness can enhance specifications; strictness can elevate standards; strictness can foster a good atmosphere; strictness can result in good products; strictness can enhance technology; strictness can come up with better methods; strictness can eliminate the market for liberalism and individualism; strictness can banish unhealthy trends and evil practices; strictness can prevent mistakes; strictness can preserve ideological and political conformity; strictness can guarantee cohesive action; strictness can guarantee unity.

At the time, I felt that all of this was empty sloganeering, for which he should have been criticized.

The occurrence of mistakes like these had their objective and subjective reasons. Objectively, we, as a nation, had been closed up for too long. As a result, little was known about what went on outside of China. We were ill-informed,

5 The Gang of Four included Jiang Qing 江青 (Mao Zedong's wife), Wang Hongwen 王洪文, Zhang Chunqiao 张春桥, and Yao Wenyuan 姚文元. In the beginning of the Cultural Revolution, Mao Zedong entrusted power to these four people. In 1974, Mao Zedong criticized them for trying to form a faction and called them the Gang of Four. On September 9, 1976, Mao died. On October 6, the Gang of Four was arrested and tried by the Communist Party's Central Committee, headed by Hua Guofeng 华国锋. They were sentenced in 1981.

like the proverbial frog looking up at the sky from the bottom of a well. Ever since the Great Leap Forward started in 1958, exaggeration ("launching satellites" or *fang weixing* 放卫星) became common throughout society, in all areas of discourse. Subjectively, Yu Qiuli and Kang Shi'en got dizzy with success and became conceited. Their positions gave them not only political authority but also "speech right," i.e., authority over the media and propaganda, and therefore they could have their thoughts published without being challenged. Of course, just because one can say something does not mean that one should. To conscientiously use one's speech right in a disciplined way takes willpower, and may be considered a virtue.

CHAPTER 14

"In Industry, Learn From Daqing" Reconsidered (Part Two)

1 Premier Zhou—the Chief Director of the Daqing Oil Campaign

Oil gushed out of Songji-3 on September 26, 1959. The following December, Premier Zhou Enlai convened a Northeast Economic Cooperation Meeting in Harbin, during which he directed Ouyang Qin and Yu Qiuli to plan and organize the oil campaign in Daqing. He stressed that this campaign must be guided by Mao Zedong thought, and that it must use the standpoints and methods of dialectical materialism to analyze and solve problems encountered during the campaign. Afterwards, Yu Qiuli decided right there to study Mao's two theses. From this time on, until the CCP Central Committee formally approved the campaign in February 1960, Zhou personally asked about and arranged important campaign matters, making sure that the campaign would get off to a quick start. After the campaign began, he visited Daqing three times, more than any of the other leaders in the Central Committee.

The first time he went was in August 1961, during Daqing's most difficult period. At the First Plenary Session of the Third National People's Congress in December 1964, he summed up the experience of the oil campaign at Daqing in his "Government Work Report." From the beginning to the end, he satisfactorily carried out his duties as the "chief director."

In 1970, when the Cultural Revolution was severely impacting Daqing, Premier Zhou gave the timely instruction that Daqing should once again study Mao Zedong's two theses. That would put Daqing back on course. With Premier Zhou's care, Daqing continued to move forward, reaching its highest production capacity of 50 million tons in 1978, which it maintained for decades. Even though Premier Zhou passed away a long time ago, he is still much respected and his memory is still cherished.

2 The Positive Influence of "In Industry, Learn from Daqing"

Aside from its lofty political implications and the noise of propaganda, the slogan "In Industry, Learn from Daqing" exercised a positive economic influence that cannot be overlooked.

 | DOI:10.1163/9789004402737_015

Petroleum production in China went from 2.5 million tons in 1960 to 138 million tons in 1990, which included Daqing's annual production of 50 million tons. This more than fifty-fold increase in 30 years resulted in two significant changes: one, China changed from an oil-poor country to an oil-rich county, the eighth largest producer in the world; and two, China changed from an oil-importing country to an oil-exporting country. In fact, oil export revenues made up two-thirds of China's total national export revenues in 1985. The oil industry paid the most taxes to the country. In 2000 alone, Daqing Oilfield's taxes amounted to half of the total from all state-owned enterprises. During the Cultural Revolution especially, when many industrial enterprises and transportation came to a halt, energy supply was also very tight. As a remedy, Premier Zhou ordered the oil industry to increase production. It is estimated that from the start of the Cultural Revolution in 1966 to 1978, the country burned through 249 million tons of oil in 12 years, averaging 20 million tons annually. One year, it burned more than 30 million tons. Thus the majority of oil workers stood fast at their posts. Work and production never stopped, even for a day, in the oilfields. Moreover, one after another oil campaign was launched, which rapidly increased oil production and shored up the national economy, which was near collapse.

Therefore, during the thirty years from 1960 to 1990, oil was the main act among Chinese industries. The subtext of the slogan, "In Industry, Learn from Daqing," was, "The national economy relies on oil." Such was the positive effect of the slogan.

3 Daqing, Emblematic of the Times

After 1959, the "deity-making" movement in the country intensified. Among the lessons from the Daqing experience, "Mao Zedong thought has been the fundamental source for our achievements at Daqing" was sequentially the first and the most frequently quoted. This "lesson" needed no testing and could not be tested. Its only purpose was building on Daqing's success to praise "the invincible Mao Zedong thought." This was not particular to Daqing but rather the prevailing style of official writing of the time, and it was also a remnant of the style used in "memoranda to the throne" used for thousands of years in Chinese officialdom. In addition, all over the nation, people were studying the PLA; China and the Soviet Union were seriously at odds with each other, so a national movement was launched throughout the country to attack Soviet Revisionism. The lessons from the Daqing experience bore the stamp of all of this. The fruits of Daqing—good and bad—were all products of that time.

4 Kang Shi'en—the Actual Leader of the Daqing Oil Campaign

Having worked for and observed Kang Shi'en for almost 30 years, I have a good understanding of the man. In his early years, he studied geology at Qinghua University in Beijing. Later, renouncing the pen for the sword, he joined the Communist Eighth Route Army fighting the Japanese invaders. In September 1949, he was appointed by General Peng Dehuai to be the chief military representative of Yumen Oilfield, then the largest one in China, and so he stepped into a leadership position in the oil industry. In 1960, he became the commander of the Daqing Oil Campaign. He lived and had his meals at the simple and crude No. 2 Courtyard and stayed there for four years, directing this campaign and personally formulating the ideas that would then become the lessons of the Daqing experience. In 1965, he was promoted as Petroleum Minister and led the oil industry forward until he retired in 1982. He understood petroleum geology, the different oilfields, and the workforce. He was the most decorated official in the Chinese petroleum industry.

In Daqing, Kang Shi'en had two effective helping hands: Deputy Commanders Zhang Wenbin and Jiao Liren. After 1965, they were successively promoted as Deputy Ministers. In the more than 20 years following the Daqing Oil Campaign, the Ministry of Petroleum planned and organized a dozen consecutive major oil campaigns and important startup projects. For nine of

FIGURE 14.1 Kang Shi'en at the ribbon-cutting ceremony of the first shipment of crude oil from Daqing; June 1, 1960

these campaigns, Kang Shi'en sent Zhang or Jiao out to direct the campaigns on-site. Thus the three of them worked hand-in-hand in the oil business for decades. They were outstanding men of action in the oil industry and well deserved to be called "the Three Fierce Generals of Petroleum (*shiyou san han jiang* 石油三悍将)."

5 The Ministry of Geology's Contributions Should Have Been Included in the Lessons of the Daqing Experience

After much consideration and reflection, I think that the Daqing lessons boil down to the two that Premier Zhou had affirmed: a high degree of revolutionary spirit and a strict scientific attitude. As for high revolutionary spirit, the arduous efforts of oil workers like Ironman Wang Jinxi were representative, and as for strict scientific attitude, there was the quiet dedication of scientific and technological personnel like Han Jingxing, Zhong Qiquan, and Feng Jiachao, as well as the scientific explorations by generations of Chinese geological workers. The Ministry of Geology, the Ministry of Petroleum, the Chinese Academy of Sciences, the Ministry of Coal, as well as the Nationalist government's National Resources Commission all played a part in Daqing's discovery. American, Japanese, and Soviet geologists also made their contributions. From 1955 to 1959, people from Ministry of Geology did a lot of hard and detailed work in Songliao Basin—but this was not reflected in the lessons from the Daqing experience.

In early 1966, a "Daqing Exhibition" was held in the National History Museum by Tiananmen Square in Beijing (I took part in its preparation as one of the chief editors from the Ministry of Petroleum). It was extremely popular. Among the hundreds of exhibition boards, only one, at most one meter in width, in the introductory section had a picture showing people from the Ministry of Geology at work, and the caption underneath only read, "In the discovery of the Daqing Oilfield, the Ministry of Geology also did much preliminary work." In the visitor's book, some of the well-informed visitors pointed out that it was unfair that the exhibition mentioned so little of the Ministry of Geology. But Kang Shi'en was mainly responsible for this, since he repeatedly considered and reviewed the exhibition's outline.

As for the Ministry of Geology's achievements and functions in Daqing, there was a rumor Mao Zedong once said in 1964, "Li Siguang gave the directions, Yu Qiuli did the actual work." This would be a fair and precise assessment, corresponding to reality.

6 "Fine Troops and Capable Generals" versus Cadres from "All Corners of the Country"

As soon as oil was found in Daqing, the Ministry of Petroleum transferred "fine troops and capable generals (*jingbing qiangjiang* 精兵强将)" from a few old oilfields in the northwest and Sichuan to Daqing for the campaign, which proved successful. This too was considered one of the main lessons from the Daqing experience. However, it was also problematic. Looking back, I feel that the decision to transfer the top people into this area from outside was both rash and biased, because it ignored the contributions of the Songliao Petroleum Bureau, which was already present, so that its staff was given a role secondary to these outsiders. This created ill-will in the workforce and made unity hard to achieve going forward. As a result, during the ensuing Cultural Revolution, there were two opposing factions consisting actually of workers from the original Songliao Petroleum Bureau versus the outsiders. Yu Qiuli and Kang Shi'en had not anticipated this. Unfortunately, this method of importing outsiders became the rule. Most of the dozen-plus oil campaigns after Daqing used this practice of replacing the original management with Daqing's "fine troops and capable generals," and the practice became even more pronounced at the height of the "In Industry, Learn from Daqing" movement. For example, in early 1978, when a new Petroleum Minister and ten Deputy Ministers were appointed, seven of the eleven came from Daqing. This new ministerial team then transferred Daqing management to other important oilfields to head up key Party Committee positions, so that they would lead these oilfields in their "In Industry, Learn from Daqing" movements and take charge of hiring. This flew in the face of the Party's usual practice of recruiting cadres from "all corners of the country (*wuhu sihai* 五湖四海)." As a result, cries of discontent rose all around among the oil workers and rifts were numerous. Therefore, in all things there is a measure, and any good thing can turn bad if taken to an extreme. It is an ironclad law of dialectics.

7 No Contemporaneous Historians Are Believable

The Daqing Oilfield and the Ministry of Petroleum published numerous pictorials and history books, some of which were of exquisite quality, because they had the money to do so after the launch of the "In Industry, Learn from Daqing" movement in early 1964 and especially after the end of the Cultural Revolution in 1976. What struck me most after reading these is that the first few pages always feature photos of the Party Central Committee leaders headed by Mao Zedong, followed by all the Petroleum Ministers (or chief executives). These photos and the publications they are in convey only a single message: the achievements of Daqing and the oil industry were all due to the loving care of the Party Central Committee and the leadership of the Ministry of Petroleum. I have always felt that such "officialese" runs contrary to the materialist conception of history, which the Party had so staunchly advocated, and which holds that "the masses create history (*renmin qunzhong chuangzao lishi* 人民群众创造历史)." In many of these works on the history of oil, the authors go so far as to disregard the facts in their efforts to laud their superiors' achievements and shift blame for errors on other people. Changing colors to suit the political climate, they willfully revise and distort historical facts. There is a saying in Chinese, "No contemporaneous historians are believable (*dangdai wu xin shi* 当代无信史)." It means that throughout Chinese history, historians of whatever period they lived in have been beholden to those in power, so that their accounts cannot be believed. It is very hard for Chinese historians to escape this curse.

8 The "Meek Little Girl" Phenomenon in the Daqing Experience

After the nine lessons from the Daqing experience were formulated in late 1963, the leading slogan became "Hold High the Great Banner of Mao Zedong Thought (高举毛泽东思想伟大红旗)." It served as a leitmotif throughout all subsequent political movements. Sophistry was rampant and political adversaries' words got plucked out of context to paint a distorted picture of their authors. For example, during the Cultural Revolution, using Mao Zedong thought as a weapon, huge numbers of people criticized the revisionism of Liu Shaoqi[1] in a big way. His bland comment, "You who work with oil must correctly handle the relationship of man, technology, and rocks," got framed as revisionist. (Heavens, how was that revisionism?) Then Deng Xiaoping's "cat

1 Before the Cultural Revolution, Liu Shaoqi was the first Vice President of the Central Committee of the Communist Party and Mao Zedong's designated successor. After the Cultural Revolution started in 1966, he was removed from power, and in 1969, he died under the harsh conditions of his imprisonment. All of his speeches were denounced as revisionist. He was rehabilitated in 1980.

theory"[2] came under criticism. After Lin Biao died in a plane crash in 1971,[3] it was his and Chen Boda's 陈伯达 turn.[4] The newly-reinstated chief at Daqing, Song Zhenming, adopted the motto, "We at Daqing shall not be fooled again" (咱们大庆再也糊弄不了了)," which received Kang Shi'en's approval. Kang repeatedly urged me to insert that statement into Daqing's annual report for 1974. When the tables turned in 1976, the words of the Gang of Four were cited to prove how they had brought calamity to the oil industry. Thus when I look back at the Daqing experience and the many ways it has been distorted, I can only think of a comment someone once made: "History is like a meek little girl, you can dress her up any way you want."

9 Daqing Becomes a "Club"

During the Cultural Revolution, Daqing became a "club." On January 8, 1967, at an oil industry meeting attended by 5,000 people, Zhou Enlai proclaimed, "Daqing is a red banner personally hoisted by Chairman Mao," and "Yu Qiuli is the general appointed by Chairman Mao." In 1969, he also said, "Kang Shi'en was named by Chairman Mao" (to go to Jianghan 江汉 Oilfield). After that, especially in the later stages of the Cultural Revolution, within the oil industry Daqing became a "club" with which one faction beat the other in the ongoing Struggle of the Two Factions. Those who criticized Daqing and wanted to overthrow Yu and Kang were all blamed for being "Against the Red Banner of Daqing."

At a public meeting denouncing Bo Yibo held at the Ministry of Petroleum in 1967, I witnessed the rebel faction criticizing two statements that he had made. After a disastrous fire at Daqing Oil Depot in the winter of 1960, Bo Yibo made the statement, "The big campaign is a big melee," and in 1964, at

2 Before the Cultural Revolution, Deng Xiaoping was the General Secretary of the Communist Party and Vice Premier of the State Council. He was removed from power together with Liu Shaoqi. In 1961, while criticizing some cadres as being only good at empty talk, not at practice, he said: "It doesn't matter if a cat is white or black as long as it catches rats!" It was derided as his cat theory and denounced as revisionist during Cultural Revolution.

3 Lin Biao, famous general of the Chinese Communist Party, one of the Ten Marshals. In 1966, he became Mao Zedong's designated successor.

4 Chen Boda had been Mao Zedong's political secretary in 1938. When the Cultural Revolution started in 1966, he was vested the power to lead the Party Central Committee's Group on the Cultural Revolution. In 1970, at the Party Central Committee meeting, he and Lin Biao proposed their "genius theory" (meaning that Mao Zedong was the genius) and proposed that the country have a president, wanting to elect Mao Zedong to be the Chairman again, an act that enraged Mao Zedong. Mao denounced Chen Boda as "political swindler."

the height of the "In Industry, Learn from Daqing" movement, Bo quipped that Mao's praise of Daqing was like "having the love of three thousand all to oneself."[5] Taking the simple facts of the matter and raising them to the higher plane of political struggle, the rebels claimed these were "Anti-Chairman Mao" and "Anti-Daqing Red Banner" statements. One of the founding fathers of the State and the Party, the wretched Bo Yibo stood there, head down, speechless, stunned beyond compare. In Daqing, the model worker Zhang Hongchi had been lauded for "creatively studying and applying Chairman Mao's Works." Before the Cultural Revolution, he was as famous as Ironman Wang. He also kept a journal, so he was called "the living Lei Feng (雷锋) of Daqing."[6] During the Cultural Revolution, with the support of the Military Control Commission, Zhang became the deputy director of Daqing's First Revolutionary Committee. However, after the withdrawal of the Military Control Commission and the reinstatement of Daqing's original leadership to power, Zhang Hongchi became an "Anti-Daqing Element" and disappeared. Thus Daqing had become a "club" for hitting your opponents, and it must be said that it was another kind of sorrow for Daqing. I now regret having identified with such activity at the time.

10 Success Is Also the Mother of Failure

Propelled by the momentum and enthusiasm generated by "In Industry, Learn from Daqing," oil production increased quickly, year after year, until 1990. This success made decision-makers overly optimistic and, as a result, each ensuing oil campaign, such as the Jianghan Oil Campaign in 1969 and the Huabei Oil Campaign in 1976, made the mistake of over-planning, over-staffing, and being tremendously wasteful. In this respect, the Tarim Oil Campaign was the worst, resulting in zero growth in on-land petroleum through the 1990s (for details, see "The Tarim Era" below). These stories of "In Industry, Learn from Daqing" also make it clear to people that if we do not handle success well, then success can also be the mother of failure. Interesting? Plausible? Is this not worth pondering?!

5 Bo Yibo was alluding to a line in "Song of Everlasting Regret (长恨歌)" by the Tang poet Bai Juyi 白居易 (772–846): "In the rear palaces dwelled three thousand beauties, / But the love for three thousand was placed on one alone." The poet was retelling the love story of the Tang Emperor Xuanzong and his concubine Yang Yuhuan.

6 Lei Feng, who was born in 1940, was a driver in the PLA who died from a work-related accident. He worked unselfishly for the public good and loved to help others. His diary recorded his short life. In 1963, Mao Zedong called for "Learning from Comrade Lei Feng," which starting a long-lasting movement.

CHAPTER 15

A Scribe—My Petroleum Career

1 Unexpected Assignment in Daqing

In April of 1960, at a military veteran officers' training class in Dalian, the instructor had a talk with me, asking me about my future plans. I merely said, "I have nowhere in particular to go, so I'll just follow the Party's arrangements." My only request was a month's leave before they sent me to the new company, so I could go to Shanghai to see my newborn daughter.

The instructor replied, "A new oilfield has been discovered in Heilongjiang Province. They need people. Would you go there?"

"What's an oilfield for?" I asked. "Is it the same as a rice field?"

"An oilfield produces gasoline. It's underground, not above ground."

"That's fine," I said. (Actually, I had no choice, even if it had not been fine.)

The instructor said, "I have no problem granting you a month's leave. However, many people are eager to go there right away in order to secure a good position."

I insisted, "When my daughter was born, I was not there. I let my wife and daughter down. Even if I were assigned to cleaning up pigsties, I must go to Shanghai to visit the two of them first."

I travelled alone to Shanghai, bringing with me a letter of introduction and 200 yuan, which included a retirement pension of 164 yuan, about two month's salary, for my ten years of military service. In Shanghai, I spent 152 yuan on a Panda brand radio—made in Nanjing, and the best in the country at the time—as a gift for my newborn daughter. I stayed in Shanghai for 28 days and took a train to Harbin in late June. In those days, the train station in Harbin was very chaotic. The transfer station for oil people was at Sankeshu 三颗树 (literally, "Three Trees"), which took me a long time to find. There was a sea of people there. Like a refugee, I wound up climbing onto a freight train. After wobbling along for almost three hours and covering less than 100 km, the train finally reached Anda Town, seat of the Daqing Oil Campaign headquarters at the time. The army veterans' reception station had set up shop there in an adobe house. I was given shelter at an inn whose only amenity was a huge heated brick bed, which could sleep more than a dozen people. The inn was filled with veterans, some bringing their wives. After some time, some would make love, as if there were no one around, which provoked loud rebukes from the others.

 | DOI:10.1163/9789004402737_016

Two days later, I received a notice, asking me to report to the nearby Publicity Division of the Party Committee of the campaign headquarters. Some others had been staying at the inn for more than twenty days and still had not been assigned a post. They must have seen my file ahead of time and believed I would be useful, so this was lucky for me.

Thus began my life at Daqing. After reporting to the Publicity Division, I learnt that the division chief had left to stay at the Sartu frontline. There was only one other colleague, Yu Shilin 于世林, who had arrived a few days earlier than I. We became good friends later on. We had nothing to do and no supervision from any one. From July to September, the two of us would spend the days strolling around on the street. There was only one street in Anda Town, which was full of oil workers. I remember we could get freshly stewed chicken for a half yuan each. In the evenings, we would go to different places to see performances. At the time, the older oilfields in the northwest and Sichuan would send performing troupes, as a show of support. After performing for drilling crews, they would come to the campaign headquarters in Anda County to perform a few times. I cannot remember what those performances were. I only remember seeing big swarms of mosquitos under the lights on the stage. Sometimes, as soon as a singer opened his mouth, a big mosquito flew right in. He had to spit it out first and then resume singing.

2 In the Land of the Blind, the One-Eyed Man Is King

By September of that year, I had been idling in the Publicity Division for more than two months. Wen Sheng 闻胜, the office manager of the campaign headquarters' Party Committee said to me, "Young Mao, you have nothing to do at the Publicity Division. I've been attending daily meetings held by Deputy Minister Kang Shi'en and Deputy Party Secretary Wu Xingfeng. The day before yesterday, Minister Kang decided that on behalf of the campaign headquarters, a campaign progress report should be written to the Party Committee of Heilongjiang Province and the Party Group of the Ministry of Petroleum. I'm shorthanded, and Secretary Wu has agreed, so would you consider temporarily being transferred to the Party Committee office to help us?"

"I have nothing to do at the Publicity Division," I replied. "If you think I can do it, I'll do it!"

Then, I started writing this report. How did I go about it? At the time, I was not senior enough to attend meetings held by a deputy minister of the state. To write such a report, Wen Sheng would take down notes of what was said at these meetings and pass them on to me. I would then organize the information and write it up. He would deliver my writing to Kang and Wu. After repeating this process three or four times, the report would be ratified by Kang and Wu. At the time, there was no printing equipment at the campaign headquarters or Anda Town. A messenger who handled top-secret documents took the report to the secret printing plant in Harbin to be printed. Then it was delivered by courier ten days before the National Day holiday. Now, I do not recall the content of the report, except for a rhyme I made about petroleum buried deep underground, and how hard it was to extract: "A cherry's delicious, the tree hard to plant, / And underground petroleum quite hard to get!" This rhyme won the praise of Kang Shi'en, who said to Wen, "That young man who helped you write—his rhyme was well-made." Naturally, I was pleased with his comment, but I was even more pleased with the fact that my writing got printed for the first time. Before that, other than articles that I had contributed to newspapers, nothing got printed, except for mimeograph copies I made myself. This was the humble beginning of my career as a scribe. Not that writing up the report required a lot of brilliance—there was just no other suitable candidate around. I was like the one-eyed man who becomes king in the land of the blind.

3 A Basket of Embarrassments

During my first four years in Daqing, I faced cold, hunger, and sleep-deprivation, but the hardest thing to endure was hunger. From the winter of 1960 until the fall harvest of 1961, I had to battle hunger at every moment, because starting in October 1960, the monthly food ration per person, consisting only of corn and sorghum, was reduced from 32 *jin* to 27 *jin*. I worked overtime every day. During the day, I went to various meetings and took notes. Sometimes I went to the various worksites to investigate noteworthy people who had accomplished some particularly crucial task. In the evenings, I would be at my desk, writing bulletins and articles, staying up until one to two in the morning. After working like this for two to three months, I finally developed dropsy. However, I came up with a number of ways to deal with an empty stomach, including a few embarrassing ones:

Embarrassing method no. 1: stealing cabbage. In November 1960, the yard of the campaign headquarters' vehicle yard next to our office was filled with Napa cabbage, stored there for winter. For a few days, around 11:30 at night, I got too hungry to continue writing. I asked my colleague, Brother Gu Ge 固革, who minded the telephone next door, to steal cabbage in that yard, suggesting that we would each grab a head and run. Gu Ge said that wouldn't do, explaining that it would be a disaster if people discovered us and reported our theft. So instead, we quickly peeled off some cabbage leaves and walked back to our office. After washing those leaves, we put them in the iron water kettle and boiled them on the coal stove. This temporarily solved my problem. But one or two hours later, after peeing outside of the office, the hunger returned. I dealt with it by going to sleep. We did this a few times, but then they moved the cabbage into the vegetable cellar, and feeling increasingly uneasy, we gave up on any further attempt.

Embarrassing method no. 2: stuffing myself at any chance. For nearly two years, the event I looked forward to most of all was always the oilfield-wide end-of-month all-level cadre meeting. A thousand people would attend these three-day events, which were one of the main campaign activities, but they also served as a means, during this period of hardship, to take care of grassroots-level cadres, improve their diet, and let them balance work with rest. During these meetings, people could eat as much of the food staples as they wanted, as long as they paid their daily grain coupon, which for us was eight *liang* (400 g). There were pork and chicken dishes as well. The most memorable dish for me was a specialty of northeast cuisine, pork stewed with bean or potato noodles and pickled cabbage. People would use washbasins as dishes and pile the food in, and if that was not enough, they would go for seconds. I was present at all these meetings, because I had to draft the final report for each one, but my chief motivation for going to these meetings was to eat my fill for three days, and to eat what we southerners prefer—rice. Frankly speaking, I would eat until my stomach hurt!

Embarrassing method no. 3: taking the best. In May 1961, I brought my wife and daughter from Shanghai to the wilderness of Daqing. This was right when the whole nation was enduring extremely difficult circumstances. In Daqing, not only was the grain ration severely reduced, there was no meat in the cafeteria for months at a time. People were forbidden to go to the little towns' farmers' markets, which were considered a remnant of capitalism and sites for speculation and profiteering. Throughout the oilfield, orders forbidding employees and their families to buy things at the farmers' markets were repeatedly issued. Even so, I privately convinced my wife to shop there, buying an egg for a half yuan, one *jin* of soy sauce for 10 yuan, a piece of beet for a half yuan in order to improve our daughter's diet. That summer, when the yellow lily flowers bloomed in the grassland, the cafeteria of our political department organized employees to pick them. Afterwards, we would pile them up in our room and mix them into our morning corn porridge. Sometimes, because they had not been washed thoroughly, there would be ants and little bugs in our porridge. At such times, we would joke that we had meat in our food. Nowadays dried yellow lily flowers are a rare and expensive foodstuff, similar to the wood-ear fungus. I regret to this day that I agreed to my wife's proposal that for her portion, she would eat only lily flowers, leaving the corn porridge for our daughter to eat. Because she ate so many lily flowers, even now, decades later, the sight of them still disgusts her.

I have many other embarrassing stories about my attempts to satisfy my hunger. For instance, every night, I would look forward to having my boss Wen Sheng sneak out two steamed buns (made with refined flour) for me from the late-night meals (usually noodles, steamed buns, wontons, vegetables, etc.) given at the meetings for senior officials. To satisfy my desire for liquor, I stole my wife's medicinal alcohol brought from Shanghai, mixed it with water, and drank it with a few friends. It gave me severe headaches and still triggers fear in me whenever I think about it.

4 My Writing Trick: Be a Willing Epigone

As mentioned above, I was involuntarily stuck with the drab clerical job of drafting documents in 1960, at the beginning of the Daqing Oil Campaign. No need to be modest, but from September 1960 to December 1963, I wrote many comprehensive articles and documentary materials for Daqing Oilfield, and as a result I received very high praise and honor. In 1961, 1,000 employees, out of the 40,000 total, were nominated for the "Five Good Red Banner Pacesetters" award. From these, 100 model pacesetters were chosen. I was one of them. In early 1964, after Chairman Mao coined the slogan, "In Industry, Learn from Daqing," the Ministry of Petroleum formed its Political Department, and I was one of the four people to be first transferred to Beijing to take part in that body.

In chronicling the Daqing spirit and "the Daqing Man," I did indeed make a meager contribution. Actually, scribes have been the companions of rulers since antiquity. Although respectable, it is a living earned by looking up at someone else's nose. Now that I have the time to think about it, I can come up with three reasons that contributed to my success in that endeavor.

First, in those days, the written summaries corresponded to the same format that had been taught to me by the political instructor at our military academy in the winter of 1950—achievements, experiences, shortcomings, and future tasks, etc. I was very familiar with this kind of formulaic writing and could write it with ease and dexterity.

Second, even though I was an ordinary person who lacked experience and broad perspective, once I was confronted with the grand scene of the Daqing Oil Campaign, I had no choice but to rely on careful observation and revision, on detailed note-taking and the drawing of comparisons. I also relied on three general sources to help formulate my writing: one was documents of the CCP's Central Committee and speeches and commentaries from the *People's Daily*, *Guangming Daily*, and *Red Banner Magazine*; another was speeches made by the leadership of the campaign headquarters; still another was the vivid stories that base-level workers and other employees could tell. To put it simply, the trick of writing consists of three words: observe, record, and compose. That is all. Nothing more.

Third, I benefitted from the appreciation of two people. One was the chief campaign commander and deputy petroleum minister Kang Shi'en. He was a quick thinker, logical, expressive, and exceptionally adept with agitprop. I remember in 1971, he took me and his secretary Wang Fuzhen 王福臻 from Jianghan Oilfield to Wuchang to attend a meeting of the Party's Standing Committee of Hubei Province. We stayed at the Hongshan Hotel, the best in the province. One night, behind closed doors, the three of us were talking about the uncivilized behavior of the Red Guards. Kang indignantly commented: "These Red Guards nowadays are no match for those Beijing students of the December 9th Movement of 1935. I remember that at the time, when we were out on the streets, we'd use just a few words and twenty to thirty people would gather to listen to us talk: 'Fellow countrymen! The northeast is in danger. Let's mobilize. Down with Japanese imperialism!'" When I wrote for him, he would dictate and I would record. We would repeat the process two or three times, and the project would be accomplished.

The second person was Wu Xingfeng, the Deputy Secretary of the Party Committee of the Daqing Oil Campaign, and later the Director of the Political Department of the Ministry of Petroleum. Born in Malaysia, Wu studied elementary education in college. When he returned to China, he joined the Communist Eighth Route Army and followed it through the Shaanxi, Gansu, and Ningxia region to fight the Japanese. Learned, reflective, and eloquent, Wu had the reputation of being "a star with many talents." He was the first to discover and make use of me. As with Kang Shi'en, when I wrote for him, he would dictate, and I would record and arrange.

Speaking of writing, Kang Shi'en once told me a story about what he wrote as part of his entrance exam for Qinghua University in 1936. He said that when he was in high school in Beiping (as Beijing was called then), he had to stop and go into hiding here and there, because he was being hunted for his involvement in the Communist-led anti-Japanese student movement. One day in Shanghai, he saw a student-recruiting notice on the street by Qinghua University. He applied. While preparing for the written exam, he heard that the head of Qinghua's Chinese Department was the literary giant Zhu Ziqing

朱自清 (1898–1948). He bought a copy of a collection of Zhu Ziqing's essays and studied the famous "Retreating Figure (Beiying 背影)," about Zhu's father seeing him off at a railway station. And then a miracle took place. Qinghua's composition title for that year happened to be "My Uncle." Therefore, he wrote in the same style as Zhu's "Retreating Figure" and received high marks.

Comparing my experience to Kang Shi'en's story, I came to a realization. If asked why my writings in Daqing won approval and praise, my answer is simply that I was a willing epigone.

5 The Daqing Exhibition's Popularity and Its Disastrous Fate

After 1964, the "In Industry, Learn from Daqing" movement continued to swell. In the summer of 1965, the Ministry of Petroleum decided to organize a Daqing Exhibition at the National History Museum, by Tiananmen Square. A dozen editors and designers, headed by a Political Department deputy director surnamed Li, were transferred from Daqing to Beijing to prepare the exhibition. As one of the main editors from the Ministry, I was in charge of the layout and the explanatory comments. The exhibition took up the whole second floor of the museum with hundreds of meters of displays. The motorized model of Daqing Oilfield was more than a 100 m^2. The scale and grandeur of the exhibition was unprecedented. We prepared for more than half a year. The exhibition opened with great acclaim after the Chinese New Year in 1966. The Ministry of Petroleum decided to open the exhibition to the general public for free.

The Red Guard Movement, which started in the early summer of 1966, soon swept across the entire nation. Given that modes of transportation such as railways were free for the Red Guards, they went all over the country to exchange experiences and establish ties. The streets of Beijing were filled with Red Guards. At the time, all parks, museums, libraries were closed, although Beihai Park, restored to its Ming and Qing royal garden glory, was set aside for Mao Zedong's wife Jiang Qing's exclusive use. Hundreds of thousands of Red Guards crowded in Beijing, with only one activity to do—reading big-character posters in Beijing's universities and colleges while waiting to see Mao Zedong when he received them every ten to fifteen days at Tiananmen Square. At this time, the Daqing Exhibition became a popular hangout for the Red Guards. Receiving tens of thousands of Red Guards from all over the country each day, it was all the rage. Of course, the Ministry of Petroleum was laughing up its sleeves. Who knew this would be the calm before the storm? What came next was a senseless farce.

In December that year, the General Office of the State Council suddenly sent the Ministry of Petroleum a notice of an impending visit from the Red Guards of Beijing Normal University and ordered Yu Qiuli to receive them. This frightened the Ministry's leadership. At his house, Yu Qiuli met with Kang Shi'en, Wu Xingfeng, and me, to discuss how we would respond. It was decided that I should accompany Yu Qiuli to the exhibition. At 10:00 that morning, when we arrived at the museum, a dozen Red Guards were waiting. Without going through the exhibition, they quickly surrounded Yu Qiuli. Shouting over one another, they hurled criticism at him, saying, "This is a counter-revolutionary exhibition," "A big anti-Maoist poisonous weed," and so on. They listed three specific crimes: (1) Why was Liu Shaoqi's photo in the exhibition? (The exhibition was prepared in early 1966, while Liu was still a Vice Chairman of the CCP Central Committee. All four Vice Chairmen—Liu Shaoqi, Zhou Enlai, Zhu De, and Chen Yun—had visited Daqing, so all of their photos were in the exhibition, whereas Mao Zedong's big-character poster titled "Bombard the Headquarters" had appeared on August 5, 1966.) (2) Why was Vice Chairman Lin Biao's photo not in the exhibition? (Lin Biao had never visited Daqing. In addition, he became the first-ranking Vice Chairman after August 5, 1966.) (3) Why did the gift shop sell Liu Shaoqi's pernicious book *How to Be a Good Communist* (*Lun gongchandangyuan de xiuyang* 《论共产党员的修养》)? (This was a fabricated charge. The gift shop belonged to the museum, not the Ministry of Petroleum.) Yu Qiuli remained silent. One Red Guard pressed sternly, "Have you seen the exhibition?" Yu timidly replied: "I have not." I was flabbergasted. I knew for a fact that Yu Qiuli had come more than once.

The meeting was hastily ended; it had lasted less than 20 minutes. Three days later, pamphlets appeared in the streets of Beijing, claiming that Chen Boda, the team leader of the CCP Central Committee Cultural Revolution Team, met with the Red Guard representatives of Beijing Normal University and publicly announced that Yu Qiuli had told lies. He criticized the Daqing Exhibition for selling Liu Shaoqi's harmful book on self-cultivation. The exhibition was then completely shuttered. To prevent the situation from deteriorating further, Zhou Enlai, considered the "fire brigade captain for the Cultural Revolution," ordered the exhibition to be recorded in the form of a documentary film, "for future criticism." The Ministry of Petroleum borrowed a director and two cameramen from Changchun Film Studio. They also had me work with them. The four of us labored for two months on the empty second floor of the museum, banishing the Daqing Exhibition to obscurity.

6 The "Four Bigs" and "Enticing the Snakes Out from Their Holes"

From the sudden rise of the Cultural Revolution in 1966 to my departure from Beijing in October 1969, I personally witnessed and participated in the entire process of the Cultural Revolution in the Ministry of Petroleum. Based on what I saw and experienced, it went through two stages.

During the first stage, which lasted in 1966 from the issuance of the May 16 Circular of the CCP Central Committee, through August 5, the date of Mao Zedong's "Bombard the Headquarters—My Big Character Poster (Paoda silingbu—wode yizhang dazibao 炮打司令部—我的一张大字报)," and to the end of the year, the Ministry of Petroleum was busy receiving Red Guards from all over the country and was entirely under the spell of the "four bigs (*si da* 四大)": speaking out freely, airing views fully, holding great debates, and writing big-character posters. Each person was required to write first ten, and later dozens, of big-character posters daily. The number of one's posters became an indicator of whether one was loyal to Chairman Mao. Sometimes people had to burn the midnight oil, writing until one or two o'clock at night. In the beginning, the content of these big-character posters was to express loyalty and praise the Party's wisdom and greatness. Later on, the content was turned toward exposing the shortcomings and mistakes of the Ministry's "not putting politics in command" and exposing and criticizing the "crimes" of those previously labeled landlords, rich peasants, reactionaries, rightists, as well as those who had worked under the Kuomintang regime. These activities were all officially arranged by the Ministry's Party Group. The Ministry's leadership at the time figured that the Cultural Revolution had to be a second Anti-Rightist movement.

At this critical moment, the Party Committee discovered one big-character poster, amongst thousands of posters, hung prominently on the wall in the main hall at the building entrance. Co-written by two deputy division chiefs and a few engineers and technicians of the Seismic Division of the Petroleum Exploration Bureau, the poster was entitled, "The Party Committee of the Ministry Failed to Put Politics in Command (Jiguan dangwei bu tuchu zhengzhi 机关党委不突出政治)." Its main idea was that the Party Committee had recently sent out a technical personnel questionnaire, which asked where one had attended school, what had been his major, and how many foreign languages one knew, instead of asking if one was a Party member or a Communist Youth League member, what was one's class origin, what awards or demerits he had received, and so on. For this reason, it accused the Party Committee of not putting politics in command. With the approval of the highest leadership in the Ministry, the Ministry's Party Committee regarded this poster as a signal to attack the Party—that it was anti-Party, an anti-socialist weed. All of a sudden, the wind had changed. The whole Ministry building was filled with big-character posters criticizing that original poster. Endless meetings, big and small, were held criticizing the technical personnel. They seemed to have become the new rightists.

However, within a month, on August 5, Mao Zedong published his big-character poster "Bombard the Headquarters," with Liu Shaoqi the target of the attack. A new upsurge took shape across the nation, criticizing the bourgeois reactionary line. Universities started to target the work groups dispatched by Liu Shaoqi and Deng Xiaoping. Under such circumstances, the tide of the Cultural Revolution turned against the Ministry. All of the big-character posters targeted Yu Qiuli, Kang Shi'en, and others in authority. Those who wrote the original poster mentioned above became the number one "heroes" who had been oppressed by the reactionary line. Accordingly, they set up a "Purification Fighting Squad" spearheading the movement in the Ministry. At this time, a few colleagues in our office also formed a "Liberation Fighting Squad," which together with the Purification Fighting Squad put forward the slogan, "Down with the Yu and Kang Dynasty (*dadao Yu Kang wangchao* 打倒余康王朝)." The Ministry bustled with noise and excitement.

The Ministry also became a chaotic mess. On all of the building's floors, big-character posters covered all the wall surfaces, making it difficult for anyone even to walk through. The building was crammed with contract workers and temporary workers, as well as Red Guards from petroleum universities and colleges across the country—all of whom had issues to take up with the Ministry. Aside from a few phones being used in the Production Bureau, nobody carried on much work. Kang Shi'en was taken by Premier Zhou Enlai to stay at Zhongnanhai for a few days. Later, he checked into a military hospital. The work was managed by Deputy Minister Tang Ke. I was assigned duty at a courtyard house (*siheyuan* 四合院) in Beijing that he used for meetings, so I commuted there every day on my old bike. Every evening, Deputy Minister Tang Ke held a meeting in the house, listening to reports on the Cultural Revolution situation within the Ministry itself; he also heard my reports on various oilfields. Actually, everyone knew that the Ministry's leadership at the time had no power at all over the oilfields, auxiliary plants, and the Ministry itself. The entire petroleum system had fallen into a state of anarchy.

On December 24, my colleague, Wei Buren 韦布仁, who was one of the key figures of the Liberation Fighting Squad, told me in a solemn voice, "Old Mao, you'd better come back to the Ministry to participate in the movement and receive education. You shouldn't be running into the city all day anymore." I hurriedly agreed, "Okay, okay." The situation had become serious, so

I decided not to go into the city the next day. In the afternoon of December 26 (I remember the day very clearly because it was Mao Zedong's birthday), in the large meeting room on the fifth floor of the Ministry, the Liberation Fighting Squad and the Purification Fighting Squad jointly held their first meeting to criticize the bourgeois reactionary line. Two to three hundred people were packed into the room, and I was crowded in among them. Kang Shi'en, who had been taken from the military hospital, was then escorted to the front of the room. I have forgotten most of the content of the speeches. However, Ren Chengyu 任成玉, an old Red Army veteran and deputy director in the Political Department, wailed, "I have failed to follow Chairman Mao's strategic plan, mistaking this movement for another Anti-Rightist Movement. Therefore, when enticing the snakes out of their holes, I made the mistake of suppressing the masses."[1] Amid the shouts of "Down with the Yu and Kang dynasty!" and "Down with Kang Shi'en!" Kang passed out, which stunned everybody. With a few helping hands, Kang's secretary Wang Fuzhen carried him out of the room, and the criticism meeting ended abruptly.

7 A Life-and-Death Struggle between Two Factions

The second stage of the Cultural Revolution in the Ministry of Petroleum started in the beginning of 1967, when a long-term factional struggle set in.

1 The phrase "enticing the snakes out from their holes" (*yin shé chu dong* 引蛇出洞) was used in the early summer of 1957, when the CCP launched the Rectification Movement. Mao Zedong encouraged the whole nation to give criticism and suggestions to all levels of leadership. At the height of the movement, Mao Zedong abruptly changed his attitude, believing some suggestions were Rightists' vicious attacks on the Party, and then the Anti-Rightist movement started, during which hundreds of thousands of intellectuals and students were arrested. It is now widely believed that the strategy of this Anti-Rightist movement was to "entice the snakes out of their holes"—first one encourages people to criticize the Party (enticing the snakes to come out), and then one clobbers them.

Similarly, in the Cultural Revolution, Mao Zedong called for the whole nation to offer criticism. In universities in Beijing, students skipped classes to take part in the revolution. All of a sudden, they were rioting everywhere, and the universities were in chaos. Liu Shaoqi and Deng Xiaoping, who were managing the Party's daily operations in Beijing, decided to send work groups to universities to stabilize the situation and restore order. Mao Zedong returned to Beijing in July and angrily denounced this action of sending work groups, deeming it a bourgeois reactionary line. Liu Shaoqi and Deng Xiaoping were sacked and labeled as class enemies. From then on, the rest of the nation followed suit, and managers in all walks of life were removed from their positions. This continued until Mao Zedong's death in September 1976. The Cultural Revolution lasted for ten years, so it is sometimes called "the Ten-Year Catastrophe."

On January 6, 1967, in Zhongnanhai, Zhou Enlai met with a few ministers and some representatives of mass groups from the Ministry of Petroleum as well as representatives of Red Guards from the Beijing Petroleum Institute. Later, my good friend and colleague Wu Xunduo 吴训铎, who was present at the meeting, told me that Zhou had asked about the ministers' thoughts concerning the revolutionary masses and Red Guards who were attempting to take charge of the Ministry—they wanted to seize leadership not only over the Cultural Revolution within the Ministry itself but also over petroleum production. A few deputy ministers replied that the leadership in both areas could be seized. Kang Shi'en, on the other hand, replied, "The leadership of the Cultural Revolution I can give up, but the leadership over production was given to me by the State Council. Without an order from the State Council, I cannot give it up." Then Zhou Enlai stated clearly, "Kang's attitude is correct. You, Tang Ke, and other deputy ministers should not take advantage of someone's precarious position. A few days ago, a deputy finance minister openly tried such a move on the Minister [Li Xiannian, a Vice Premier]. You better watch out!"

I remember many interesting stories about attempts to seize power around this time. For example, the Red Guards from the various universities in Beijing, incited by the Central Cultural Revolution Group,[2] stormed into institutions and companies throughout the city in search of the seals that government entities used for official documents. They thought that once they got these seals, they had seized the power to lead. Thus groups of Red Guards could be seen parading around the city, carrying bags of seals, chanting Mao Zedong's slogan, "Revolution is no crime. It is right to rebel." One does not know whether to laugh or cry in such a situation.

In the afternoon of January 8, 1967, at the Beijing Workers' Stadium, approximately 5,000 people attended the "Beijing Petroleum Employees Meeting to Criticize the Bourgeois Reactionary Line." A gigantic banner of "Down with the Yu and Kang dynasty" hung in the stadium. Songs of Chairman Mao's quotations were broadcast, and wave upon wave of chanted slogans such as "Down with the Yu and Kang dynasty" filled the air. The emcee of the meeting announced at the start that Premier Zhou had been invited to give a speech. However, ten minutes before the meeting started, people working at the meeting suddenly took down the "Down with the Yu and Kang dynasty" banner, and the chanting of the slogan ceased. According to my good friend Zou Jiazhi 邹家

2 Composed of mostly radical followers of Mao Zedong, including his wife Jiang Qing, the Central Cultural Revolution Group (中央文革小组) acted as the de facto top power organ during the first few years of the Cultural Revolution. It was formed in May 1966 and disbanded in 1969.

智, prior to the start of the meeting, Zhou Enlai met backstage with representatives of the Red Guards and the Ministry masses. Holding Zou's hand, Zhou Enlai said, "You know Kang Shi'en best. Criticize his shortcomings and mistakes, but you can't overthrow him." Afterward, Zhou walked to the podium. First he personally led the thousands of people through the singing of "The East Is Red" and "Sailing the Seas Depends on the Helmsman." The gist of his speech can be boiled down to two phrases: "Daqing is a red banner personally erected by Chairman Mao," and, "Yu Qiuli is the general appointed by Chairman Mao." At this same meeting, when people shouted the slogan, "Down with Liu Shaoqi," I saw Zhou Enlai immediately turn his back to the crowd, and he did not raise his hand to acknowledge the chant. I took that to mean that he did not agree with it. At the time, Liu Shaoqi had been stripped of his title of Vice Chairman of the CCP, but he was still a member of the Politburo.

In the aftermath, there developed two factions within the Ministry. One faction was composed of the Purification Fighting Squad, which had been labeled as counter-revolutionary only a few months prior, and the Liberation Fighting Squad, which had initiated the slogan "Down with the Yu and Kang dynasty." This faction now advocated criticizing but not overthrowing Yu and Kang, but it was an extreme minority, counting only a hundred or so people. The other faction still wanted to depose Yu and Kang. With new college graduates as its main force, this faction was larger, with five or six hundred people. Over the next few years, the Cultural Revolution in the Ministry centered around the issue of whether to protect Kang or overthrow him. Yu Qiuli had been transferred to the State Planning Commission as Deputy Commissioner in early 1965, so he no longer figured in the debate. Most of the disagreements between the two factions were non-violent and carried out with brush and ink; they aired their differences on big-character posters. Since the Ministry was located in Beijing, where the atmosphere was relatively civil, only two or three fistfights took place in the Ministry. My colleague Wu's nose and glasses were broken when he was punched in the face, but it was no big deal.

In late spring and early summer of 1967, the Military Control Commission (军管会) took over the Ministry. The first two commissioners were Red Army veterans, a lieutenant general and a major general. Gradually, military men were assigned key posts throughout the country. As a result, there were too few top officers to fill the Military Control Commission positions, and soon lower-ranking military personnel started assuming these outside roles. This was indeed reflected in what happened in the Ministry. After 1968, the next military commissioner in the Ministry had been a mere vice president of some military academy, a colonel, and the deputy had been the rear service division chief of that same academy. This demonstrates the lack of available military officers after military management was put in place in enterprises across the country.

At first, the Military Control Commission appeared to be fair, calling for a "grand union" of the two factions in the Ministry, but as time went on, they showed their true colors. It became more and more obvious that they leaned toward the majority "overthrow Kang" faction. Thus, while that faction gained strength each day, things were becoming more difficult for the minority "protect Kang" faction. In the fall of 1968, in a meeting attended by 100,000 people held at the Beijing Workers' Stadium, Mao Zedong's wife Jiang Qing publicly announced that "Yu Qiuli is He Long's henchman."[3] This boosted the "overthrow Kang" movement to a new high.

In March 1969, at a National Planning Meeting, Zhou Enlai requested Kang Shi'en to take charge of the oil campaign in Jianghan, Hubei. The Ministry's deputy military control commissioner attended the meeting on behalf of the Ministry. He objected to Zhou, "Kang Shi'en followed the reactionary line and is still in detention. The masses won't understand (the move)!" Zhou replied, "Chairman Mao specifically appointed Kang Shi'en to Hubei. You mean Chairman Mao doesn't represent the masses?" From this, one could see the Military Control Commission's position on the Cultural Revolution in the Ministry.

8 Words of Regret

During the Cultural Revolution, ministers and bureau chiefs were thought to be "the faction in power (*dangquanpai* 当权派)" while heads of divisions and below were considered part of the "revolutionary masses (*geming qunzhong* 革命群众)." The revolutionary masses could form fighting groups on their own and write big-character posters to expose ministers and bureau chiefs, but they could not write posters aimed at people inferior to division heads. Otherwise, they would be labeled as "masses fighting masses." In the first year of the Cultural Revolution, none of the fighting groups dared to accept me, since they were afraid that their opponents would attack me as "the spy" of the minister. I was happy to enjoy the leisure, a member of the "bystander faction" (*xiaoyaopai* 逍遥派). I consider myself lucky for being in Beijing. Had I been in Daqing, I would have been physically abused, confined, and possibly tortured to death since I was born into a landlord's family (more of this below), was the head of a division, and one of the most-favored Red Banner Pacesetters. For

3 Yu Qiuli had joined the Red Army in 1929 and served in He Long's division until 1949.

this, I am thankful that I was resident in Beijing—the Cultural Revolution here was still much more "civilized" than in Daqing.

However, I need to confess that I had my own opinion of the Cultural Revolution in the Ministry—I was not a true bystander. I sympathized with the "protect Kang faction," and those heading the faction were well aware of it. In the latter part of 1967, I finally joined the Liberation Fighting Squad, the nucleus of the protect Kang faction. Considering my status, they assigned me to write "big criticism" posters, thus leaving me outside of the fights between the two factions. I wrote posters criticizing Liu Shaoqi and Deng Xiaoping. No one could object to that! Honestly speaking, at the time, I supported all of Mao Zedong's theories and directives from the bottom of my heart and tried, in a thousand and one ways, to interpret them so as to criticize Liu and Deng. One article that I wrote criticized Liu Shaoqi for sending work groups to implement his reactionary line, and it was broadcast as a model composition in a nearby high school, and so it indirectly harmed young people! I regret that I, with too little schooling or understanding of history, just blindly echoed others' views, that I was so thoroughly poisoned by them!

Now I should pause here to say something about my family history and explain why I was considered to have been born into a landlord's family. My hometown, Heiyudang 黑鱼垱, was a small rural village near Changde 常德, Hunan Province. There were four brothers in my father's family: the eldest, Mao Yifeng 毛翼丰 (1896–1986), was a nationally prominent textile executive in Chongqing and considered a hero for having defended Yufeng Cotton Mill (豫丰纱厂) during the liberation of Chongqing from marauding stragglers from the Nationalist army; the second older brother and my father Mao Jingwu 毛敬梧 (1901–1958) stayed in the village to farm, and the fourth brother followed the eldest brother to Chongqing when he was a child. My father worked his plot of land himself and hired temporary help only during harvest. Before the Revolution, when I was 13, I too had gone to Chongqing in order to get a better education, and I stayed with my uncle's family. In December 1949, less than a month after Chongqing was liberated, I left National Central Industrial College (国立中央工业学校) and joined the PLA.

Immediately after the Hunan countryside was liberated in September and October 1949, everyone in the Hunan countryside was divided into different classes and designated as landlords, rich peasants, middle peasants, poor peasants, and so on according to their assets, particularly in land and real estate. Those who were classified as landlords and wealthy peasants first had to give up their movable assets (*fen fucai* 分浮财) and then, under Land Reform, their land and houses without compensation to the poor peasants. At the same time, they became the political targets of the "dictatorship of the proletariat," because they were considered class enemies who had obtained their wealth by exploiting the poor peasants. In my village, the classification was decided by the local Party branch.

They first classified my second older uncle as a landlord. They locked him up in a small barn, interrogated and tortured him, and denied him food. He died in a couple of months. The village Party branch was then under pressure to find one more landlord, and even though there were a few other families in the village, members of the Zhao clan, who had been richer than my father, the secretary of the village Party branch still chose to designate him as the other landlord, because the committee members of the village Party branch were all Zhaos.

I do not know whether my father had known he would also become a target, but before that happened, he informed the village Party leader that he was going to Chongqing to visit his son in the PLA. Starting in April 1950, he walked all the way to Chongqing, a distance of 600 to 700 km. When he finally reached me in Chongqing in June, he tried to convince me to leave the PLA. I refused. However, he knew that things had also become very chaotic at home, so he decided to remain in Chongqing. First he stayed with his older brother, then with my older half-sister, and for the next few years he worked as a coolie carrying luggage for passengers at the port of Chaotianmen 朝天门.

During his absence, my step-mother was left on her own with my little sister, who was still a child, and she could only depend on sewing for people in exchange for food and rice. One night, in 1955, she hung herself, and the next morning my sister woke up to the sight. Then my father was exposed by his own niece as a "runaway landlord" and had to return to Heiyudang in the fall of that year. Afterwards, he was always under supervised labor and suffered tremendous discrimination. From my military officer's salary, I would wire him and my sister around 20 yuan per month (more than one-third), but as soon as he received it, the village cadres would come to his house asking for it, and they would never pay him back. Those who got the money would leave happily, and those who did not would curse him out. On one such occasion in 1958, the village production team leader, along with others, demanded more money from him. Because my father did not give him any, he accused my father of stealing his grain. In fury and despair, my father tied a big stone to his waist with an apron, walked into the ditch behind his house, and drowned himself. My sister at the time was 13.

Since I was the son of a "class enemy," I too was treated differently, though to a lesser extent, and there was little I could do to avoid it. For years, as I went about my own life, such matters were always weighing heavily on my mind, and I had to be very circumspect in all that I said or did. In this regard, I was probably typical of most people of my generation.

9 A Scribe's Impression: the Longer the Article, the Easier It Is to Write

In early June 1972, when the United Nations Conference on the Human Environment was held in Stockholm, Sweden, Premier Zhou Enlai decided to send a Chinese delegation led by Deputy Minister of Fuel and Chemical Industries Tang Ke and Deputy Commissioner of the State Planning Commission Gu Ming. A dozen experts from the Ministry of Foreign Affairs, the Ministry of Health, and environmental specialists under the State Construction Commission made up the rest of the delegation. The Ministry of Fuel and Chemical Industries was responsible for Tang Ke's grandiose speech.

One month earlier, I had been transferred back to Beijing from the Jianghan Oil Campaign in Hubei. I worked at the Political Work Group of the Ministry of Fuel and Chemical Industries (formerly the Political Department of the Ministry of Petroleum). Kang Shi'en, the Minister of Fuel and Chemical Industries, asked me to write the speech. Another senior writer had worked on the speech for more than a month, but they were not satisfied, and time was running out. I was nervous, not knowing whether I was up to the task.

After I got the assignment, Kang Shi'en held a seminar and decided on the main content of the speech. Experts from the Ministry of Foreign Affairs provided me documents on our country's foreign policy, the Ministry of Health provided documents on family planning policies, and the State Construction Commission provided documents on environmental protection policies. I studied these documents carefully. I also looked up relevant political commentaries and articles in the recent CCP Central Committee's newspapers and magazines. Gradually the outline of the speech was formed, which included the following: first, a firm denunciation of the American imperialists for initiating the Vietnam War, "creating unprecedented severe damage to the environment of mankind" and of the "superpowers [that] in order to seek world domination ... pollute and poison the environment of their own countries and other countries"; second, promotion of the achievements of our country's economy and development, with a clear statement of our environmental policy; third, a description of our family planning policy; fourth, clarification of our nuclear policy: "... we must resolutely fight the superpowers' nuclear monopoly, nuclear threat, and nuclear blackmail," etc.

It took me ten days to write this speech. Afterwards, the delegation gathered in Qianmen Hotel for a week. Key members discussed the speech and had me modify it. The final version, which was around 5,000 words in length, was submitted to Premier Zhou for approval. I read the copy that he had read and noticed that he had read it very carefully, using a pencil and punctuating each sentence. He revised the paragraph denouncing the United States for starting the Vietnam War by adding or deleting a few sentences. Finally, he approved the speech. I felt consoled and moved. This was the biggest chunk of writing that I had done during my career as a writer. Receiving Premier Zhou's approval and permission was no doubt an honor. On June 10, the speech was delivered by the leader of the Chinese delegation Tang Ke at the United Nations Conference on the Human Environment, the first such held by the UN. On June 11, the whole speech was published in *The People's Daily*. The Stockholm Declaration that resulted from the conference was a milestone, with important ramifications for improving and protecting the human environment around the world. The occasion also marked the first time that a Chinese governmental delegation attended a UN conference after the People's Republic of China was admitted to the UN in 1971.

These past events now seem as transient as fleeting clouds. When I recall how I wrote that speech, all I did was to carefully study relevant documents at the time, copy and quote key words and phrases, follow a prearranged outline, write, and edit. It did not require any of my own ideas or creativity. It was not even necessary for one paragraph to connect well with the next. My approach was to add words like "Mr. Secretary-General, delegates ..." between paragraphs. This made things seem concise and was a bit of an empty show of strength. In any case, I was a mere copier—this was the pathetic truth of my career as a scribe.

10 The First Session of the Fourth National People's Congress Held in Secrecy

From January 13 to 17, 1975, the first session of the Fourth National People's Congress was held in Beijing. I had the privilege of attending the entire meeting. Here are a few of the scenes that are impressed most deeply in my memory.

Scene One: Taking the secret assignment and riding a mysterious chartered plane. Right after New Year's Day, my boss notified me and three other colleagues to report to the First Guest House of the State Council. I was also enjoined not to let my wife and family know or to make any phone calls. I was supposed to say only that I would be away for business. The four of us were terrified, and it was not until we checked ourselves in at the guest house that we learned we were going to work in the secretariat of the National People's Congress. Before going, we went through two days of training, with the Deputy Director of the Organization Department of the CCP Central Committee lecturing—mainly on the work of the secretariat and the rules for maintaining secrecy. We had two tasks. One was to deliver Zhou Enlai's "Report on the Work of the Government (政府工作报告)" and Zhang Chunqiao's[4] 张春桥 "Report on the Revision of the Constitution (修改宪法报告)" to the provincial and autonomous regional governments, so that local representatives could use them in preliminary meetings. The second task was to accompany provincial and autonomous regional delegations back to Beijing and take part in the small group discussions of these delegations. Under the leadership of each delegation head, we would be responsible for writing the delegations' bulletins—this was our main task. There were two rules: one was to keep it all a secret, the other was to go to these small group discussions only with our pens, not our mouths. We were absolutely not allowed to express our own views, and so on.

One colleague and I were assigned to Hunan Province, and I was the group leader. The other two colleagues were assigned to Hubei Province. At 6:00 in the morning of the third day, the four of us were sent in a military car to the Shahe 沙河 military airport outside of Beijing's Deshengmen 德胜门. The car came to a stop at the airstair of a Soviet Tu-154 plane. The captain, a regimental deputy, saluted us, which was flattering but made me feel uneasy. Fortunately, it was still around dawn, so the dimness hid my embarrassment. I hurriedly boarded the plane.

This was the first and only time that I rode in a special military plane, so naturally I felt great. This special plane was 30 to 40 m long, and the interior was arranged roughly as follows. Behind the cockpit was the seat for the senior officer. There were two rows of spacious couches facing each other, with a tea table in the middle, similar to the arrangement found in the cushioned first class seats of a train. A curtain separated this space and the officer's bedroom, which featured a queen-size bed. The aisle passed around the left side of the bed. From what I could see, the bed did not look very comfortable, and

4 Zhang Chunqiao (1917–2005) joined the CCP in Yan'an in 1938. He was one of the secretaries of the Shanghai CCP Party in charge of propaganda and cultural work. Because he often published ultra-left views, Mao Zedong thought highly of him. When the Cultural Revolution started, he was named Deputy Director of the Central Cultural Revolution Group, and his political career took off. He became a member of the Politburo Standing Committee, Vice Premier, member of the Central Military Commission Standing Committee, and Director of the PLA General Political Department. As a member of the Gang of Four, he was arrested in October of 1976 and was imprisoned until 1998.

the officer would not have been able to sleep there. I thought to myself that it must be there for display. After that, there was a cabin, similar to economy class on an ordinary plane, for the senior officer's entourage. There were seven or eight rows with four seats in each row, and the aisle in the middle. Behind this cabin was the lavatory. After the four of us sat down behind the seat for the senior officer, a young female soldier brought over four cups of tea and a bowl of fruit. Engines humming loudly, the plane took off. An hour later, the plane landed in the military airport in Wuhan, and those two colleagues disembarked. The plane took off again, and 40 minutes later, it landed at Changsha's military airport. Again, the captain and his crew saluted us goodbye. I warmly shook hands with them to show my appreciation. They probably had no idea of our stature or what our business was. Knowing this, I laughed to myself.

Scene Two: The many mysteries of the two-thousand-year-old corpse. At Changsha airport, a Deputy Secretary General of the Hunan Provincial Party Committee took us in two military cars to the First Guest House of the Provincial Military District. The place had been cordoned off and was guarded by a platoon. By then, more than 80 Congressional representatives of the province had gathered there. The two of us took our documents and stayed in one standard room, and we had a single task: distributing and collecting documents in the mornings and evenings, with no mistakes allowed. Moreover, we were not to distribute these documents directly to the representatives, but rather give them all to the Deputy Secretary General. In the evenings, he would return the documents to us, with the correct copies. When the provincial representatives read the two documents in groups, we did not participate. These so-called "preliminary meetings" went on for five days, and my colleague and I were bored to death. Fortunately, the Deputy Secretary General assigned someone to us who drove us by car to show us places in Changsha. We visited the Hunan First Normal College, Orange Isle in Xiang River, Yuelu Academy, and so on—places where Mao Zedong had spent his youth. I felt rather moved when I remembered his agitating poems such as "When we hit midstream, the waves held back our speedy boat."

Most memorable was our visit to the Western Han dynasty (206 BCE–9 CE) tombs at Mawangdui 马王堆. The museum had been closed to avoid sabotage from the Red Guards. When the two of us went to visit, those who accompanied us introduced us to the museum manager as two leaders from Henan Province. The manager humbly showed us around and explained the exhibit to us in detail. I was most curious about the identity of the ancient corpse, because there had already been many reports about it, and much speculation about the corpse's identity. The most widely believed story was that she must have been a palace woman in the imperial court, and that after the emperor had relations with her, she gave birth to a boy. Once the empress learned of it, she claimed the boy as her own and made him crown prince. The young woman was placed under house arrest and was about to be killed, but with the help of an old eunuch, she escaped and was saved by a villager. She married the villager and sold tofu for a living. Years later, when the prince became the new emperor, the news reached the former palace woman, and she wanted to go to the capital to meet her son. Following imperial customs, however, the ministers dissuaded the woman from meeting with the new emperor and instead made arrangements for her and her husband to live in the kingdom of Changsha, far away from Chang'an, and they were given noble titles. The court took good care of her, treating her like the empress dowager. When she died, she received a lavish burial ... People loved to tell this old tale of palace intrigue, and some believed the story.

I was among those at the time who did. During our visit, five items caught my eye. One was an exceedingly light silk dress, as gauzy as the wings of a cicada. The second one was the T-shaped silk banner emblazoned with a bird-in-the-sun totem. These exquisite items were very well made, so they had to have been imperial items—not even the nobles could have them, much less commoners. The third were her coffins, one inside another, beautifully painted. Her body was preserved with special preservatives, something not normally achievable by nobles far from the capital. The fourth was a small bowl of grape seeds, together with wheat grains, rice grains, and other fruits on display. Grapes at the time were grown in the Western Region (present-day Xinjiang) and presented as tribute items from those kingdoms. They would not have been readily available in remote Changsha. However, grapes were part of the burial, which meant that they were very likely sent over from the imperial palace. I asked the manager of the museum about these items, but the answers he gave me were not very satisfactory. In another exhibition room, her husband's coffin was on display, and even though it was bigger than usual, made of thicker wood, there was otherwise nothing special about it. It was not beautifully painted and much inferior to the woman's. Again, I asked the museum manager. He said that perhaps men and women were treated equally at the time. I just smiled, since it was neither here nor there. Another unique item that I saw was a slice of lotus root in a bowl half full of water. The museum manager told me that it showed that Changsha had not had a 2.0-level earthquake since the burial, otherwise, the lotus slice would have sunk to the bottom of the bowl.

In 1980, I read in a newspaper in Beijing that this two-thousand-year-old corpse was the wife of the prime minister of the Kingdom of Changsha. I still have my doubts about this.

Scene Three: Important officials at the National Congress, and the unexpected death of an 83-year-old farmer. Before the preliminary meeting started, the Deputy Secretary General gave me a list of the representatives' names, and I noticed that most of them were important provincial Party, governmental, and military officials. In an arrangement like this, where people are given both legislative and administrative powers, it would be difficult to have legislators and administrators countercheck and counterbalance each other, in the spirit of modern constitutionalism. Besides these, there were also a few "models" who were included as representatives of the common people. Among them was an 83-year-old peasant Ma Lao'er 麻老二, a member of the Miao minority group. He was identified as an activist who "studied and applied Chairman Mao's works in his life." He lived in his shed deep in the forest in mountainous western Hunan. He had never been out of the mountains before. Before the meeting, the county sent a military car to pick him up and drive him to Changsha, and it took three days to do so. He was completely illiterate, very lean, and his face was full of wrinkles. In the depth of winter, he wore a pair of coarse cotton pants with broad bottoms. He wore the old military overcoat given to him by the country unbuttoned. He was in high spirits. I asked him if he was cold, and he replied, "No." He kept saying, "I can go to the provincial capital and eat good food every day, go to Beijing to meet Chairman Mao. I have to thank the Communist Party and thank the big benefactor Chairman Mao ..." He did not take part in the small group meetings. He stayed in his room, accompanied by a deputy of the Organization Department of the Provincial Party Committee. He took all three meals of the day in his room. Most pitiful, however, was the fact that in the evening of the fourth day, before the delegation went to Beijing, he started coughing violently. A large mass of bloody mucus blocked his airway, and he died in an instant. The people with him did not know what to do. The following morning, the Deputy Secretary General announced the news. Without saying anything, I felt a tightness in my chest. I thought, had he not been taken from his shed, had he been left to live his life peacefully in the deep forest, the old man would probably have lived a few more years. Why on earth did they take him over to "participate in state affairs"? How could he even have participated in state affairs?

Scene Four: The People's Congress goes through the motions, and I write the briefs in fear and trepidation. After the Changsha preliminary meeting concluded, the Hunan delegation and a few staff members took a special train to Beijing and checked into the big Xiyuan Hotel, west of the zoo. I remember that on the third evening, the opening ceremony for the Congress was held in the Great Hall of the People. To get there, we were secretly bussed over to Jingxi Hotel in order to take the subway (still not in operation for the public).

It was very quiet on Tiananmen Square, with little trace of people and traffic. To this day I do not understand why such a grand event, so beneficial to the country and people—convening the People's Congress—had to be done in such secrecy.

I recall that after the several thousand representatives took their seats, we staff sat in an area on the right-hand side. I sat in the ninth row, rather close to the front. At 8:00 PM, as the tune "Greeting the Guests (迎宾曲)" was played, Zhu De, the Chairman of the Standing Committee of the National People's Congress, who was then almost 90 years old and in poor health, walked slowly to his seat with the assistance of two attendants. After people on the rostrum were seated, the old man stood up. His attendant helped him put on his reading glasses and handed him notes prepared in advance. He announced the opening of the Congress. The national anthem was played. Premier Zhou Enlai was invited to give the "Report on the Work of the Government."

At this time, from the right side of the rostrum, Premier Zhou strode out at a vigorous pace, but he was very lean and looked ill. The whole audience quickly stood up, with prolonged applause. Premier Zhou waved his hand a few times, signaling people to calm down. He started delivering the report, which I knew ran more than 10,000 words. However, he read only the first few hundred and the last hundred or so words, and then, amidst another enthusiastic standing ovation, he left.

Afterwards, I felt pity for these two old men. One was almost 90 years old and already senile, and the other was in his seventies, very ill. Why would they bother to come to such an occasion and go through the motions of acting out the pre-scripted parts of a play? Was it necessary? Was it not too hard on them? Why should they put themselves through so much trouble?

During the meeting, my only job was to write the bulletin of the discussions of the Hunan provincial delegation. Four bulletins were issued: two on Premier Zhou's "Report on the Work of the Government," and two on Zhang Chunqiao's "Report on the Revision of the Constitution." Each bulletin was approximately 2,000 words long. Before the meeting, a Deputy Secretary General of the National People's Congress held two meetings and repeatedly stressed that we should not write bulletins similar to those written by the Huabei regional delegation from the Lushan Conference in 1970.[5] The Deputy Secretary General

5 During the Lushan Conference, Lin Biao and Chen Boda proposed the "Genius Theory" and proposed to reinstate the position of the "Chairman of the People's Republic of China." Lin Biao's four generals spoke at the meeting of the Huabei regional delegation. Their speeches were printed in the bulletins from the Huabei delegation. However, when he read these speeches, Mao Zedong was enraged and criticized Lin Biao and Chen Boda.

also told us, "To avoid the Lushan Conference situation, this meeting will make some changes. The CCP Central Committee will send competent cadres to form the secretariat of each delegation. You will write the bulletins, and the chief of each delegation will examine them and decide on the final versions ..." He kept repeating himself. I got the hint.

While writing these bulletins, I was extremely tense, afraid of making any mistake. I tried very hard to gather each representative's speeches and did my utmost to sing the praises of Mao Zedong. Meanwhile, I included idioms of Hunan local speech such as "Men are in charge outside of the house and women are in charge inside of the house" and *Xiang meizi* for "Hunan girl." My bulletins received good reviews from the secretariat. Now, decades after the event, I often think that this meeting, which involved a few thousand people laboring for weeks, was all a drama for one person's enjoyment. Everyone acted according to one person's whim. Was this a socialist democracy? Was this how people should take part in politics?

Scene Five: A down-to-earth banquet, and Premier Zhou's swan song. Immediately after the Congress's closing ceremony, at 7:30 in the evening, all the representatives and staff had a banquet in the dining room on the second floor of the Great Hall of the People. I was naturally very happy that I could attend such a state banquet. I still remember the scene. Several hundred large round tables were arranged in the immense dining room, and twelve people were seated at each table. For staples, there were steamed buns, twisted rolls, stuffed buns, rice, and cake. There were ten warm dishes, seven with meat, and three with vegetables. The most memorable dish was served in a huge earthenware pot placed in the middle of each table, filled with Napa cabbage, tofu, bean noodles, seaweed shreds, slices of pork bellies, meat balls, and jumbo shrimp rarely seen in the market at the time. This dish was delicious but light and delicate, not very oily, which was more typical of Huaiyang cuisine (the cuisine of Premier Zhou's hometown). If the food in the pot were finished, it would be refilled. After the banquet, each person was given a big orange and a big tangerine. Such fruits were not sold on the market. I did not eat them but took them back home the next day for my two daughters. This was the only state banquet that I ever attended, and it was the banquet with the most people that I have attended. To this day, two memories from the banquet still remain as fresh as new. First, even though two to three thousand people attended the banquet, all of the thousands of dishes were served warm. I had great respect for the people who had planned and worked for this banquet. Second, it was truly sumptuous, suitable for southern, northern, eastern, and western tastes, so that in its own way, it was the kind of feast that would have appealed to the tastes of any ordinary person. Compared to the way current officials wine and dine on public funds, I cannot help but feel that yesterday was better than today.

After the feast, at 10:00 o'clock, we rode big buses back to the hotel and passed by Tiananmen Square and the wide West Chang'an Avenue. The night was quiet and hazy under dim moonlight. At 7:00 PM the next day, CCTV broadcast the news of the People's Congress and announced Zhou Enlai's proposal to realize the four modernizations (industrial, agricultural, defense, and science and technology) by the end of the twentieth century. This became Zhou Enlai's swan song in the Chinese political arena. He gave it all to the country until his dying day! I have profound respect for him.

CHAPTER 16

Advancing on the Crest of Victory: Switching the Battleground to Bohai Bay Basin

1 Southward to Bohai Bay Basin

In February 1964, when the national "In Industry, Learn from Daqing" movement had just started, Yu Qiuli at an oil industry managers' meeting in Beijing announced that a big oil exploration battle would be launched in the Bohai Bay Basin 渤海湾盆地. Starting that year, the focus of oil exploration thus became Bohai Bay Basin.

Bohai Bay Basin, also known Huabei Basin, has a total area of around 195,000 km^2. It is situated south of the Yan Mountains 燕山, east of the Taihang Mountains 太行山, and north of the Tai Mountains 泰山, so it stretches over Liaoning, Hebei, Shandong, Shanxi, and Henan Provinces as well as Tianjin and Beijing Municipalities. Two-thirds of the area is on land and one-third extends out into the Bo Sea (Bohai) at an average depth of 17 m. This basin is large and its sedimentary rocks thick. It has good oil-generating conditions. Geologically speaking, it has key exploration areas such as the Jiyang Depression (济阳坳陷, i.e., Shengli Oilfield) in Shandong Province, Huanghua Depression (黄骅坳陷, i.e, Dagang Oilfield) in Tianjin, Lower Liaohe Depression (下辽河坳陷, i.e., Liaohe Oilfield 辽河油田) in Liaoning Province, Dongpu Depression (东濮坳陷, i.e., Zhongyuan Oilfield 中原油田) in Shandong and Henan Provinces, Jizhong Depression (冀中坳陷, i.e., Huabei/Renqiu Oilfield华北/任丘油田) in Hebei Province, and so on. This area has great potential for oil, with prospective reserves estimated to be around 40 billion tons. It has long attracted the attention of Chinese and foreign petroleum experts. In China, geologists such as Huang Jiqing, Sun Jianchu, Xie Jiarong, Li Siguang, and Weng Wenbo had all written about the great oil and gas prospects of Bohai Bay Basin.

2 Cooperation between the Ministries of Geology and Petroleum

The Ministries of Geology and Petroleum worked hand-in-hand on oil exploration in this vast Bohai Bay Basin. As early as November 1960, the two ministries held a meeting in Tianjin concerning Bohai Bay Basin exploration. Deputy Geology Minister Kuang Fuzhao 旷伏兆 presided over the meeting,

 | DOI:10.1163/9789004402737_017

Meng Jisheng, Director of the Petroleum Survey Bureau, and other relevant experts of the Ministry of Geology from Beijing, Tianjin, Hebei, Henan, Shandong, and Shanxi attended the meeting. From the Petroleum Ministry's side, Director of the Petroleum Exploration Bureau Shen Chen and geologists Qiu Zhongjian and Cai Bijian also attended the meeting. This meeting drew on their collective wisdom and useful ideas and served as a prelude to the oil and gas exploration in Bohai Bay Basin. After much discussion and study, consensus was reached at the meeting that the Dongying (东营) area in Shandong Province, northern Dagang area in Tianjin, and Huanghua area in Hebei Province should be the key areas of exploration.

3 The Central Committee Approves the Oil Campaign

On January 15, 1964, the Party Group of the Ministry of Petroleum decided to organize the Huabei Oil Exploration Campaign. On January 21, Yu Qiuli reported the decision to Vice Premier Bo Yibo. Bo consented and requested a written report. The next day, the Party Group of the Ministry of Petroleum formally submitted to the Central Committee of the Chinese Communist Party "A Concise Report on Organizing the Huabei Oil Exploration Campaign (关于组织华北石油勘探会战的简要报告)." On January 25, the Central Committee approved the report, stating, "This is another important campaign following the Daqing campaign in Songliao Basin. We hope that relevant local governments and agencies give it their assistance. The National Economic Commission is instructed to coordinate the assistance."

To ensure the campaign's success, the Ministry of Petroleum emulated the Daqing model and formed a campaign headquarters and campaign working committee. Kang Shi'en took on the role of chief commander and secretary of the working committee. The campaign was conducted in Dongying, Shandong Province and in Dagang, Tianjin Municipality. Zhang Wenbin and Yang Wenbin were appointed commanders of Dongying and Dagang respectively. From then on, a large-scale search for oil started in the Bohai Bay Basin.

4 China Gets Its First Thousand-Ton Well, and Shengli Oilfield Secured

Bohai Bay Basin's Jiyang Depression is situated in the Yellow River delta, in northern Shandong Province. It has the sea to the east, Hebei Province to the north, and Jinan, Zibo, and Weifang to the south. It also includes a vast neritic zone. Its total area is approximately 30,000 km^2. After many years of survey and exploration, it proved to be a rich, complex oil and gas area with many oil and gas types and many oil deposit systems. Shengli Oilfield is in this depression.

The site of the well Hua-8, determined in late 1960, was located in the town of Dongying Village (now City) in Guangrao 广饶 County, Shandong Province, and like Daqing's Songji-3, it is the well that led to the discovery of Shengli Oilfield. It was drilled by the 32120 Drilling Crew starting on February 26, 1961. On March 5, with much excitement, geologists An Peishu 安培树 and Shuai Defu 帅德福 of the Huabei Oil Exploration Division ran to the Ministry of Petroleum to report good news. They presented Yu Qiuli with a little bottle that was tied with red silk and contained glistening oil sand. It had been found in the well at a depth of 1,194 m. Yu Qiuli exclaimed, "This little treasure is much more valuable than gold!" He immediately showed it to the other deputy ministers and bureau chiefs in the ministry. Meanwhile, he sent someone to inform the cafeteria to prepare a meal. He invited the two geologists for lunch and made a toast to them. That afternoon, he sent the Drilling Division Chief Deng Lirang of the Ministry's Petroleum Exploration Bureau and geologist Xie Qinghui 谢清辉 to work at the drilling site. He then transferred the chief geologist of Qinghai Petroleum Administration Bureau Wang Shangwen to Dongying to supervise. It was smooth sailing afterward. On April 14, the well tested successfully, producing 10 tons of oil.

In 1964, after the CCP Central Committee formally organized the oil campaign, large groups of people were transferred from Daqing. The number of campaign employees grew to more than 20,000 people. In addition, several hundred sets of drilling, transportation, and mechanical equipment were sent. As the exploration progressed in the Dongying area, the importance of a cluster of structures around Tuozhuang 坨庄 gradually became obvious. More than 20 rigs fiercely attacked this cluster. On January 25, before the Chinese New Year, good news arrived. The Tuo-11 well started gushing out 1,100 tons per day! It was the first thousand-ton well in Chinese history! Deputy Minister Zhang Wenbin, who acted as the commander on site, delivered the good news to the Ministry of Petroleum via telephone. Yu Qiuli and Kang Shi'en were ecstatic. They immediately invited those geological people present to a Mongolian restaurant to celebrate with a meal of instant-boiled hotpot mutton. As someone said, the oil workers get a big holiday when they find a high-yield gusher.

Soon after, nine oilfields in succession were found near Tuo-11. Since Tuo-11 is located in Shengli Village, these oilfields in the Dongying area are jointly referred to as the Shengli Oilfield. A sizable new oil base now appeared on the Yellow River delta.

5 Gang-5 Gushes Oil, and Dagang Oilfield Secured

Dagang Oilfield is situated on the coast, in the southeastern part of Tianjin Municipality and in Huanghua County, Hebei Province. Geologically, it belongs to the Huanghua Depression in the Bohai Bay Basin. Covering an area of 19,000 km^2, it has Shengli Oilfield to its southeast, Hebei Renqiu Oilfield to its southwest, and the sea to its east. The Ministries of Geology and Petroleum began doing oil exploration work in Huanghua Depression in 1955.

Based on the achievements of the Ministry of Geology, the CCP Central Committee formally approved the Huabei Oil Campaign in January 1964. The Ministry of Petroleum ordered Daqing oil workers to move south, through Shanhaiguan Pass, to Dagang and Huanghua. These workers, more than 7,700 people strong, led by commander Yang Wenbin, brought four or five thousand items of equipment and materials weighing more than 20,000 tons. They had braved Daqing's extreme winter cold, which ranged between −20°C to −30°C. They marched and gathered at the seashore near Tianjin and started the oil campaign. Since this group arrived in Dagang in January 1964, it was called the "641 Plant." As with all Chinese industrial projects during this time, it was cloaked in secrecy. It was not until twelve years later, after the Cultural Revolution ended, that it was formally named Dagang Oilfield.

In 1964, right before National Day, Kang Shi'en made a one-day stop in Dagang on his way to Dongying. After repeated comparisons and analysis with geological personnel, he decided to have Gang-5 well drilled on the shore south of northern Dagang. Before that, even though exploratory wells showed some oil and gas, there had been no major breakthroughs. Some campaign troops were about to withdraw. Kang repeated, "We need to finish drilling this well no matter what. If we were to let go now, it would be like 'failing to build a mound for want of one final basket of earth'—it would be a great pity." Soon thereafter, the geologist Wu Huayuan 吴华元 took relevant personnel to carry out reconnaissance on the well position. On December 1, a miracle occurred. Gang-5 had a blow-out of high-yield oil and gas, producing 20 tons of crude oil per day and natural gas on the order of 30,000 to 40,000 m^3 per day. The blow-out brought the campaign back from the brink of death. Some oil workers were already at the Tianjin Railway Station ready to head west to Sichuan. Once the blow-out occurred, they were immediately notified, told to get a refund for their tickets, and return to Dagang to rejoin the campaign. Like Daqing's Songji-3 and Shengli's Hua-8, Gang-5's blowout announced the birth of Dagang Oilfield. From then on, Dagang Oilfield became the second most important oil industry base in Bohai Bay Basin, after Shengli Oilfield.

This story illustrates that there was an aspect of inevitability to Dagang Oilfield's discovery, but it almost did not happen. It was inevitable, because starting in 1955, the Ministry of Geology and the Ministry of Petroleum had been working there for ten years. After a year-long oil exploration campaign they had nothing to show for their efforts and were about to abandon the project. Just at that moment, by persisting with the drilling, they found Gang-5, which marked the birth of the oilfield. One might call it luck, but ultimately one must recognize that the science behind it made it inevitable.

In early February 1965, the Ministry of Petroleum held its annual managerial conference on plants and oilfields in the Peace Hotel on Wangfujing Street, Beijing (by then, Yu Qiuli had been appointed as the first Deputy Commissioner of the State Planning Commission). Minister Kang Shi'en presided over the meeting. He got straight to the point: "The Bohai Bay Campaign has made breakthrough advances. Zhang Wenbin has struck oil in Dongying and Yang Wenbin has struck oil in Dagang. For finding the Shengli and Dagang oilfields for the country and bringing good news to this meeting, let us show them our warm congratulations!"

6 Wang Tao Strikes a Pose: Comparing Bohai Bay Basin with Venezuela's Lake Maracaibo

In February 1965, the good news about Shengli and Dagang Oilfields lifted the morale of employees working in the Ministry of Petroleum's office building. Everyone was happy and excited. I attended the managerial meeting described above. At the time, I was a division director in the Ministry's Political Department. I was part of the meeting's secretariat, in charge of writing conference bulletins. I remember at a plenary session, one young and handsome man stood before a geological map, comparing our country's Bohai Bay Basin to Venezuela's Lake Maracaibo, extremely rich in oil and gas resources. His voice loud and sonorous, his speech excited and rousing, and his vision grand, he made Kang Shi'en, who was presiding over the meeting, laugh wholeheartedly and lavish praise on him. The two of them would speak in turns, one adding to the other, as if one were singing and the other joining the chorus, to constant applause. I was sitting in a corner, dumbfounded. I asked around and learned that his name was Wang Tao 王涛, that he was an outstanding geologist trained in the Soviet Union. I felt then that he would be a new star in our oil industry, with a great future. As I had expected, he was soon appointed Dagang Oilfield's deputy commander and chief geologist.

Fifty years later, when I read Kang Shi'en's biography, my recollections were confirmed. Kang Shi'en said:

> At this meeting of the leading cadres from the oilfields, Wang Tao, trained in and newly returned from the Soviet Union and then the Director of the Exploration Division of the Ministry's Petroleum Scientific and Research Institute, made a report on the prospects of Bohai Bay Exploration. Putting forth a bold vision and an interesting comparison, he systematically analyzed China's Bohai Bay and Venezuela's Lake Maracaibo and believed that the two shared many similarities in terms of size, sedimentation type, and geological evolution.[1]

This story provides a concrete illustration of the general state of mind among people working in the oil industry that had taken root with the "In Industry, Learn from Daqing" movement. While this no doubt created a positive outlook and momentum, as I look back on it today I think that if you are a geologist, no matter what your own personal mindset, opinions, or fantasies might be about the potential richness of a particular basin, that has no bearing at all on whether those ideas are correct or not. One cannot object to a geologist having opinions in this regard, because everything will eventually be tested in the fullness of time and practice. However, once you become a decision-maker, what is really needed is the ability to think objectively, keep a level head, listen to all opinions, size up the times, weigh all the pluses and minuses. Above all, one must be willing to make adjustments as situations develop. Otherwise, one will be prone to chasing after unrealistically high targets. Many of the strategic blunders made later on in the Chinese oil industry attest to this point.

1 Kang Shi'en zhuan bianxie zu, ed., *Kang Shi'en zhuan* (Beijing: Dangdai Zhongguo chubanshe, 1998), 258.

PART 4

The Jianghan Era (1970–1979): Against the Current, China Becomes a Big Oil-Producing Country

During this decade, Jianghan became the focus of the petroleum industry's campaign efforts. Despite the adverse effects of the Cultural Revolution, it built up momentum and achieved great successes. In 1965, throughout all of China's 31 provinces and regions, there were only 11 oil and gas fields, but by 1975, that number had increased to 21. In 1978, when national annual production of petroleum reached 109 million tons, China became the eighth largest oil producer in the world, and its average annual oil production increase was 7.4 million tons. Oil exports increased steadily. The "CCP Central Committee Document No. 13 (Zhonggong zhongyang di shisan hao wenjian 中共中央第十三号文件)," dated June 4, 1975 (issued when Deng Xiaoping led the Central Committee), pointed out: "It is especially pleasing that the oil industry has been in the lead, developing at a high speed." During the 1970s, China's rapidly developing oil industry made outstanding contributions that helped support the country at a time when its economy had virtually collapsed. It stabilized society and helped the country through hard times. Even so, the damage to the oil industry caused by the Cultural Revolution was severe and long-lasting.

CHAPTER 17

The Jiangan Oil Campaign: Zhou Enlai's Moves to Protect the Oil Industry

1 Zhou Enlai's Plea

In May 1969, after the CCP's Ninth National Congress put forth the theory of continuous revolution under the dictatorship of the proletariat, the Cultural Revolution was expected to continue. I, as an "oil soldier" staying in the office of the Ministry of Petroleum in Beijing, had grown weary of doing nothing all day long. However, at this time, oil was found in Qianjiang 潜江 County, Hubei Province, and the State Council appointed Kang Shi'en to lead the campaign. I longed to go to the front lines, so, with the help of Kang's secretary Wang Fuzhen, I secured a position in the campaign in Jianghan Oilfield. My wife left Beijing for the Ministry's May 7th Cadre School in Qianjiang in May 1969. After getting my two young daughters and aged mother-in-law settled in Beijing, on October 4, I traveled alone and reported for duty at the campaign headquarters in Qianjiang as the head of the secretarial group in the political department. My job was to draft important reports and documents for Kang Shi'en, and I would continue doing this for the next four years. In 1972, when the Jianghan Oil Campaign ended, I was transferred back to Beijing. Therefore, I was both a participant and eyewitness of the Jianghan Oil Campaign. As with Daqing, I want to leave the world with my impressions of the place and period.

The Jianghan Basin in Hubei Province, which has an area of approximately 28,000 km^2, has Wuhan to the east, Yichang 宜昌 to the west, the Dahong Mountains 大洪山 to the north, and the Dongting Lake 洞庭湖 to the south. The Yangtze and Han rivers run through this basin, which borders Sichuan and Guizhou provinces. Geologically speaking, it is an ancient salt lake basin, with good oil- and gas-producing conditions. It was, for this reason, a key inland area for oil and gas exploration. Exploration of the basin started in 1958, and after many years of geological surveys and regional exploration by both the Ministry of Petroleum and the Ministry of Geology, exploratory wells drilled Qianjiang County, Hubei Province in 1966 produced commercial oil flow.

In July 1967, at the beginning of the Cultural Revolution, the Ministry of Petroleum submitted a "Report on Organizing a Big Oil Campaign in Jianghan Region (关于在江汉地区组织石油大会战的报告)" to the State Council, which stated, "This region has many structures. Although its individual structures

 DOI:10.1163/9789004402737_018

are small, generally ranging from 10 to 20 km^2, it is possible to find many of relatively small oilfields … Therefore, we are prepared to gather our best forces in Jianghan Area and immediately start the campaign." However, the Cultural Revolution followed instead and swept across the whole nation, so that the Ministry's plan to organize the Jianghan Oil Campaign was shelved for three long years.

In the latter part of 1968, employees of the Ministries of Geology and Petroleum that continued to work in the basin discovered more new oil-containing structures and drilled two exploratory wells that turned out to be gushers. On February 2, 1969, the Ministry of Petroleum's Military Control Commission submitted its "Report on Organizing Oil Exploration and Campaign in the Jianghan Region of Hubei (关于在湖北江汉地区组织石油勘探会战的报告)" to the party Central Committee and the State Council. On February 6, the report reached the hands of Chairman Mao Zedong, and the next day Mao summoned Premier Zhou Enlai to his study to discuss the matter.

In March 1969, during a National Planning Meeting in Beijing, Premier Zhou proposed the organization of a relatively large-scale oil campaign in the Jianghan region. He appointed Kang Shi'en, who had still not been released, to be in charge of planning for the project. At the end of April, Kang Shi'en returned to Beijing after conducting twenty-plus days of investigation and research with his group in Jianghan. On May 18, 1969, he met with Li Fuchun, Li Xiannian, Su Yu 粟裕, You Qiuli, and others of the State Council business group, and he reported on the petroleum geological status of the Jianghan region and the plans to organize a campaign. On June 26, 1969, the State Council formally granted permission to start the oil campaign.

On the eve of Kang Shi'en's departure from Beijing for Jianghan, Zhou Enlai received him in Zhongnanhai. At their parting, Zhou held Kang Shi'en's hand and earnestly entreated him, "You must remove any form of interference and firmly fight this campaign to find oil in our nation's interior. You will need to think of ways to preserve the backbone forces of the oilfield and keep up their good traditions and good working style."

Afterwards, when Kang Shi'en remembered the meeting, he said that at the time his eyes teared up and that he solemnly nodded, indicating that he would remember the Premier's request.

I am mentioning Zhou Enlai's words now because they illustrate that, as one of the country's top leaders, he had great foresight and was taking special care of the oil industry. In that turbulent era, Zhou Enlai was carefully protecting the oil workers that made up the backbone of the industry, so that they could continue building the momentum of that industry's development.

2 Sent by the State Council, but Needing a Military Escort

To guarantee that the oil campaign would be carried out in an orderly fashion, the State Council and the Party's Central Military Commission decided that the Jianghan Oil Campaign would be led and directed by Wuhan Military Region and Hubei Province. The Ministry's Military Control Commission would be in charge only of production, development plans, and supplies. On July 25, the State Planning Commission and State Construction Commission issued "A Notice to Support the Jianghan Oil Exploration Campaign (关于支援江汉石油勘探会战的通知)" to relevant municipal and provincial agencies. On August 1, the Jianghan Oil Campaign Headquarters was established. Han Dongshan 韩东山, the deputy commander of Wuhan Military Region, was appointed campaign commander, while Kang Shi'en and Zhang Xianyang 张显杨, deputy chief of staff of Wuhan Military Region, were named as deputy commanders. Kang Shi'en was also deputy director of the provincial Revolutionary Committee and member of the provincial Party Standing Committee. Putting the Wuhan Military Region in collective command with Kang Shi'en was an important decision, because under the turbulent circumstances of the time, only the senior military officers in command of major military regions had enough authority and power to prevent various disruptions, and this enabled them to steer the vessel of the oil campaign forward.

To prevent the influence of factions and guarantee that the campaign would proceed normally, Han Dongshan, Kang Shi'en, and Zhang Xianyang decided to implement a few unusual measures: (1) the campaign workers would be organized into regiments, battalions, and companies like a military establishment, thus doing away with the trend of setting up a "Union of Leadership, Technical Personnel, and Workers," which tended to become hotbeds for radicals; (2) they restored the regular meetings of the Communist Party and the Communist Youth League; (3) they instituted the "Three Key Points (*san da jiang* 三大讲)," which emphasized talking about reality, duty, and unity. Meanwhile, it eliminated all factional struggle and reduced the impact of the slogan, "It is Right to Rebel." Some people much later on have commented that arranging oil workers in the military way was "ultra-leftist," but they do not understand the difficulty of accomplishing anything at this time. More than a decade later, General Zhang Xianyang visited with the Jianghan Oilfield employees, and when asked about the matter, the distinguished old general roared, "How could we not have had campaign troops arranged in a military way at the time? If we had had a Revolutionary Committee and let those son-of-a-bitch radicals come in, they would have spent all day rounding up and criticizing the leading cadres, and then how could the oil campaign have been completed?"

Jianghan Oil Campaign received the close attention of the State Council and Premier Zhou. Once it started, drilling went fast, and news of success kept pouring in. What was achieved in August and September 1969 equalled that of the previous few years combined. Eleven exploratory wells showed oil and gas, among which five had high-yield oil flow. Early in the morning of October 16, 1969, Han Dongshan and Kang Shi'en traveled to Beijing to report on these two months of the oil campaign to the Central Committee and the State Council. The leaders of the State Council, after having worked through the previous night, listened to Kang Shi'en's report for the whole day. Vice Premier Li Xiannian said, "The Premier is very interested in the development of Jianghan Oil Campaign. He called twice last night, asking if you [i.e., Han and Kang] had arrived, and he had another instruction: 'Please ask Xiannian, Desheng, Dengkui, Qiuli, and Su Yu to listen to their reports, the sooner the better.' The Premier also put emphasis on the four words 'the sooner the better.'"

This campaign was conducted under the extremely chaotic circumstances of the Cultural Revolution. For example, Kang Shi'en was 54 when he went alone to Qianjiang, and in 1969, when he was attending the National Day celebrations in Wuhan, he was barred from entering the gate of the Provincial Revolutionary Committee (i.e., the provincial government building), no matter how much he tried to persuade the guards. The guards did not believe him to be a person of importance because he travelled in a battered old Soviet-made jeep. He was finally let in when General Han Dongshan personally came out to verify his identity. Afterwards, Han sent him an ordinary Soviet Volga sedan from Hubei Province, even though Kang had not requested it. According to Kang's secretary Wang Fuzhen, in the beginning of the oil campaign, the Ministry of Petroleum's Military Control Commission treated Kang Shi'en as an "unliberated capitalist-roader." They stripped him of his secretary and monitored him everywhere.

During the first few months of the campaign, Kang Shi'en often made long-distance midnight phone calls to Yu Qiuli in Beijing. At the time, long-distance phone calls were difficult to make; callers had to shout to be heard. Sometimes, his calls would last for one or two hours as he talked about one well after another. During this time, one of the two staff officers from the Ministry's Military Control Commission would often eavesdrop outside of the window of the one-level house where Kang lived, in order to find out what these two "capitalist-roaders" were talking about. Once the eavesdropping was discovered, Han Dongshan and Zhang Xianyang were enraged. Zhang suggested that Kang choose a secretary from Wuhan Military Region, but

Kang politely declined, saying that he originally had his own secretary. Han Dongshan then promptly notified the Ministry's Military Control Commission to restore Kang's secretary Wang Fuzhen and send him to Jianghan Oilfield. Meanwhile, they negotiated with the Ministry's Military Control Commission to transfer the two eavesdropping staff officers back to Beijing.

Looking back, we cannot really blame the two officers for acting irrationally, for they had been swept up by the trends of the time. Even so, this incident shows the resolve of the Wuhan Military Region leadership to do away with all kinds of interference or problems in order to let the Jianghan Oil Campaign proceed normally.

There were many other instances in which the Wuhan Military Region had to protect the Jianghan Oil Campaign. In 1969, among the oil workers who had come from different parts of the country, there were numerous political factions. A minority among them claimed to be rebels. They wanted to organize "great criticisms" and looked for "targets" everywhere. Whenever Kang Shi'en went to a work site, some of these people from the rebel factions would say, "Your arrival is so timely, for you are the biggest capitalist roader in the Ministry of Petroleum. In the past, we couldn't criticize you even if we wanted, but today we'll have an on-site criticism meeting." When the upright Zhang Xianyang learned of this, he was furious: "These people are complete outlaws. Kang Shi'en was sent by the State Council to be in charge of the oil campaign. In the future, if such things happen again, I won't be so civil." Meanwhile, he told Kang, "Old Kang, to tell you the truth, when it comes to fighting a war we here are okay, but when it comes to finding oil, it's all up to you oil experts. Just do your job and be bold about it. Commander Han and I will escort you and protect you along the way." This encouraged Kang Shi'en and the oil managers who participated in the campaign.

Once when Kang Shi'en and Zhang Xianyang were carrying out an inspection of the 1832 Drilling Crew, a few rebels surrounded Kang Shi'en at the drilling site and insisted, "Here you are, Kang Shi'en. You insist on productivity theory and all that [i.e., without paying attention to class struggle], so today we'll give you a five-minute study class." On hearing that, Zhang Xianyang was furious. With his arms akimbo, he shouted, "You're sabotaging Chairman Mao's 'Grasp Revolution and Promote Production,' so I'll have you arrested first and then put *you* through a study class!" When they saw that behind this awe-inspiring old general stood a pistol-bearing bodyguard, the group of rebels had no choice but to dejectedly slink away.

At the time, most of those taking part in the campaign from the Wuhan Military Region were officers, all holding formal positions with various work units. Han Dongshan stated repeatedly at leadership meetings: "In terms of business, we need to rely on comrades who know business. Military managers should not interfere and poke our noses into these affairs. We are here mainly to support the work of oil managers like Kang Shi'en."

The fact that the Jianghan Oil Campaign was successfully carried out was a result of the efforts of both the military and the people. The Jianghan Oil Campaign signaled the petroleum industry's determination to advance, not retreat, during the Cultural Revolution, and it came about as a result of Chairman Mao and Premier Zhou's resolve, the careful arrangements of the State Council and the Central Military Commission, and the protection of the Wuhan Military Region.

3 Preserving the "Backbone Forces of the Oil Industry"

Recall that before Kang Shi'en left for the Jianghan Oil Campaign, Zhou Enlai advised him, "You will need to think of ways to preserve the backbone forces of the oil industry." To carry out Premier Zhou's instruction, Kang Shi'en made good use of the opportunity provided by the Jianghan Oil Campaign in order to protect, preserve, and foster a group of managers who would become a reserve force for the future development of the petroleum industry. Per Kang Shi'en's arrangement, many of those "capitalist roaders," "reactionary technological authorities," and "class enemies of all descriptions," many of those overthrown managers and engineering and technical personnel at all levels in the Ministry of Petroleum and its various agencies and oilfields who had ever been "bombarded" or put "under fire," as well as Daqing's outstanding 1202 and 1205 Drilling Crews, were gathered at Jianghan Oilfield. They were all assigned to different positions so they could work hard for the country's oil industry development. I shall illustrate the significance of this move with the following stories of some of the individuals involved.

(1) Zhang Wenbin. In 1952, on Chairman Mao's order, an entire military division was transferred into the civilian petroleum industry. Zhang Wenbin was that division's political commissar. In the fifties, he was the chief of the Xinjiang Petroleum Bureau. In 1960, he was transferred to be Deputy Commander and Deputy Party Secretary in Daqing. In 1964, he became the commander of Shengli Oilfield and was promoted to Deputy Minister of Petroleum. He was soon transferred and became the commander of the second oil campaign in Sichuan Province. From the time of the Daqing Oil Campaign, he had been one of Kang Shi'en's most important assistants.

Soon after the beginning of the Cultural Revolution in 1966, Zhang was detained and tortured, both psychologically and physically. According to

Wang Fuzhen, in early 1970, soon after the Jianghan Oil Campaign started, Vice Premier of the State Council Li Xiannian instructed Wang Xing 王星, the director of the Ministry's Military Control Commission, to take some people to work and be of service in Jianghan Oilfield. A few days later, at a meeting in Wuhan presided by the Wuhan Military Region Commander Zeng Siyu 曾思玉 and the Region's Political Commissar Liu Feng 刘丰, Kang Shi'en proposed transferring Zhang Wenbin to participate in Jianghan Oil Campaign. Tang Xianzhen 汤贤珍 from the Military Control Commission in charge of personnel responded, "Zhang Wenbin shouldn't be transferred—he was both a stubbornly unreformed capitalist-roader and a coward during the War of Resistance Against Japan."

On hearing this, Liu Feng asked, "Where did you get this material of yours from?"

Tang answered, "It was mentioned in a big-character poster that the masses of the Ministry posted."

Liu Feng loudly scolded him. "What year are you living in? You're still using the masses' posters to determine a manager's character? Ridiculous! I could testify that during the War of Resistance Against Japan, when I was a regiment commander, Zhang Wenbin was the bravest battalion political instructor. He was young and capable. He was good with his pen as well as his gun. If you don't believe me, you could go ask so-and-so and such-and-such—now they're respectively the commander of such-and-such army and the political commissar of this-and-that army!"

With that, the officer from the Ministry's Military Control Commission was made speechless. He could only agree to contact Sichuan Province right away to transfer Zhang Wenbin. Afterwards, Yu Qiuli personally stepped in, asking his old comrade General Zhang Guohua 张国华, Commander of Chengdu Military Region and Director of Sichuan Province Revolutionary Committee, to get Zhang Wenbin out of detention at Sichuan Petroleum Administration Bureau and put him in charge of production in Daqing Oilfield. Later, when the oil pipeline in the northeast urgently needed to be built, Zhang was transferred to Shenyang to be in charge of the project. Zhang eventually became Deputy Minister and Deputy Party Secretary in the Ministry of Petroleum. He made outstanding contributions to the development of the oil industry.

(2) Jiao Liren. In September 1949, when Yumen Oilfield was liberated, Jiao became its deputy military representative. He later became the Director of Yumen Petroleum Bureau. In 1960, he was transferred to Daqing to become the Deputy Commander. He was also Kang Shi'en's primary assistant. He became the Director of the Exploration Bureau at the Ministry of Petroleum in early 1965. Soon afterward, he went to Shengli Oilfield as the Commander.

The Cultural Revolution came and greatly impacted Jiao. He had still not been rehabilitated in 1970, so he was still someone to keep an eye on. However, he was well respected by the employees of Shengli Oilfield, and people cordially called him "Master Jiao." In the latter part of 1970, at a critical moment in the Jianghan Oil Campaign, Han Dongshan saw that Kang Shi'en was under duress and proposed that an assistant be assigned to him. Kang said that he would like to have Jiao Liren from Shengli Oilfield to come, even though it might be difficult to arrange. Han agreed to bring it about. Very soon, Jiao became the Deputy Commander for Jianghan Oil Campaign, assisting Kang with production. In 1972, after the Jianghan Oil Campaign, he was transferred back to Beijing, where he assisted Kang with daily production. He became Deputy Minister and Deputy Party Secretary of the Ministry of Petroleum in 1978, continuing to contribute to the oil industry.

(3) Zhang Fuzhen's exoneration. In 1970, the long-distance oil transport company in Dunhuang County, Gansu Province was transferred in toto to Jianghan Oilfield's auto repair plant; this included the employees, amounting to 1,600 people. Among them was Zhang Fuzhen's widow, Tian Shuzhen 田淑珍, who brought her husband's case with her to Jianghan.

Zhang Fuzhen had been the Division Commander of the original Petroleum Division. In 1952, on Chairman Mao's order, he, the Political Commissar Zhang Wenbin, and his whole division of officers and soldiers were transferred to the oil industry. He was appointed manager of the industry's best-funded transportation company, which had many hundreds of large imported oil tanker trucks. In his early years, after he joined the Communist Party, he did underground work for the Party in the senior Kuomintang general Zhao Shoushan's 赵寿山 army. He became a division commander. Later, he led his division to revolt and cross over to join the PLA's 57th Division. After he moved to the oil industry, he became one of the managers most respected by employees. He was hardworking, responsible, kind to people, and beloved by most.

During the Cultural Revolution, he was, without rhyme or reason, branded as a traitor because of his time in the Kuomintang army. On the night of September 14, 1968, he "fell" into a deep water well on site and died. He continued to be branded a traitor after his death. When his wife Tian Shuzhen moved to Jianghan Oilfield, she requested an investigation into her husband's mysterious death.

All of the Jianghan Oil Campaign employees were concerned about Zhang's exoneration. Kang Shi'en was particularly so, and he talked to General Zhang Xianyang about it. On May 2, 2000, Li Shaoliang 李绍亮, who had earlier been the chief of the Jianghan Petroleum Bureau, told me that in early 1972,

General Zhang specifically went to the Campaign Logistics Department, which oversaw the auto repair plant, and summoned three associate managers. Two were from the Wuhan Military Region and one was Li Shaoliang. General Zhang was attentive, astute, thoughtful, and thorough, despite his gruff demeanor. He first questioned Li, who responded, "Although he did not know me, I knew Zhang Fuzhen, because he was the commander of the oil division. After reading his material, I think if he had been an undercover Communist in the Kuomintang army, he couldn't have been a traitor to the Communist Party."

General Zhang then queried one of the military control commissioners. He replied, "I read the materials. I felt that the accusations were too harsh, and there was no hard evidence."

General Zhang then asked the other military control commissioner. He responded, "Even though there was no hard evidence, the accusation of treason had been handed down by the Lanzhou Military Region. It might be inappropriate to overturn it."

The hot-tempered General Zhang pounded the table and got up. Pointing to the man, he shouted, "What the fuck are you talking about? Can I accuse you of being a traitor based on empty talk?" That commissioner was at a loss for words.

Zhang then decided to assemble a few trusted military officers and oil industry managers to readdress Zhang's case. Under General Zhang Xianyang's personal attention, the case was soon rectified. Soon afterward, Tian Shuzhen asked to have a memorial meeting to publicly restore Zhang's reputation. When General Zhang heard about this, he helplessly replied, "Even though I, Zhang Xianyang, took a risk by standing up to those guys, please tell them that any memorial meeting must be put off until a later time."

There are hundreds of stories similar to the above three in Jianghan. I choose to tell these stories because they marked a turning point. Afterward, Jianghan Oil Campaign worked vigorously to protect and preserve its leadership, exonerating the wrongfully-accused, and freeing and reinstating many managers and core technical personnel. Key managers from more than 40 petroleum enterprises were released from detention and immediately set out for Jianghan to resume their positions. In the afternoon of May 3, 2000, Li Yong 李勇, a former chief of Jianghan Oilfield Bureau who was also from the oil division told me that a few years earlier, some old comrades gathered historical data on the oil division. According to their statistics, there had been more than two thousand managers, a third of the oil division at the time, who had been released from detention and reinstated in Jianghan. It is worth mentioning that this type of rectification at Jianghan Oilfield happened during the extremely chaotic Cultural Revolution and preceded the 1978 downfall of the Gang of Four by seven or eight years, and that it was essential for the oil industry's continued growth. Had it not been for Jianghan Oil Campaign's success in reinstating these managers and technical personnel, there would have been no large-scale growth in oilfields throughout the nation after 1972, and there would have been no annual national oil production exceeding 100 million tons in 1978.

4 Keeping Up the Oil Workers' "Good Traditions and Good Working Style"

Abiding by Zhou Enlai's advice to keep up the oil workers' "good traditions and good working style," the Jianghan Oil Campaign employees, under the adverse circumstances of the Cultural Revolution, successfully fought one arduous battle after another.

The Jianghan region in ancient times had been a big lake called "Yunmeng daze (云梦大泽)," meaning "Great Marsh of Clouds and Dreams." It sits on a low, marshy alluvial plain. The Yangtze, Han, and other rivers and streams crisscross its expanse. In addition, due to abundant rainfall, it is flooded nine years out of ten. After the Liberation of 1949, it was designated as a flood discharge area for Jing River 荆江.

A hundred thousand-strong force of oil workers and hundreds of thousands of tons of oil campaign equipment and materials suddenly poured into this low, marshy area with its network of rivers and streams. Roads became a high priority. Jianghan Oil Campaign Headquarters made two decisions: one was to immediately organize crews to build roads and expand the wharf at the bank of the Han River, thereby solving the transportation problem; another was to motivate employees to carry forward the spirit of the Daqing Oil Campaign and to manually pull and carry equipment if need be in order to put it quickly into position.

When the famous Daqing 1202 Drilling Crew arrived in Jianghan, they did not ask for food or shelter but instead, imitating the model set by Ironman Wang, asked where their drilling position was and made it clear they were eager to start work. When they heard that trucks were in short supply and roads were bad, they each carried their own luggage, food, and tools weighing dozens of kg, and walked more than 30 km on muddy roads to the drilling site. Once there, they dropped their luggage and immediately started fixing the road and leveling the drilling site. When the rig and 13 items of equipment arrived, they pressed on and finished installing them in one day. When Kang Shi'en learned

FIGURE 17.1 Daqing 1202 Drilling Crew takes part in Jianghan Oil Campaign (1970)

of this, he rushed to the site to visit them and subsequently promoted their work ethic to all.

When the Guang-5 well near Guanghuasi 广华寺 was near completion, rain poured for days, making it impossible for any vehicles to reach it. By order of the Campaign Headquarters, more than 2,600 employees, working four days and nights and traveling more than five km, pulled and carried more than 6,000 m of casing, tubing, and 5,400 bags of concrete to the drilling site, completing the project on time.

I once took part in the casing transport. One casing was more than a dozen meters long and weighed 500 kg. Eight or nine of us carried it along the muddy dirt road for about a hundred meters or so to the drilling site. Because I am tall, I carried more weight on my shoulder than the others, but it would have been embarrassing to mention it, so I just kept it to myself.

When the construction workers built the oil depot at Guanghuasi, there was a truck shortage, but they did not wait for or rely on others' help. Instead, they carried more than 1,200 tons of building materials to the site from two kilometers away. The foundation for the oil tanks needed 200 m^3 of concrete, but there was no water source or mixer on site, so the workers carried more than ten thousand buckets of water from a ditch hundreds of meters away and manually mixed and poured the concrete, in order to complete the construction project on time.

5 "Resolutely Fight This Campaign to Find Oil in Our Nation's Interior"

After more than three years of the oil campaign, Jianghan Oilfield achieved relatively good results with respect to production and construction, and it realized its projected targets. Its achievements included the following:

- A better understanding of the geological situation of Jianghan Basin was obtained. Its oil reserves were found to exceed 50 million tons. New oilfields such as Xingang 新港 and a group of oil-bearing structures were found. Jiannan 建南 Gas Field in Enshi 恩施, in the western part of Hubei Province, was also found.
- Following to the "Find One, Drill One, Develop One" principle, oilfields such as Wangchang 王场 and Guanghuasi were developed. By 1972, the annual crude oil production capacity reached one million tons.
- Jinmen Oil Refinery (荆门炼油厂), capable of refining one million tons of crude oil per year, was built. After several expansions since then, it has become an important petrochemical enterprise capable of treating tens of millions of tons of crude oil annually.
- The auto repair plant, oil machinery plant, and oil instrument plant were transferred respectively from Dunhuang in Gansu Province, Dushanzi in Xinjiang, and Daqing in Heilongjiang Province. In two years, that is to say by 1972, all reached their pre-transfer production capacities.
- Beneath Wangchang Oilfield, a large, highly saline underground lake was discovered. The nation's largest salt chemical plant was built there, using salt water a few thousand meters underground as raw material. This produces and exports large quantities of high-quality salt chemicals each year.

The complete development of Jianghan Oilfield realized Premier Zhou's call to "resolutely fight this campaign to find oil in our nation's interior." In those days, from the "war preparation" perspective, the function of this oilfield could not be replaced by the big oilfields along the coast or frontier.

In February 1972, after the State Council had granted partial transfer of the Jianghan Oil Campaign troops, Kang Shi'en and Zhang Xianyang, the Wuhan Military Region deputy commander and Jianghan Oil Campaign commander, held a meeting in Beijing Hotel to discuss how to carry this out. In the two months following the meeting, more than 1,000 bureau- and division-level managers and engineers, more than 60,000 workers, more than 900 sets of equipment, and more than 100,000 tons of oil industry materials were transferred to the Shengli, Dagang, Henan, Daqing, Changqing, and Jilin Oilfields, and to the newly founded Planning Institute of Petroleum Exploration and

Development in Beijing. This formally brought to an end the inland oil campaign, which had lasted three years.

6 Blunders of the Jianghan Oil Campaign

At the end of 1963, my colleagues and I summed up and publicized the Daqing Oil Campaign as being conducted "in a difficult time, in a difficult place, and under difficult conditions." I want to say that ten years later, the Jianghan Oil Campaign was carried out "in an extraordinary time, an extraordinary place, and under extraordinary conditions." The "extraordinary time" refers to the turbulent Cultural Revolution. The "extraordinary place" refers to Jianghan, a flood-discharge area filled with lakes, rivers, and marshes. The "extraordinary conditions" refers to Jianghan's extremely hot summers, with temperatures exceeding 40°C.

Achieving the above-mentioned success under these three extraordinary circumstances was indeed a feat. However, in retrospect, there were many blunders on the part of the leadership.

Blunder no. 1: The plan was grandiose, and too many people were transferred. At the time, more than one hundred large-scale oil rigs capable of drilling more than 3,000 m were transported to Jianghan from other major oilfields in the nation. Oil workers were transferred from Daqing, Liaohe, Shengli, Sichuan, Xinjiang, Qinghai, and Yumen. A petroleum machinery plant, an auto repair plant, and a petroleum instrument plant were all transferred there. Meanwhile, they made a request to the Party Central Committee's Military Commission for 25,000 army veterans to be assigned to Jianghan Oilfield to participate in the campaign. The underlying intention to house, maintain, and train these oil workers in order to shield them from the harms of the Cultural Revolution was very obvious, and in that regard it was quite effective. However, in hindsight, the waste it caused was also obvious—it was "using an ox-butcher's knife to kill a chicken."

Blunder no. 2: The construction of Jinmen Oil Refinery, which started in 1970, was done entirely in accord with Mao Zedong's instructions to "prepare for war" and Lin Biao's instruction that infrastructure be "dispersed, close to mountains, and in caves." Thus, instead of being built in the oilfield itself, the refinery was built in hilly Jingmen 荆门 County more than 100 km away, and with individual refining units dispersed in different valleys a few kilometers apart (normally, the distance between single units should be no more than a few dozen meters). This resulted in great waste, as it required extensive piping.

Once it was built, it cost much more water, electricity, steam, and fuel to run. It was difficult to manage, with potential safety issues.

Blunder no. 3: The construction of the oilfield still adhered to the Daqing model. It had scattered, isolated worksites that were entirely self-sufficient, small but complete. This made it wasteful and hard to manage. Had the oil town been built in Wuhan, the capital of Hubei Province, or in the historical city of Jingzhou 荆州 about 50 km away, or in nearby Qianjiang, it would have been wonderful!

CHAPTER 18

An Oil Industry Ready to Launch Out in All Directions

1 The Origin of "a Dozen Daqings"

In 1970, when the Cultural Revolution was at its height, the State Council was developing and discussing the country's fourth Five-Year Plan. The Military Control Commission of the Ministry of Petroleum, not knowing the petroleum business, had to invite Kang Shi'en back to Beijing from the Jianghan Oil Campaign in Hubei Province, so that he might take charge of compiling the fourth Five-Year Plan for the petroleum industry. Upon his return to Beijing, Kang Shi'en organized a group of experts to study data related to the oil industry and compare it with the latest information on foreign oil industries. He also had lengthy discussions with Yu Qiuli and with numerous geologists. At last, they proposed the outline for the oil industry's fourth Five-Year Plan.

This outline made clear the purpose and main theme from the very beginning: "Our nation's oil and natural gas resources are very rich. In the twenty years since 1949, the overall exploration situation is that the more we explore the terrain, the better the situation; the more we explore resources, the richer they become; the more we explore, the more confidence we have; and the more we explore, the brighter the future."

The outline also stated:

> There are more than 4 million km^2 of continental sedimentary rock in our country, and every province has sedimentary basins where oil and gas can be found. How great are the prospects for developing our nation's oil and gas resources? First, let us draw a comparison with the United States, which produces the most oil in the world. The American continental sedimentary area is similar to our nation's. They have been extracting oil for more than a hundred years. They have drilled [a total of] 2 billion m and have found 41 billon tons of petroleum reserves and more than 10 trillion m^3 of natural gas reserves. In the past twenty years, we have thoroughly explored only less than ten percent of our sedimentary rock area, drilled [a total of] 17 million m, and found 3.6 billion tons of petroleum reserves. This fully demonstrates that our country could, within a relatively short

 | DOI:10.1163/9789004402737_019

> period of time and with a relatively smaller amount of work, catch up with and surpass advanced levels in the world.

The outline goes on to state clearly:

> Our country has more than 4 million km^2 of sedimentary area. According to a formula now commonly understood worldwide, it is entirely possible to find on average 10,000 tons of petroleum reserves for every km^2. Therefore, we could find at least more than 40 billion tons of petroleum reserves. This is a solid foundation for the development of our nation's petroleum industry.

In other words, as early as 1970, Kang Shi'en, using the United States as a comparison, proposed that if the 40 billion tons of oil reserves were found, then, applying the one percent recovery rate formula that was commonly used at the time, one could set an annual production target of 400 million tons. This target was the origin of the "dozen Daqings" proposed seven or eight years later.

2 Relying on Mao's "Two Theories," Reestablishing Order in Daqing Oilfield

During the Cultural Revolution, even though oil production continued to increase in Daqing Oilfield, there were many hidden production problems. Due to Cultural Revolution-related interference and sabotage, managers who were familiar with production were purged at all levels; rules and regulations could not be maintained, and normal production procedures were disrupted. Scientific and technological work stopped. Although many new wells went into production, water injection decreased, which resulted in excessive rates of reservoir voidage, as well as decreased reservoir pressure, decreased oil production, and a drastic increase of wet wells. By the first quarter of 1970, according to an investigation that Daqing Oilfield conducted on 583 wells, daily production at each well had dropped 7% on average. Production at 61% of the wells had decreased. Meanwhile, the water content of the wells was as high as 6.78%, more than twice the normal rate.

Faced with this severe situation, in 1970, Ironman Wang Jinxi, who was the deputy director of the Revolutionary Committee of Daqing Oilfield, went to Beijing himself to submit to Premier Zhou Enlai the report on the decline of production in Daqing Oilfield. Zhou Enlai commented on the report, "Daqing must return to fundamentals of development by relying on Chairman Mao's two essays" (i.e., "On Practice" and "On Contradiction"). Zhou Enlai's written instruction became a turning point in the development of Daqing Oilfield. It was also by such measures that the top leaders such as Mao Zedong and Zhou Enlai protected the oil industry during the Cultural Revolution.

To carry out Premier Zhou's instructions, a large group of managers such as Song Zhenming and Chen Liemin were reinstated to managerial positions. They mobilized the masses and conducted large-scale underground investigations, formulated reasonable technical policies, gradually restored the system of personal responsibility for each line of work, and soon turned the situation around. Crude oil production was stabilized and increased. By 1973, after three years of adjustments, Daqing Oilfield's crude oil production had increased more than 50% compared with 1970.

3 Making a Big Effort during the National Crisis, Fully Developing Daqing Oilfield

By the early 1970s, the Cultural Revolution had truly led to "great chaos under heaven." Industrial and transportation enterprises came to a partial or complete halt, and the national economy was near collapse. Coal, which accounted for 80% of the nation's fuel supply, mattered a great deal to the economy and to people's livelihoods, but its production and rail transport were repeatedly in a state of emergency. This situation worried Zhou Enlai, the "head butler" of eight hundred million people. He tried time and again to find a solution. Kang Shi'en's secretary at the time, Wang Fuzhen, told me, "Premier Zhou Enlai often asked Kang Shi'en over to talk about work late at night, when it was quiet, and once they got started they'd talk for hours. When he returned, Kang Shi'en would tell me, with anguish, that even though Zhou was the premier of such a big country and had to handle countless state affairs every day, he was still forced to become a 'coal dispatcher.' To guarantee that some of the big cities and big power plants would have enough coal, many times the premier would personally use the phone to dispatch the trains and ask about their transportation status, station by station, along the line. To ensure smooth passage for coal trains through railway hubs like Xuzhou and Zhengzhou, he would sometimes mobilize army forces to protect them." Such was the reality of China's industry and transportation at the time. It sounds absurd, but it is true.

As for the oil industry, even though it was impacted by the Cultural Revolution, it still forged ahead against the wind, thanks to the protection of Zhou Enlai and Mao Zedong. Production increased at old oilfields such as Daqing and Shengli, and new oilfields were discovered at Jianghan, Changqing, Liaohe, and elsewhere. In 1965, China produced 11.31 million tons of crude oil, but in 1970, the figure reached 30.65 million tons. This oil production played an important role in maintaining the national economy and guaranteeing people's livelihoods.

Even so, the decline of coal production and the increase in oil consumption could not be stopped. This was a matter of importance to social stability and national security. After weighing the pros and cons of the situation, Zhou Enlai decided in the fall of 1972 to immediately develop Lamadian Oilfield in Daqing. Lamadian Oilfield is situated at the northern tip of Daqing. Discovered during the Daqing Oil Campaign in 1960, it is a high-yield oilfield, with a thick reservoir, good physical properties, and proven reserves of 480 million tons. It was the oil industry's "nest egg," left there as a strategic reserve oilfield. In the early 1960s, the Ministry of Petroleum's Party Group had held a special meeting on it. Yu Qiuli said that if it were to be developed in the future, the Ministry's Party Group would need to discuss and approve it by a unanimous show of hands.

But now this was a time of national crisis. Since Premier Zhou had made the decision, Kang Shi'en carefully organized and made the arrangements for a campaign to develop Lamadian Oilfield; he transferred Jiao Liren, who had just returned to the Ministry of Petroleum from the Jianghan Oilfield, over to Daqing to oversee the campaign. In early 1973, braving the extreme cold, Daqing oil workers gathered at Lamadian Oilfield. Following the slogan, "It Is Right to Work Hard for Socialism, It Is Meritorious to Work Hard for Socialism, It Is an Honor to Work Hard for Socialism," they labored away, keeping Cultural Revolution-related interruptions and sabotage at a distance, and vigorously carried out production and development. Everywhere else there was "stopping production to carry on revolution" and violence and fighting among factions, but here, "working hard for socialism" prevailed.

After two years of development, Lamadian Oilfield was put into full production in 1975 and had an annual output of 10.992 million tons. This surpassed design capacity and contributed approximately one-fourth of Daqing Oilfield's total production at the time. Lamadian Oilfield's successful development spurred an increase in crude oil production at Daqing Oilfield, so that in 1976, Daqing's total production reached 50.30 million tons, up from 46.259 million tons the previous year.

4 Shengli Oilfield: Ten Times the Reward

In June 1972, in order to deal with disruptive meddling from the rebel faction in the Revolution Committee, Kang Shi'en, with the agreement of Shandong Province, had Shengli Oilfield's Revolution Committee changed to Shengli Oilfield Campaign Headquarters. He then transferred a group of managers and several thousand drilling and oilfield construction people from Jianghan Oilfield to Shengli Oilfield. This greatly strengthened Shengli Oilfield.

While they stabilized production in the old subsidiary oilfields, they found and developed more than 20 new ones, such as Gudao 孤岛, Dongxin 东辛, Yihezhuang 义和庄, and Shanghe 商河. By 1978, there were 46 oilfields of different sizes in Shandong. Many moving stories come from this time.

In 1968, Gudao was still a big stretch of beach where the Yellow River met the sea, but it became a vast watery expanse whenever the Yellow River flooded. That year, the 3293 and 3297 Drilling Crews were drilling two exploratory wells, which proved to be key wells that discovered Gudao Oilfield. One day in June of that year, geologists Liu Xingcai 刘兴材 and Zhang Zhongxin 张忠信 went by car to the drilling site. Their original task was to tell these two drilling crews to retreat from Gudao before the July flooding and ensure the safety of crew members and equipment, given that an earlier drilling crew had been besieged by flood for weeks. As soon as they got out of their car, they were surrounded by workers, who escorted them to a simple work shed. At the time, the leaders of drilling crews had been sidelined, and the heads of the two opposing political factions simultaneously asked them, "Could the two of you decide? Do these two wells have oil or not?" Interestingly, those two heads who otherwise held different views on just about everything, shouted in unison, "If there is oil, and the two of you ask us to leave, we'll turn against you and won't be nice about it!" One should keep in mind that when so many factories had long ceased production because of revolutionary activity, these workers were still willing to risk their lives to look for oil for the country.

After looking over the geological data and records collected by the drilling crews, Liu and Zhang concluded that there was oil. Once they returned to their base, they reported that the workers from two opposing factions strongly requested to not only finish drilling these two wells and testing for oil before the flood season, but also to drill two more exploratory wells. This plan, which won full support from the sidelined Shengli Oilfield Campaign commander Jiao Liren (he was still not released at the time), was ultimately ratified. Everything was carried out per the workers' request. Soon, the two wells produced commercial oil flow, and the two new wells were finished in succession. According to the data collected from these four wells, Gudao Oilfield was estimated to have 200 million tons of geological reserves. It was the biggest oilfield found in Shengli during the Cultural Revolution and remains one of Shengli's main oilfields.

On September 21, 1973, the 32192 Drilling Crew, while drilling Luo-5 well, had a powerful blowout. The head driller, Wu Yutian 吴玉田, ignored the danger and kept minding his post, despite lethal levels of hydrogen sulfide fumes. He kept "moving the string"[1] and died as a result. His dead hand still tightly gripped the brake lever to protect the safety of equipment and the exploratory well.

Once Shengli Oilfield's exploration was in full swing, competitions between workers became popular. The 3252 Drilling Crew were champions of the work competition. Workers of this drilling crew, embracing the "Ironman" spirit, made bold improvements to drilling technology. In 1973, they drilled 102 wells and set the national record of drilling 151,420 m in a year. The Ministry of Fuel Industries awarded them the honorary title of "Ironman Heroic Drilling Crew," and the crew leader Yang Zhiyu 杨志钰 was designated a model worker.

From 1966 to 1978, Shengli Oilfield was fully developed and continued to grow, its annual production increasing 27.9% on average. In 1978, it produced 19.46 million tons, ten times that of 1966. This period was aptly dubbed "ten years of endeavor, ten times the reward." Shengli Oilfield became the second largest oilfield in the country.

5 Liaohe Oilfield: Number Three

Liaohe Oilfield is situated in the delta of Liao River, in the midwestern part of Liaoning Province. It is surrounded by mountains on three sides, and on the fourth it extends south to Liaodong Bay. The Liao, Daling 大凌, and other rivers all flow into the sea from here. The place used to be a no-man's land, full of wild reeds, and was known as the northeast's famous "Southern Wilderness." After 1949, irrigation systems were built, the wilderness was reclaimed, and it became the main center in Liaoning Province for farming rice and reeds. Liaohe Oilfield is conveniently located near industrial zones like Shenyang, Anshan, Jinzhou, Yingkou, and so on. Geologically, Liaohe Oilfield is called the Lower Liaohe Depression; it is part of Bohai Bay Basin and has a total area of 24,000 km^2. The area has good conditions for oil and natural gas generation, transport, and collection. Oil exploration in this basin started as early as 1955, when the

1 "Moving the string" (*huodong zuanju* 活动钻具) is part of the process for recovering the drilling pipe.

FIGURE 18.1 Cluster wells in Liaohe Oilfield

First Survey Crew of the Ministry of Geology worked here. By 1966, they had drilled 13 exploratory wells, seven of which showed either oil and gas or had commercial oil flow.

But in the latter part of 1966, the Cultural Revolution swept over the country, and by the end of that year, the offices of the Ministry of Petroleum had succumbed to total chaos. Under these difficult circumstances, in March 1967, first deputy commissioner Yu Qiuli of the State Planning Commission decided to let the Ministry of Petroleum organize a 600-person group of oil explorers from Daqing Oilfield to go into the reed paddies in the lower Liao River. They duly pitched their tents or built plank houses, and got to work on the oil campaign. Perhaps at the time the rioters nearby were too busy with internecine power struggles and factional fighting to notice that out in the reed swamp there were a group of "idiots" who were only interested in finding oil and not at all interested in revolution.

After three years of hard work, by the end of 1969, they had drilled 36 exploratory wells in this basin. They drilled a total of 90,000 m, obtained commercial oil flow on nine structures, and had basically put the oil-bearing area of Huangjindai 黄金带, Yulou 于楼, and Rehetai 热河台 Oilfields under control. In November 1969, they were drilling Huang-5 well in Huangjindai Oilfield, and while they tested for oil, a blowout occurred. Oil spurted as high 50 m, and the rumbling could be heard from 15 km away. The blown-out natural gas formed a long plume in which a single spark could have resulted in a disastrous fire. The blowout at Huang-5 well shocked the State Council. Premier Zhou Enlai instructed those in charge to spare no cost to control the blowout and save the oil and gas field. Seeing how dangerous the situation was, people in the surrounding area, coordinated by the local government and army, extinguished all potential spark and fire sources. The Ministry of Petroleum sent a group of experts headed by Qin Wencai 秦文彩 to assist at the site. As many as 100,000 soldiers and people in the area worked together for 30 days and nights. A team of rescue commandos, risking their lives, got on the rig and tried to switch off the gate. After 26 hours of continuous struggle, the blowout was stopped. This incident signaled the presence of a high-yield oil and gas field. But there would be more.

In February 1970, the Military Control Commission of the Ministry of Petroleum submitted a request to pool manpower and resources to continue expanding oil exploration in Liaohe Basin. On February 24, the State Council approved the request. The Ministry of Petroleum gradually transferred personnel and equipment from Dagang Oilfield to Liaohe to start the oil campaign there. Thus in 1978, Liaohe Oilfield's oil production reached 2.54 million tons.

In May 1984, when CCP Secretary Hu Yaobang 胡耀邦 visited Liaohe Oilfield, he wrote in calligraphy, "I hope you will open your minds, broaden your vision, continue to march forward, and endeavor to be 'the Country's Third Largest Oilfield.'" In 1986, when Liaohe Oilfield's annual oil production reached 10 million tons, it indeed became China's third largest oilfield. In 1995, it hit its peak production of 15.52 million tons. In 2014, when it produced 10.76 million tons, it still ranked ninth in the nation.

CHAPTER 19

The Sudden Rise and Fall of Huabei Oilfield

1 A Concerted Effort of the Ministries of Geology and Petroleum

Starting in September 1975, I worked in Huabei Oilfield for eight years, until I was transferred back to Beijing at the end of 1982. Since I personally experienced its vigorous start and sudden decline, it is my responsibility and duty to document them.

Huabei Oilfield is situated in the vast region south of Beijing and Tianjin, west of Cangzhou 沧州, east of the Taihang Mountains, and north of Shijiazhuang and Hengshui 衡水. Geologically speaking, it is called the Central Hebei Depression, with an area of 56,000 km^2. It borders Tianjin's Dagang Oilfield in the east and Henan Province's Zhongyuan Oilfield in the south. Not far to the southeast is Shengli Oilfield in Shandong Province. These oilfields all belong to the oil- and gas-rich Bohai Bay Basin. Since Huabei Oilfield's main oilfield was discovered in Renqiu County in Hebei Province, it is also called Renqiu Oilfield.

The exploration of Huabei Oilfield started in 1955. The Ministries of Geology and Petroleum did a large amount of geological investigation, geophysical prospecting, and some drilling. In 1958 and 1972, more exploratory teams were sent out. These early endeavors had little success, though much valuable data was obtained. The fourth big attempt occurred in June 1973. Per Kang Shi'en's instructions, the Ministry's Planning Institute of Petroleum Exploration and Development invited experts from the Ministries of Geology and Petroleum to conduct research and decide on the positions of the first group of wells on the four geological structures in Renqiu, Gaoyang 高阳 County, and Xian 献 County in central Hebei Province. This move yielded instant results. In June 1974, 3269 Drilling Crew of the Second Drilling Company of Dagang Oilfield, while drilling Jia-1 家一井, experienced a blowout at the Tertiary oil-bearing stratum. They completed the well and conducted well testing. It produced 63.3 tons of crude oil and 4,379 m^3 of gas each day. This was the first exploratory well in the plains of Central Hebei that produced a commercial oil flow.

2 China's First Buried-Hill Oilfield

In September 1974, when the Ministry of Geology Hebei Bureau's 3505 Drilling Crew successfully drilled the Jimen-1 well (冀门一井) at the Shimenqiao

 | DOI:10.1163/9789004402737_020

石门桥 structure in Renqiu County, they obtained an oily core measuring 0.92 meters, but the core for the first time was found in a Paleozoic stratum. This was an unprecedented discovery in China's oil exploration history.

The discovery of oil at Jimen-1 in a Paleozoic stratum got the attention of the Ministry of Petroleum. Under Kang Shi'en's direction, in October 1974, more than 100 geological personnel from the Petroleum Geophysical Prospecting Bureau, Dagang Oilfield, and Planning Institute of Petroleum Exploration and Development met at a petrochemical industry guest house in an eastern suburb of Beijing. Yan Dunshi 闫敦实, the vice president of Planning Institute of Petroleum Exploration and Development, presided over the effort. Petroleum Geophysical Prospecting Bureau experts such as Wang Shangwen, Meng Ersheng 孟尔盛, and Yuan Bingheng 袁炳衡 participated. Without using computers or calculators, people studied decades of seismic and drilling information from central Hebei Province. After several weeks of difficult research, they produced a general plan that included the drilling of Ren-4, the fourth exploratory well in Renqiu County.

To facilitate this mission, they transferred 3269 Drilling Crew, which had drilled Jia-1, the first oil-producing well in central Hebei, to the site. Under the guidance of Yan Dunshi and Xian Xuefeng 咸雪峰, the deputy manager of Second Drilling Company, the crew repeatedly found oil and gas shows at Cenozoic strata. But they did not let this distract them, nor did they run tests while drilling. Since the Ministry of Geology's Jimen-1, only five or six km away, had shown oil and gas at Paleozoic strata, they drilled down to that point. They entered the Paleozoic strata at 3,162 m. Throughout the drilling, geological personnel and workers cooperated closely on carefully recording data. Among countless cuttings, a worker found a few Paleozoic cuttings that contained oil. This was a good sign and a surprise. At 3,177 m, there was severe mud loss. Drilling was finished and the well tested at 3,200 m. On July 3, 1975, oil from the Paleozoic stratum gushed from Ren-4. In September the same year, the well received acid treatment. Producing as much as 1,014 tons of crude oil per day, it became the nation's first high-yield buried-hill oil well.

Experts explain that 500 million years ago, during the Paleozoic Era, the site of Huabei Oilfield might have been covered by sea, at the bottom of which lay very thick carbonate rock. Due to crustal movement, the bottom of the sea rose and became land, and the bulging parts became hills. Long-term erosion created cracks and caverns on the surface of the carbonate rock. In the Cenozoic Era, the crust moved downward, and the carbonate hills of Paleozoic Era were gradually covered over by Cenozoic sedimentary strata. The hills that were formed by Paleozoic Era strata and buried underneath Cenozoic Era strata are called "buried hills," and the oil and gas reserves formed in these buried hills came to be called buried-hill oil and gas reserves. Huabei Oilfield is one such large-scale buried-hill oilfield. Its discovery and development opened up new potential search areas for oil and gas and enriched our understanding of petroleum reservoirs.

3 Four Achievements in a Year: a Vigorous Start

Amidst the whirlwind of the Cultural Revolution, production and construction of Huabei Oilfield started, swiftly and vigorously. Within 10 to 15 days after Ren-4 began producing oil, a few nearby wells also started gushing, all producing more than 1,000 tons per day. The highest yielding well, Ren-7, in its early stages produced 4,620 tons of oil per day. On October 26, at the Ren-6 site, more than 3,000 people attended the opening celebration for Ren-4 and Ren-6. Kang Shi'en came to the event from Beijing for the express purpose of congratulating the workers. He gave a silk banner embroidered with "A Drilling Crew of Iron and Steel" to the 3269 Drilling Crew responsible for drilling the two wells.

Deputy Petroleum Minister Zhang Wenbin was initially this campaign's commander and party secretary. After 1977, Deputy Minister Jiao Liren assumed those positions. In the spring of 1976, tens of thousands of people were transferred to the plains of central Hebei Province, where all of a sudden, drilling machines roared, traffic grew dense, and the clamor of voices could be heard. People worked hard, even in windy or rainy weather. By September that year, 14 high-yield oil wells had been drilled and daily crude oil production had reached 30,000 tons. At the same time, around Renqiu County, a group of medium and small Paleozoic and Cenozoic oilfields were found. Near Beijing, in the Langfang 廊坊 area of Hebei Province, a natural gas field was found. In Renqiu, two big oil gathering stations were built. In three months, workers laid a 100 km pipeline capable of transporting 5 million tons of oil annually from Renqiu to Cangzhou. Workers also built in six months a 120 km pipeline capable of transporting 10 million tons of oil annually from Renqiu to Beijing Fangshan 房山. These large oilfield construction projects, using the most advanced technology, achieved the best results for their time. By the end of 1976, Huabei Oilfield proved to have oil reserves of 500 million tons, developed an annual oil production of 10 million tons, and produced 5.97 million tons of crude oil. Within one year, the oil reserves were proven, the oilfield was developed and built, and the state's investment repaid. In 1977, it produced 12.3 million tons, and in 1978, it reached its peak of more than 15 million tons, contributing to the nation's annual crude oil production of over 100 million tons.

4 Production Drops Sharply: a Sudden Decline

From my present vantage point, I can see that many blunders were made in Huabei Oilfield, some very serious. Even though some mistakes can be ascribed to circumstances largely out of people's control and could not be helped, in many other cases the blame may be placed squarely on our own shoulders. There were three major blunders, one minor one.

The first blunder was building the plan on overly-optimistic expectations. At the time, in the later stages of the Cultural Revolution, the national economy was near collapse, so the oil industry stepped up production to stave off complete calamity. For years, people in the oil industry had been expecting high-yield oil wells with daily outputs of 10,000 tons and high-yield oil and gas fields like those in Iran and Venezuela. Once a high-yield oil well with 1,000 tons daily output was found, people thought they had dug up "a big golden baby." In addition, after the downfall of the Gang of Four in October of 1976, the idea of building a dozen Daqings producing an annual 500 million tons in total seized hold of the entire oil industry. Under such circumstances, the plan that year was to make Huabei Oilfield into "one Daqing," with an annual output of 50 million tons. In reality, Huabei Oilfield had been built in 1976, and that year it produced 5.97 million tons of crude oil. It reached its peak of more than 15 million tons in 1978. But since 1980, the production has continued to slip, and by the end of the twentieth century, it was producing only 3 million tons annually. The adverse effects of this grandiose planning were profound. I also made the mistake of over-planning with the oilfield's vocational education, an area that I was in charge of.

The second blunder was employing too many people. Before the discovery of Huabei Oilfield, the area was a subsidiary unit of Dagang Oilfield and had been drilled by the Second Drilling Company. After oil was found in Huabei Oilfield in October 1975, half of Dagang Oilfield's workforce, more than 40,000 people, were transferred to Huabei Oilfield. All of the company's 50 large-scale rigs were shipped to Renqiu. In early 1976, the Ministry of Petroleum transferred teams from oilfields such as Yumen, Changqing, Xinjiang, Lanzhou, Dunhuang, Sichuan, Shengli, Daqing, Jilin, Jiangsu, Liaohe, and between 1977 and 1978, it signed on large numbers of new workers (this did succeed in bringing the sons and daughters of many oil employees back from the countryside after the Cultural Revolution). Between 1978 and 1982, the oilfield's vocational school admitted large numbers of students. As a result, by the end of 1982, the oilfield employed more than 120,000 people, making it the lowest in terms of oil production per capita among oil and gas fields in the country.

The third blunder was drilling too fast. We could politely call this "over-drilling," but it would be more accurate to describe it as "killing the hen to get the egg out." Every year, oil production fell short of the plan. As a result, at the end of each year, the chokes were unplugged in order to produce more oil and "complete the whole year's plan above quota," because this was "a serious political duty affecting the whole situation." When this unprecedented high-yield carbonate rock oilfield was found, little careful scientific research and testing were conducted. People just unplugged the chokes and let the oil erupt. At the end of 1975, one after another 1,000-ton daily high-yield well appeared. However, the 100 km pipeline from Renqiu to Cangzhou was still under round-the-clock construction, so the crude oil could not be transported. Even so, the oilfield had to contribute to the fulfillment of the annual national production plan, so orders were given to dig earthen pools in the ground near these high-yield wells that could only store a few thousand m^3 of crude oil. Daqing had such earthen pools back in 1960—they would just let the oil gush into them—but afterward the oil would be sold cheaply to nearby villagers. One night, I accompanied Deputy Minister Zhang Wenbin on a walk, and we passed by the earthen oil pool at Ren-6. Zhang only sighed, "Just for the sake of reputation and fulfilling the national plan, this is indeed too wasteful."

This resulted in the premature depletion of the new oilfield. It is a lesson in what happens when scientific research is not done. A fine "plump golden baby" had in three or four years turned into "an old man" sapped of vitality. What a sudden rise and decline!

The fourth blunder was that the construction of the employee communities still followed the "little village and town model" established by Daqing in 1960. Employee housing, services, and amenities were built right in the middle of the oilfield—a practice that acquired the fine-sounding name of "moving command to the frontline." As a result, one second-tier unit would be widely separated from the next, and communities sprawled across dozens of kilometers, resulting in tremendous waste and no end of trouble for the future. Had these oil communities been built in nearby towns and cities such as Shijiazhuang, Baoding, Langfang, Cangzhou, it would have been much better for these cities, as well as for the oil workers! How come we, at the time, had not thought of doing a thing like that? It would have been good for the country, good for local governments, and good for the oil industry!

The achievements and blunders of the Huabei Oil Campaign can all be traced back to the Daqing Model. Its achievements stemmed from carrying out Daqing's two fundamental practices: instilling a high revolutionary spirit and following strict scientific practices. Its blunders were due to the campaign inheriting the mistakes of the Daqing model, such as being overly-confident after victory, over-planning, following an outmoded management style, construction of village-style oilfield communities, and so on. Such errors highlight the fact that the oil industry can resume marching forward only by modifying the Daqing Model.

CHAPTER 20

Diary of the Deputy Executive of Huabei Oilfield

1 My Appointment

In August 1975, 3269 Drilling Crew of Dagang Oilfield's Second Drilling Company drilled the Ren-4 well, which produced 1,000 tons of oil per day. The following month, Petroleum Minister Kang Shi'en ordered the formation of the five-member Central Hebei Oil Campaign Leadership Group, which was led by Ren Chengyu, the deputy director of the Ministry of Petroleum's Political Department. I was one of the members. In October, six days after National Day, the group led approximately 30 people from Dagang Oilfield straight to a wheat field, south of Renqiu Town, putting up shanties and cotton tents that served as headquarters for the oil campaign. A few days later, Director Ren asked me to lead a work group of four people within 3269 Drilling Crew, in addition to my usual task of attending decision-making meetings. I carried my bedding and lived in the drilling crew's cotton tent. For more than a month, while taking part in all the work related to campaign planning, I also experienced first-hand the oil workers' hardships and joys.

Meanwhile, the campaign developed very fast, as 1,000-ton per day wells kept popping up. To strengthen leadership, the Ministry assigned Deputy Minister Zhang Wenbin in January 1976 to serve as commander and secretary of the Party Committee of Huabei Oil Campaign. Deputy Director Ren Chengyu was named deputy commander and deputy secretary of the Party Committee, in charge of political and ideological work. Deputy Minister Zhang had been deputy commander and deputy party secretary of Daqing Oil Campaign in 1960. He was also my old boss. As soon as he assumed this post, he asked me to draft a "Report on Organizing an Oil Campaign in Central Hebei Province (关于组织冀中地区石油会战的报告)" to the State Council on behalf of the Ministry of Petroleum, Hebei Province and Tianjin Municipality. I remember that the report was divided into the following parts: a foreword, an account of the current good situation of Huabei oil exploration, the tasks and planning for organizing the campaign, details concerning the transfer of campaign troops, and the main measures for completing the tasks. Deputy Minister Zhang headed the discussions on the outline twice. It took me a week to write. Once the report was written, it was first sent by courier to Tianjin for the municipality's approval. Then Zhang and I travelled by night on very bumpy country dirt roads for six hours from Renqiu County to the city of Shijiazhuang. On the second

 | DOI:10.1163/9789004402737_021

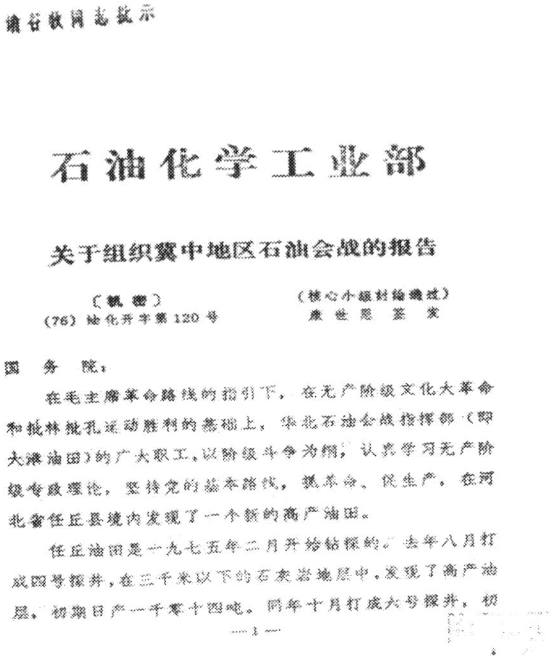

请谷牧同志批示

石油化学工业部

关于组织冀中地区石油会战的报告

（机密）　　　　　　（核心小组讨论通过）
(76)油化开字第120号　　　　康世恩 签发

国务院：

在毛主席革命路线的指引下，在无产阶级文化大革命和批林批孔运动胜利的基础上，华北石油会战指挥部（即大港油田）的广大职工，以阶级斗争为纲，认真学习无产阶级专政理论，坚持党的基本路线，抓革命、促生产，在河北省任丘县境内发现了一个新的高产油田。

任丘油田是一九七五年二月开始钻探的。去年八月打成四号探井，在三千米以下的石灰岩地层中，发现了高产油层，初期日产一千零十四吨。同年十月打成六号探井，初

—1—

FIGURE 20.1 Huabei Oil Campaign report (January 1976)

day following our arrival, we went to Hebei Hotel to deliver the report to Ma Li 马力, the first deputy secretary of the Party Committee of Hebei Province. With fear and trepidation, I read the report to him in five or six minutes. We waited for Ma's comments and instructions. After a moment of silence, Ma said calmly, "I think it is well-written. However, in terms of criticizing Comrade Deng Xiaoping's right-deviationism, I suggest the word 'comrade' be used only once, when first mentioning Deng Xiaoping, but it should be deleted in all other places. That would be more appropriate. I totally agree with the rest of the report." Later, I sighed that the Communist Party's upper-level leadership, when faced with all kinds of perverse acts of the Cultural Revolution, had become as silent as a cicada in cold weather, keeping quiet out of fear. That same night, I went back to Beijing to report to the Ministry of Petroleum. I remember going to the hospital alone to report to Deputy Minister Sun Jingwen, who was in charge of business in Beijing. After hearing the report, he signed right away, without any comments or requests to edit. The report was submitted to the State Council on January 28, 1976. Two days later, the State Council ratified the report. The Huabei Oil Campaign thus formally started.

This report was the last piece of important writing I did as a scribe for the oil industry. Afterward I became a member of the oil industry's leadership, and so this marked my transition from "child" to "parent"!

2 In Charge of the Political Department: Two Good Deeds and One Stupid One

In accordance with the report ratified by the State Council, Huabei Oil Campaign Headquarters was officially formed in February of 1976. Based on military protocol and following the Daqing model, it was under the collective leadership of the Party Committee and had a Production Office and a Political Department (in China after 1949, the political departments in industrial enterprises were the most important departments). Within the oilfield's Political Department, there were seven or eight divisions, including the Secretary Division, Propaganda Division, Organization Division, and Personnel Division. I was appointed as a member of the Party Committee and first deputy director of the Political Department. The director never came, so I was in charge of the operation, and I doubled as the secretary of the Party Committee of the Second Drilling Company. Thus I was mainly in charge of the daily operations of the Political Department; I visited the Second Drilling Company only once per week as its secretary of the Party Committee.

During 1976 and 1977, Huabei Oilfield was bustling with all kinds of activities. The campaign was in full swing. During my days in the Political Department, I did two good deeds and one stupid deed.

The first good deed was that I hired more than a thousand technicians to work in the oilfield, since the Political Department was in charge of personnel management and hiring. As a result of the ten-year Cultural Revolution, there was a severe shortage of technical staff on the campaign teams. Also, ever since the new China was founded, there had been restrictions placed on where people could live, so-called population movement restrictions. Therefore, in Beijing, Tianjin, Shijiazhuang, and Baoding, there were many technicians in different enterprises who could not bring their families to live with them, i.e., wives who lived in the country could not live with their husbands in the cities. Some couples had thus been separated for many years, even decades. With the support of Hebei Province, drawing from the Daqing experience, we allowed wives with agricultural residence cards to move to the oilfield and transferred their husbands who worked in cities to the oilfield, thereby uniting couples. The Political Department I headed was responsible for this work. I remember that from 1976 until the first half of 1977, more than a thousand technicians swarmed to the oilfield from the nearby big cities. Many among them were experts from the nation's top scientific and research institutions, such as a researcher of an atomic energy institute or someone who got his associate Ph.D. from the Soviet Union. All of these technical experts later on became standouts in the educational, scientific, cultural, and medical fields in Huabei Oilfield.

My second good deed was to organize and preside over the Fat Pig Reward Celebration Meetings. In the first half of 1976, news of victory kept pouring in. Every ten or fifteen days, we would be notified of a new well producing over 1,000 tons of crude oil per day. At the time, one of the key tasks of the Political Department was to organize a big on-site celebration party every time this happened. Hundreds of representatives from dozens of subsidiaries and a hundred or so drilling crews would go to the given oil well with drums and gongs to participate in the exciting event. I designed the format of these parties. First, thirty or forty workers from the drilling crew would be invited to the makeshift podium made up of eight trucks. Deputy Minister Zhang Wenbin would adorn each worker with a red silk sash emblazoned with a big red silk flower. Then the crew leader would read the report on the good news. A few workers would then go up to the podium to make pledges. Finally, Deputy Minister Zhang would give a speech. The climax of the celebration would come after Zhang's speech. With an iron pole, two workers would carry a live pig weighing more than 50 kg to the site, as reward to the drilling crew. This was the only reward we could give to the meritorious workers at a time when the whole country was still under the influence of the absurd theory of "criticizing material incentive." At such times, people would shout and jump for joy. To this day, I remember this moving sight, which truly fit a famous saying in the oil industry at the time, "The day a well produces oil is a grand holiday for the oil workers!" On one such occasion, I made a rhyme to commemorate the event:

> Who says campaigning's hard?
> I say it's the greatest joy.
> Every day there is good news,
> The more we do, the greater the satisfaction.

The stupid thing that I did was to have weekly meetings devoted to "Criticizing Deng Xiaoping's right-deviationism." It is common knowledge that in the first half of 1976, as arranged by Chairman Mao and the CCP Central Committee, criticizing Deng Xiaoping and right-deviationism was the work focus for the whole nation and Party. Even though production and construction in the

Huabei Oilfield was in full swing, the Political Department that I headed dared not treat "Criticizing Deng" lightly. After much deliberation, I thought of a way that would not interfere too much with production, and that was to have a weekly "Criticizing Deng" meeting at the only large meeting place in town—the movie theater. A few people from offices throughout the oilfield, a total of some three hundred people, would attend the meeting. Each time, a few people were chosen to speak, read newspaper articles and documents of the CCP Central Committee in order to "fire long-range missiles" at Deng Xiaoping. These meetings did not last more than an hour. There would be only three people on the podium: Deputy Minister Zhang assuming command, another speaker, and I who presided over the meeting. After each meeting, a brief report on the event would appear in the oilfield's internal newspaper, to show that "Criticizing Deng" was a key matter for the Huabei Oil Campaign. After the last of five or six of these meetings was dismissed, Zhang said to me in a low voice, "Old Mao, I think we've held enough of these meetings. Let's not do it anymore." Although I arranged for these meetings out of deference to the Party and a sense of duty, I did so half-heartedly. Even so, in the end, I still did it. I have to admit that it was a stupid thing.

3 A Skipped Investigation

On April 5, 1976, the day after Qingming Festival (Tomb-Sweeping Day), the Tiananmen Incident, when Deng Xiaoping was removed from office, sent shock waves through China. More than a million people gathered in Tiananmen Square, to mourn the death of Premier Zhou Enlai and to protest. The CCP Central Committee had the crowd dispersed by force. Soon, investigations were conducted throughout the nation with regard to the participants, on orders from the Central Committee; it was carried out at all levels. I learned the news from sources in Beijing and other oilfields. I also knew that the Party Committee of Hebei Province summoned relevant people to implement this work. Out in Huabei Oilfield, I would be in charge of this work. However, since I had not received any notice from Hebei Province, I pretended to be deaf and dumb and did not lift a hand. In the oilfield, someone asked me why the oilfield did not engage in such investigations, and I stalled him off with the explanation that I had not received any notice from the province.

After twenty some days, after the climax of the investigations had passed, Hebei Province notified me to report on our investigation. I replied, "We did not receive an investigation notice." After checking, they found that they had indeed not sent us a notice and told us to immediately investigate. I had no choice but to summon managers from second-tier political departments to carry it out, and most of them knew I was just going through the motions. Only one division manager made a great show of being earnest. He said, "Four days after April 5, I found a torn up photograph of Tiananmen Square in the garbage heap behind our office. Should we investigate it?" I said noncommittally, "It's a garbage heap, isn't it? Just sweep it up and dump it all!" With that, I conveyed my attitude: one thing more is not as good as one thing less. Why should we look for trouble?

The investigation of the Tiananmen Incident of 1976 in Huabei Oilfield ended without ever having started. Not that I disliked and resisted the perverse acts of the country's leadership at the time—I was merely being "lazy," taking advantage of the strange combination of circumstances and leaving a mistake uncorrected.

4 A Battle in a Bird's Nest: a Wonderful Drama

During the eight years I worked in Huabei Oilfield, I experienced an internecine struggle within the leadership, a so-called "battle in a bird's nest" or "nest-fighting" (*woli dou* 窝里斗). Even though its impact was inconsequential, this phenomenon concretely and vividly reflected the abnormal state of mind that prevailed within the leadership of the oil industry during those years.

In this particular instance of "nest-fighting," there were four background issues. First, after October 1976, criticizing the Gang of Four reached its climax nationwide; investigating people and things related to the Gang of Four occurred in all workplaces. Second, after the fall of 1976, since no new oil or gas fields were being discovered in Huabei Oilfield, some people in the oilfield leadership fell into the mindset of, "We can endure hardship together but not share happiness together." Third, Deputy Director Ren Chengyu of the Ministry's Political Department had created a scandal by having an extramarital affair. Even though he had been transferred out of the oilfield, I remained negatively associated with him in some people's minds because we had come from the Ministry to the oilfield together. Under these circumstances, I came under attack by some people in the Second Drilling Company. Fourth, some people in the Second Drilling Company were also upset with me for having employed or promoted two men, Jiang Yang 姜阳 and Zhang Lanzhi 张兰志, who, as it turned out, did not get along well with others.

Jiang Yang, the Second Drilling Company's deputy manager and deputy Party secretary, was a recent (1970) graduate from Beijing Petroleum Institute.

In 1966, when the Cultural Revolution first started, he was the head of Beijing Petroleum Institute's student rebel faction and was personally received by Premier Zhou early in 1967. On Premier Zhou's suggestion, he joined the faction protecting Kang Shi'en. His father-in-law, the Vice Minister of Public Security Yu Sang 于桑, and Director Ren Chengyu were old friends from Yan'an days. After graduation, he was assigned to work in Dagang Oilfield, but to train him further, Minister Kang and Director Ren transferred him to Huabei Oilfield and appointed him to be a member of the oilfield's Party Committee, as well as deputy Party secretary and deputy manager of the Second Drilling Company. This appointment angered some people in the oilfield and the Second Drilling Company, who objected: "A young lad fresh out of college—what virtues and capabilities have let him skip over three levels [workers, section level, division level] to deputy department chief?" Worse, after he arrived in the oilfield, he did not change his rebel ways. I arranged for him to work with a drilling crew, but he rarely stayed with it. Instead he would often go to Beijing in his chauffeured jeep to maintain his contacts. There was nothing I could do about it.

As for Zhang Lanzhi, I had promoted him from deputy to section head of the secretariat of the Second Drilling Company. Some people believed Zhang had deep personal ties with me, but this was a misunderstanding. Actually I did not know him at all prior to the assignment. At the time, the hiring and firing on the section levels throughout the oilfield was decided at meetings of the oilfield's Political Department, so when the Second Drilling Company submitted its assignment list to the leadership, I asked if there was a section head. It turned out there was not, so I said, "If there is no section head, then let's just name Zhang Lanzhi to that role." Afterwards, Zhang was able to attend meetings of the Party Committee of the Second Drilling Company and take notes. But Zhang, as it turned out, had come with his own set of baggage, unbeknownst to me at the time I had appointed him. Zhang was disagreeable and inconsiderate, traits which earned him the ire of his colleagues.

The fact that Jiang's hiring and Zhang's promotion were the root causes of the accusations against me did not emerge until later, when a group from the oilfield's Party Committee investigated. But some of the leadership in the Second Drilling Company aimed to take me down also because I was an outsider, a secretary of the Party Committee from the Ministry. I became the target of this particular nest fight. I hadn't seen this coming.

Now as I recall this time, it really seems like a live drama, and the characters, scenes, and plot come clearly before my eyes, so I shall try to record the drama accurately:

Dramatis personae:

Kang Shi'en: the Minister of Petroleum.
Jiao Liren: Deputy Petroleum Minister, commander of Huabei Oilfield (after Zhang Wenbin's transfer out), also the deputy commander of the Daqing Oil Campaign, and my old boss.
Mao Huahe (me): deputy director of the Huabei Oil Campaign Political Department and Party secretary of the Second Drilling Company.
Xian Xuefeng: Second Drilling Company manager and deputy Party secretary.
Jiang Yang: Second Drilling Company deputy manager and deputy Party secretary, recently graduated (1970) from Beijing Petroleum Institute.
Zhang Lanzhi: Second Drilling Company section head of the secretariat.

Time: Summer, 1977.

Scene: One June evening during the oilfield's summer wheat harvest, in 1977, in Deputy Minister Jiao's office.

It is the first day of the managers' meeting in the oilfield. As I am in charge of the Political Department, I am busy with the meeting. That evening, to welcome the Party secretary of the Bohai Petroleum Bureau, the four of us drink a bottle of Wuliangye 五粮液, a prized hard liquor. Afterwards, I go to see a film in the makeshift meeting room, which is made of reeds, though there are no good films to be seen—just newsreels and documentaries. Just as the film starts, an announcement is broadcast: "Director Mao, please attend a meeting at Minister Jiao's office immediately." The weather is drizzly and muggy, and I feel a little dizzy from the liquor, but I get up and go straight to Deputy Minister Jiao's office, two hundred meters away.

On entering the room, I notice that a few of the oilfield deputy Party secretaries headed by Jiao are sitting there poised and waiting for me. I also notice that Deputy Secretary Ren, who had an affair, is absent. Jiao voluntarily explains that Ren is busy with the summer wheat harvest—obviously a pretext. What harvest takes place so late at night? Anyway, those sitting there are all older and more senior than me. I figure they must be there to interrogate me, so I brace for combat.

I realize that in all fairness, Jiao has high regard for me from our time at Daqing and Jianghan Oilfields, but he is unable to fend off the charges against me from some people in the oilfield, especially those in the Second Drilling Company. As soon as I go in, Jiao calls out: "Young Mao, come sit by me. Let's have a little chat. I have a few things I need to check with you." Then the meeting starts. Jiao first has a clerk read an anonymous letter sent to Minister Kang. It relates how, starting in midwinter of 1976, Huabei Oilfield held meetings for over half a year to criticize Director Ren, and that in one such meeting, the Red Army veteran fainted on the spot. Ren had performed outstandingly for the Huabei Oil Campaign, but by punishing Ren, the Huabei leadership was "killing the donkey as soon as it finishes pulling the millstone."

Jiao asks me, "Do you know anything about this letter?"

"No, I don't," I reply.

Jiao: Was it written by Jiang?

Mao: Judging from the tone, it might have been written by Jiang, but whether or not he did write it, I don't know.

Jiao: Have the two of you talked about Ren's affair in the past?

Mao: No. [*Raising voice*] Minister Jiao, I know of Ren's affair—this has been causing a big fuss the past few months. Because he and I both came from the Political Department in the Ministry, it naturally affects me. But I remain unperturbed, because having an affair is something he did on his own, without any collaboration on my part, so I have nothing to hide or cover up.

Jiao [*trying to smooth things over*]: Young Mao, there's no need to get upset. We just need to clear this matter up. [*Getting to the point, though already knowing the answer*] How did Jiang happen to become deputy manager of your Second Drilling Company?

Mao [*playing along*]: He graduated from Beijing Petroleum Institute, with a good background. How Jiang came to the oilfield, I have no idea. He and I are only related through our work. Everyone knows that as the Party secretary of the Second Drilling Company, I have no qualifications or power to appoint my deputy Party secretary and deputy manager, so I can't be blamed for it. For those who want to investigate the matter, they could go ask Minister Kang and Director Ren directly. As for Jiang's poor work performance, I assume the responsibility for not helping him enough.

The Party secretaries nod in agreement.

5 Asserting Sincerity through Sophistry: the Drama Continues

The second issue had to do with the assignment and promotion of the disagreeable Zhang Lanzhi to the section head of the secretariat of the Second Drilling Company. In his capacity, he could attend meetings of the company's Party Committee and take notes. I, as the company's Party secretary, did not routinely attend meetings of its Party Committee—these were held by the company manager Xian Xuefeng. One such meeting of the Party Committee that took place a year earlier, in June 1976, focused on how to "criticize Deng Xiaoping." You Shuchang 尤树昌, a deputy secretary who had a grudge against me, pointed out that at the beginning of that year, when at Kang Shi'en's instructions I had carried out a "big inspection of political work," I had really been following Deng Xiaoping's directions on "Rectification as Guiding Principle."[1] It was true, but I had only done so per Minister Kang's orders. Xian Xuefeng chimed in without equivocation: "Minister Kang Shi'en doesn't have any serious disease. He is on vacation in the south simply because he wants to avoid criticizing Deng Xiaoping." Given the situation at the time, if you were alleged to have carried out Deng Xiaoping's orders or avoided Chairman Mao's initiative to criticize Deng, it would have severe political repercussions. The following day, Zhang Lanzhi informed me of the allegations Xian had made about Kang. When I asked him further about this, he said that he had the meeting memos as evidence. A few days later, I had the opportunity to ask Xian. He admitted that he had made an indiscreet remark. This incident partly reflected Zhang's fault-seeking tendencies.

Thus when Deputy Minister Jiao Liren asked me about Zhang's appointment, I told him about the above mentioned Party Committee meeting of the Second Drilling Company, which Zhang had recorded, and Jiao and his group were taken by surprise. Without much thinking, I said that I was totally responsible for Zhang's appointment and employment. As for Zhang's imprudent and antisocial behavior, I was only responsible for not disciplining him enough. At last I said, "When I promoted Zhang as the head of the section, I could not predict that he would not perform well later. When Chairman Mao promoted

1 At the beginning of the Cultural Revolution in 1966, Deng Xiaoping was removed from office and sent into the countryside for supervised labor. In 1974, when Zhou Enlai was gravely ill, Mao Zedong proposed to reinstate Deng Xiaoping, appointing him Vice Premier of the State Council and Chief of Staff of the military, in charge of daily operations of the Party, Government, and the Army. Deng's proposal that all work should have "Rectification as Guiding Principle" met with extreme opposition from the Gang of Four, who said that it was a reversal of the "right-deviationist verdicts." Mao Zedong agreed with them, and in the spring of 1976, Deng Xiaoping was removed from office again. A national political campaign of "Criticize Deng, Counterattack the Right Deviationist Reversal-of-Verdicts Trend" was launched. In September 1976, Mao Zedong died; the Gang of Four was arrested the next month; and Deng Xiaoping was reinstated in 1978. Soon, he became the most powerful person in China.

Lin Biao, he had no idea that Lin Biao would try to assassinate him."[2] That was a heavy remark, hitting the ground with a bang! Deputy Minister Jiao hurriedly mediated, "Young Mao, don't talk like that …" Obviously, I was resorting to sophistry, but it demonstrated how upset I was: even though I had promoted a section head by mistake, there was no need to denounce me publicly for it.

This meeting lasted less than an hour and ended inconclusively. After the meeting, I returned to my own office, which was also my bedroom, feeling dismal. It was a sultry night. I lay on my bed, in a confused state of mind, incapable of sleeping a wink. I thought about how for more than a year, having volunteered to come to Renqiu for the oil campaign, I had left my wife and daughters behind and had not even visited them after the big Tangshan earthquake. I also lamented that after working day and night wholeheartedly to find oil, in the end, others had groundlessly linked me with my superior's extramarital affair and other things. I felt very wronged.

I had the presentiment that the oilfield Political Department Director Ren Chengyu's affair would be the focus of criticism at the next big leadership meeting, which would be held in a few days, and that I would also be criticized due to my association with him. This was normal procedure for "nest fights." I knew that the previous afternoon, Minister Kang Shi'en had come from Beijing; he had held a meeting that evening for the oilfield's Party Committee secretaries, hearing reports from a few work groups prearranged by the oilfield Party Committee. The work group that went to the Second Drilling Company was headed by a deputy of the Personnel Division of the oilfield's Political Department. This man had spoken to the Second Drilling Company without talking with me once, but he had talked only with those who were at odds with me. His report infuriated Kang Shi'en, which resulted in my interrogation and the above scene.

6 A Sudden Change of Situation: the Drama Ends

Around 8:00 the following morning, the first day of the three-day preparatory meeting before the big oilfield leadership meeting, I rushed to the leadership assembly of the Second Drilling Company at the oilfield's hotel. As soon as I

2 On September 13, 1971, Lin Biao, his wife and son tried to flee China for the Soviet Union by plane. The plane crashed and they perished. Afterwards, notes on a coup d'etat plot developed by his son Lin Liguo, called "Project 571" (5-7-1 were homophones of *wuqiyi*—armed uprising) were found in Lin Biao's residence, which discussed how to murder Mao Zedong.

sat down, Jiang Yang, the company's deputy manager and deputy Party secretary, came over and said to me, "Secretary Mao, I need to talk with you about something." The two of us went to another room. Jiang said, "Last night, at 9:00, I visited Minister Kang at the hotel. We spoke for five hours, about the oilfield, about exposing and criticizing Director Ren, and about things that have happened at the Second Drilling Company. When I finally left him at 2:00 in the morning, he saw me off outside of his room and tried to console me."

"Did you tell him about the Xian Xuefeng incident?" I asked.

He replied, "I did. Kang was very upset."

I said, "O.K. Let's see how the big meeting proceeds in the next few days." Since I did not know further details, I did not tell him about the drama of my interrogation the previous night.

About an hour later, at 10:00 in the morning, Kang's secretary Zhu Binggang 朱秉刚 called me: "Director Mao, Minister Kang has asked for you to come over." I went to Kang's hotel room. Deputy Minister Jiao and the director of the Ministry's Personnel Department were also there. As soon as I walked into the door, Kang said to me, "Young Mao, tell us about the situation of exposing and criticizing the Gang of Four in your Second Drilling Company." Jiang's disclosure had given me time to think it over. I addressed the issue directly, repeating what I had said the night before. Kang was infuriated by Xian Xuefeng's remark, which had accused him of going to vacation in the south to avoid criticizing Deng Xiaoping. As I have mentioned before, at the time the Daqing Red Banner had become a club, and Kang considered himself as the representative of Daqing. Boiling with anger, he stood up, banged on the desk and roared, "Whoever is opposed to Daqing, I will fight him to the end. I'll see who can bite my balls off!" In all my years of working with Kang, I had never seen him blow up like that, shouting obscenities so hard that my ears rang. He was considered the only true intellectual in the leadership of the oil industry. I was shocked!

Kang immediately instructed Jiang to give the first speech at the big formal leadership meeting, which would take place three days later. The content of his speech would be to expose and criticize the Gang of Four and relate it to the Second Drilling Company. It was a pity that Kang returned to Beijing that afternoon to attend a related Party Central Committee meeting. Otherwise, there would have been a big scene in three days' time. For the time being, my duty was to help Jiang draft his speech. It took two days and nights to put together. I thought to myself that it should not lack fire, nor should it resemble the chaotic diatribes of the Cultural Revolution. I weighed each and every word.

When the formal leadership meeting started, more than 700 people were attending. When the chairman of the meeting announced that Jiang Yang would be the first speaker, looks of surprise came over the faces of some insiders and dozens of people from the Second Drilling Company. This week-long leadership meeting suddenly changed course. Under the pretext of exposing and criticizing the Gang of Four in the workplace, the brunt of the criticism was directed instead at the Second Drilling Company manager Xian Xuefeng and deputy Party secretary You Shuchang. Even though I managed to skirt danger and survive, I was extremely sad, weighed down with anxiety.

7 Unspoken Rules of the Nest Fight

The nest fight was the perduring evil fruit of Chairman Mao's highest order of "class struggle as the key guideline" (阶级斗争为纲) and statements such as "Once one grasps class struggle, miracles are possible (阶级斗争一抓就灵)." It became the normal state of affairs for all levels of leadership, especially during the Cultural Revolution. Its unspoken rules, in general, were the following:

First, seize on political topics and "lines" (*luxian* 路线), then on extramarital affairs, and then corruption, in that order. At the time, since everyone was poor, there was hardly any bribery, but there was taking more than that to which one was entitled. However, that was considered an "unhealthy tendency," a trifle.

Second, fully take control of the initiative of the struggle. The usual way of accomplishing this was to link any incident to the target of the struggle by reversing black and white and "pointing to a deer and calling it a horse."[3] Later on, when I had given the above nest-fight sober thought, I came to the conclusion that Manager Xian and Deputy Party Secretary You had been telling the truth, though inappropriately for the time and place, which was difficult for me to accept. I should not have made trouble for no reason, nor should I have reported the incident to Minister Kang. I greatly regret having acted so.

Third, uphold Mao Zedong's statement that "the pleasures of fighting people are endless (与人奋斗，其乐无穷)." Mao Zedong also said: "It is the mentality of emperors and kings that there is no party outside of the party; it is very strange if there are no factions within the party (党外无党，帝王思想；党内无派，千奇百怪)." It is an objective truth that there are always at least

---

3 This phrase alludes to an incident recorded in Sima Qian's *Records of the Grand Historian*. After Qin Shi Huang, the "First Emperor," unified China, during the short reign of his son the Second Emperor (r. 210–207 BCE), Zhao Gao 趙高, the prime minister, presented the young Emperor a deer. To test his own power and his ministers' loyalty to him, Zhao Gao called the deer a horse. Those who did not agree with him were executed.

FIGURE 20.2 The author at Huabei Oilfield (2013)

two factions within parties and societies; it has been thus in ancient or modern times, in China or abroad. Factions are a motive force in the development of societies. However, during the Cultural Revolution, people had no breadth of feeling or tolerance and treated everyone with opposing opinions as their enemies, and they would stop at nothing to discredit them and bring them down. This poisoned the political environment for the whole of society for a long time.

8 Witnessing the Birth of Longhuzhuang Oilfield

After the nest fight, the oilfield-wide campaign to expose and criticize the Gang of Four activity ended. The production frontline, including more than 100 drilling crews under the supervision of five different drilling companies, were hard at work on the tens of thousands of square kilometers of the Central Hebei Plain. Deputy Minister Jiao told me, "Although you have been appointed as the Party Committee secretary for the Second Drilling Company for more than a year, you have not worked there much. Now the production frontline is very busy. I suggest you go to work in the company to assume that position, while still keeping your deputy directorship in the oilfield's Political Department." I agreed right away, because I had grown weary of my days in the Political Department. Even though it seemed glamorous and busy, I was but "a stagehand in a theatrical show." To encourage me, Jiao instructed the campaign headquarters to assign me a rare Beijing jeep with a chauffeur named Ni Xiancai 倪显才. The young man ended up working for me for six years. Soon thereafter, I went to the Second Drilling Company to work.

At the time, the Second Drilling Company had 32 large-scale rigs, more than 9,000 people, and over 400 large trucks, half of which were new imports. It was one of the finest and strongest group of workers in the oil industry. It was formed in 1961 in Daqing, and then it had taken on missions in Dagang Oilfield, Sichuan Oilfield, and Jianghan Oilfield. In 1972, it returned to Dagang Oilfield, drilling deep wells in the oilfield's outskirts on the Central Hebei Plain. The Second Drilling Company had discovered Huabei Oilfield, and after the Huabei Oil Campaign started, it became the main player. Coming to this company to do something concrete, I felt both lucky and a heavy responsibility.

In 1977, soon after I assumed my position at the Second Drilling Company, I attended a meeting in which Minister Kang came from Beijing to listen to reports on oil exploration in the northern part of the Huabei Oilfield. There were seven or eight rigs drilling in Nanmeng Town 南孟镇, in Ba 霸 County (present-day Bazhou City 霸州市). At the time, to the east of this there was a structure that geologists called "Nanmeng East (南孟东)." When the geologists talked about "Nanmeng East," I chimed in, "I have been to the site many times. This place belongs to the Longhuzhuang Commune 龙虎庄公社 of Yongqing County. I suggest we change 'Nanmeng East' to 'Longhuzhuang.'" On hearing that, Kang immediately responded, "Young Mao's suggestion is very good. The name of Longhuzhuang ['Dragon and Tiger Village'] is very lucky. You should change it." I was thus very pleased that an oilfield was named at my suggestion. At this meeting, it was also decided that the Second Drilling Company would transfer four drilling crews, including the famous 3269 Drilling Crew, to conduct a "small campaign" at this site, and I volunteered to work on-site to organize and lead it. Starting in October 1976, 3269 Drilling Crew was my fixed contact spot. With my own bed in the crew's tent, I stayed there for 20 days each month. At the time, the crew's well was called Long-30 ("Dragon Well No. 30"). One day, I went to Long-29, situated two or three km away, to attend their evening meeting. The meeting went on past 11 o'clock. After the meeting, the crew leader and the deputy manager of the Second Drilling Company Gao

Bingzhong 高炳中 asked me to stay the night. I thought that I might as well spend the night there, too. I lay down at the trailer shared by Gao Bingzhong and the crew leader. The rig was humming away nearby. Around 4:00 in the morning, one driller called outside the trailer window, trying to wake up Gao Bingzhong: "Commander Gao, the crew leader asked you to hurry to the drilling site. They found gas kick in the well bore and the mud pit is full of oil bloom."

This of course was the best news that we oil workers could dream of. Gao went over to the location immediately. Even though I did not utter a sound, I was too excited to sleep. After more than half an hour, I pricked up my ears and heard footsteps rushing to the trailer window, and someone hurriedly called out, "Party Secretary Mao, the well is blowing out. Go there right away!" I ran over. A black oil stream six or seven centimeters in diameter mixed with some pungent white natural gas whistled and poured into the pre-dug earthen oil pond. Dozens of workers shouted and jumped for joy under the light of dawn. This was truly what oil workers call a holiday!

Before six o'clock in the morning, I used the drilling crew's special line to telephone Deputy Minister Jiao at the campaign headquarters in Renqiu. He was extremely pleased and kept telling me that my phone call was timely. He said, "I will bring Ministry leaders to congratulate the drilling crew right away." A week later, a celebration meeting was held at the site. Minister Kang was invited over from Beijing to bestow the crew workers with red silk sashes emblazoned with red silk flowers. Per custom, the drilling crew was awarded a fat pig—nothing more.

This marked the birth of Longhuzhuang Oilfield. Unfortunately, after a few more wells were drilled, it became apparent that this was a rather small oilfield.

9 Busy Grabbing Oil in Baiyang Lake

In October 1977, soon after the small campaign at Longhuzhuang, another drilling crew in our Second Drilling Company found a new oilfield near the famous "Pearl of Huabei"—Baiyang Lake.[4] Since there was once an anti-

4 Baiyang Lake (Baiyangdian 白洋淀), the largest natural lake in northern China, located in the middle of Hebei Province, is approximately 162 to 155 km from Beijing and Tianjin respectively and 12 km from Huabei Oilfield. Rich in fish, geese, ducks, etc., it is a popular tourist destination known as the "Pearl of Huabei." The new development of Xiong'an 雄安, to which many of the second-tier administrative offices in Beijing will be moved, is on its shore.

Japanese guerrilla detachment called the Yanling 雁翎 Guerrillas, Deputy Minister Jiao named the oilfield Yanling Oilfield and decided to transfer 23 drilling crews to campaign there; seven of the crews were from the Second Drilling Company. Once I received the order, I took the managers of the Second Drilling Company to live in a tent on-site, working around the clock for seven days and seven nights. Sometimes I was so sleepy that I would put on a padded jacket and sleep on the floor for two or three hours. When I was hungry, I would ask the chef to cook a bowl of noodles to eat. I worked and directed on site. Within seven days of the start date, I had seven deep-drilling rigs from places such as Longhuzhuang Oilfield, 70 or 80 km away, transferred to the site. They were quickly assembled in place and put into operation, breaking the record for transferring and assembling large-scale rigs. Yanling Oilfield bustled with activities.

Because of the tight schedule, workers suffered from extreme fatigue. During his shift, a head driller is in charge of the rig switch, and his role is similar to that of a car driver or a plane pilot. There is a traditional saying in the oil industry: the head driller of a crew has three kinds of lives in his hands: people's lives, the rig's life, and the well's life. During this time, one head driller became blurry-eyed with sleep and misread the weight indicator. This led him to move the brake handle incorrectly and, as a result, the derrick collapsed to the ground—a rare accident that cost hundreds of thousands of yuan. Luckily, no one was hurt.

This serious accident shocked Deputy Minister Jiao. He notified me to return to the campaign headquarters to give a report. Around 11:30 that night, soon after I left my tent, I noticed a "Road Closed" sign at the main road. If I detoured, it would take more than one hour. While I was hesitating, an old worker on the roadside asked, "Where did you come from?" My chauffeur Ni Xiancai answered, "Headquarters." The worker responded, "You must be Minister Jiao!" Ni noncommittally said, "Yeah." The worker raised his hand, signally us to pass and said repeatedly, "Hello Minister Jiao!" Feigning Jiao, I waved my hand and said, "You have been working hard. It is very cold at night—wear more clothes. Thank you." I figured that since it was dark and no one could tell who I was, I might as well pretend that I was Xiao just this once in order to save time. The following day at the meeting, Jiao lightly scolded me for the accident. What else could he do? After the meeting, I confessed to him that the night before I had pretended to be him. He responded, laughing, "Who would have thought that a scholar like you could do such a thing." Indeed! Why couldn't a scholar do such a thing?!

10 A Pair of Oily Gloves Saves Me

At the end of September 1977, one drilling crew in our Second Drilling Company was relocating north a dozen or so kilometers from Ren-9 to Ren-11, in Renqiu County. There was nothing but dirt roads. The rig was a large one, capable of drilling more than 3,000 m deep. The managers of the Second Drilling Company wanted to set a record for the move, completing all of the tasks in one day: moving, installation, starting drilling, and cementing 40 m long steel piping on the surface. I got up at 3 o'clock in the morning and went to Ren-11. The site was bustling with noise and excitement. More than a hundred workers were unloading equipment from dozens of large trucks and installing it at the new well. The head of the installation crew passed me a pair of gloves made of thick cloth. While the equipment was hoisted into position, I occasionally lent a hand. Soon, my gloves became oily. At this moment, a clerk from the Huabei Oil Campaign Headquarters arrived, asking me to go have a meeting with Minister Kang Shi'en at his hotel room. I rode there in the clerk's car, still holding the oily gloves …

I learned later that Minister Kang Shi'en and Deputy Minister Jiao Liren had come the previous day to Huabei Oilfield by car from Shengli Oilfield in Shandong. While passing Cangzhou, Hebei, they had lunch at our Second Drilling Company's base in Cangzhou. At the time, our Cangzhou base was very chaotic, with earthquake shelters from the Tangshan earthquake all over the courtyard, puddles of dirty water, and trash everywhere. I had arranged a deputy Party secretary to receive them. When Kang Shi'en and Jiao Liren asked about production, he replied, "Our company has the most rigs, one-third of the rigs among the five drilling companies in the whole oilfield. Lately, we have had frequent accidents, so as a result, four rigs are not drilling at present. Some accidents could scrap the wells. To scrap a well will cost hundreds of thousands, even two or three million yuan …" Kang Shi'en and Jiao Liren became upset at hearing this.

I was told later on that Kang Shi'en and Jiao Liren held a meeting at the hotel, listening to geologists' report on the oilfield's oil exploration. Someone mentioned a certain well, drilled by the Second Drilling Company, where drilling had stopped due to an accident. At this point, Kang Shi'en said, "What's going on? Go fetch Young Mao, see what he has to say."

When I walked into the meeting room, Jiao Liren saw me right away and called me to sit beside him. Seeing that I had sweat on my forehead and oily gloves in hand, Kang Shi'en seemed cooled down a great deal. He asked, "What have you been doing?"

I answered, "We tried to move and drill on the same day at one of our wells. I went to the site early in the wee hours of the morning."

"Not bad. You went to work. You went to take part in the labor."

"We want to break a new record."

"Why does your company have so many accidents?"

"I don't know. We will look into it and improve."

"Taking part in the labor is good. You should also take charge of reducing accidents. That will do. You go back to what you were busy doing!"

Kang Shi'en and Jiao Liren were no longer upset. I, on the other hand, had a narrow escape from criticism because of a pair of oily gloves. Later, when I brought up the incident with a friend of mine who is also an old oilman, he remarked impertinently, "I've been drilling oil wells for more than thirty years. Nobody has the fucking ability to guarantee no drilling accidents." What he meant was that because drilling accidents happen often, there is no need to over-investigate. However, at the time, I dared not to tell that to Kang Shi'en and Jiao Liren.

For years, I thought often of an allegedly true story about the Kuomintang army. When Chiang Kai-shek was still the principle of Huangpu Military Academy, he frequently inspected student dorms early in the morning. One day, he saw a student sitting at his desk, reading Sun Yat-sen's "A Program of National Reconstruction (建国大纲)." Chiang Kai-shek commended him highly and placed him in important positions later on. Chiang Kai-shek had no idea that that student had actually gambled the night away and returned to the school after daybreak. He could not sleep anyway, so he sat and read, waiting for the sound of the reveille. He hit the mark by a fluke, grateful for Chiang Kai-shek's appreciation of him.

That student caught the same lucky break as I did.

11 Building a Tent School for 4,000 Students in 64 Days

In the spring of 1978, Huabei Oilfield leadership again underwent a reorganization. The chief was still Deputy Minister Jiao. I was relieved of my position as the secretary of the Party Committee of the Second Drilling Company and promoted to be the deputy commander (later called deputy bureau chief) of the campaign headquarters, in charge of personnel, education, and public health of the whole oilfield. In the early summer of 1978, representing the oilfield, I attended the Ministry of Petroleum's National Petroleum Industry Vocational Education Meeting in Beijing. The meeting assigned to Huabei Oilfield the responsibility of enrolling 5,000 students in its technical school for the 1978 academic year.

After careful consideration, I made three proposals to the oilfield's Party Committee and Deputy Minister Jiao. One was to divide the original education division into two divisions: the Education Division, managing the kindergarten through high school education of employees' children, and the Training Division, managing employees' on-the-job training, the two petroleum technical schools (one with a two-year high school program, the other with a three-year high school program), and the petroleum university. My second proposal was that we map out a vocational education plan for the following five years. My third proposal was to build the new two-year high school campus of Huabei Petroleum Technical School, with an enrollment of 2,500 students, and establishing vocational extensions of the school at the oilfield's eight chief branches, which would enroll another 2,500 students for that year. The oilfield's Party Committee and Deputy Minister Jiao gladly accepted my proposals. I remember clearly that when I submitted the draft to Jiao, he ratified it with just one minor change. He inserted "one of" so as to make my phrase "Turning Huabei Oilfield into one of the important vocational education bases in the petroleum industry."

To implement the urgent task of enrolling 2,500 students for the main campus of Huabei Petroleum Technical School for that year, Huabei Oilfield's Party Committee asked me to serve concurrently as Huabei Petroleum Technical School's principal and Party Committee secretary. On the fifth day after National Day that year, I had a trailer moved from the Second Drilling Company, had someone make a wooden plaque, "Huabei Petroleum Technical School Preparatory Office," and got to work. First, near the headquarters we got 70 acres of vacant land where Renqiu County's crematorium used to be. Then, we transferred more than two hundred administrative and technical personnel from five drilling companies, of varying work types and positions, put them into preparatory groups, and divided up the work among each group. Those technical personnel later became teachers at the school. To meet the tight schedule, we worked day and night. I attended group meetings every evening and checked on the progress of each project. Every week, I held a dispatch meeting for the oilfield's related companies on raw materials supply, transportation, water, electricity and telephone installations, etc., coordinating and facilitating forces. According to the school record, between October 21 to December 25, 1978, in 64 days, 442 cotton tents were set up as dormitories for teachers and students, as well as 72 movable prefabricated classrooms, and 5 reed tents (each could hold hundreds of people) as cafeterias. Pebble stone paths were created throughout the school, and a dozen km of water pipes, electricity lines, and communication cables were laid. Meanwhile, 50,000 textbooks and more than 2,000 items of teaching equipment were purchased. In such a brief time, this two-year petroleum technical school was completed. It offered 47 majors in areas such as drilling, petroleum geology, and petroleum equipment maintenance. On December 25, 1978, the school's first group of 2,103 students, half of whom were children of oilfield employees, arrived for registration. On December 28, on the newly leveled playground, a grand opening ceremony was held. Oilfield employees were delighted and cordially referred to the school as "the Tent School." That such a large school was built in such a short amount of time was unprecedented in the oil industry. It received rave reviews.

Soon after the tent school started and students went through militia training, the school had winter break. On February 12, 1979, the new semester started. As the school's principal and Party Committee secretary, I asked the associate principal to take charge of instruction quality and establishing rules for instruction. I was most concerned with one thing: the prevention of fire, gas poisoning, and frostbite so as to guarantee the safety of the students and teachers. I repeatedly said at school meetings that even though the few hundred tents and close to a hundred wooden rooms looked magnificent, on windy winter nights, the thousand coal-burning stoves were a fire hazard. Any fire would be like "Burning the Camps,"[5] with grave consequences. Therefore, I issued the following orders: At check-in, provide strict education on fire and coal gas poisoning prevention to all inhabitants. Each tent should have a trained person in charge of safety and assigning responsibilities to individuals. Organize nightly patrol teams to check on each tent every two hours. From school leadership to each classroom, establish strict nightly patrol systems on each level, making sure the nightly patrol teams did their jobs. Provide an enamel latrine pot for each girl students' tent for the girls to use at night (boys' tents were not provided with this). The latter task was carried out by Zhang Mingliang 张明亮, the school's office manager, who personally delivered more than 200 pots for the girls' tents. At the time, my greatest fear was getting a phone call in the

5 "Burning the Camps" is an episode from the historical novel *Romance of the Three Kingdoms*. The three kingdoms were Wei, Shu, and Wu in the years 220–280 CE. The story goes that Liu Bei, the King of Shu, took his million-strong army to attack Wu and camped on the banks of Yangtze River. Tents in the camp stretched for 700 *li*, or around 350 km. When Lu Xun 陆逊, the young general of Wu, attacked, a key strategy was setting fire to the tents, which burned out of control, and he thus defeated Liu Bei. Liu Bei retreated to Baidi City and died from unhappiness.

middle of the night. Luckily, not one fire accident or incident of coal gas poisoning occurred. In the three years from December 1978 to the fall of 1981, before the school's relocation to its new address, more than 20,000 people studied and lived in this tent school. There was not even one serious group safety accident. I was very pleased and greatly relieved.

12 Building a Technical School for 5,000 Students

Once the tent school was in operation, in the spring of 1979, I split its leadership into two groups, and I continued to spend my time leading one, the group devoted to building the new technical school. In this regard, there were two main tasks in the beginning.

The first task was to choose the new site. With the support of Renqiu County, we looked at five possible sites, which were narrowed down to two after careful consideration. The first was the ancient town of Mozhou 鄚州, some 20 km from the oilfield. Mozhou, a town that with two or three thousand years of history, was the hometown of the famed ancient doctor Bian Que 扁鹊. During the Ming and Qing dynasties, it was a gambling town. Situated beside Baiyang Lake, it was a beautiful site. Our surveying team walked on the ancient city walls, imagining how splendid it would be to build a petroleum technical school inside the city! The problem was that it was too far from the oilfield, therefore inconvenient. The second site became the present location of the school—Yabazhuang 哑巴庄. It also borders Baiyang Lake and is a mere two or three km from the oilfield. Legend has it that Yabazhuang was named personally by the Qianlong emperor of the Qing dynasty on his way to visit the lower Yangtze River to the south. After leaving the south gate of the Forbidden City in Beijing, he travelled on a royal path made of yellow sand. As he passed this place, he saw no one and heard no sound of chickens crowing or dogs barking. It was so quiet that he said, "Let's call it Yabazhuang," i.e., "Mute Village." I reported on our choice of Yabazhuang to Deputy Minister Jiao.

He said, "Young Mao, someone suggested that you should build the new school on the site of the tent school. This would spare you a lot of trouble."

Without hesitation, I replied, "No. First of all, the old site is too small. Secondly, imagine a school with a few thousand students crowded next to the offices of the oilfield's headquarters. During school recesses, the streets would be filled with kids. That would be too boisterous!"

After thinking about it for a moment, Jiao said, "That makes sense. Let's build it at Yabazhuang." We then bought more than 130 acres of land for the school site from Changyangdian Commune 长洋淀公社. Later, we bought more than 160 acres of land for the school's agricultural use.

Now the second task was to see to the construction of the school. With the backing of Deputy Minister Jiao, I held many meetings on the school's overall planning, which involved the oilfield's Infrastructure Division, Planning Division, and Finance Division. According to this plan, which required the completion of the 190,000 m^2 school within two to three years, the oilfield's Design Institute drew up concrete drawings for the school. We had to hire construction teams. At the time, there was little new construction nationwide. Through connections, the oilfield's Infrastructure Division contacted the office of the PLA Capital Construction Engineering Corps in Beijing. Its Commander Li Yuan 黎原 personally led the 00212 Division's more than 1,000 officers and soldiers, with their fine, complete construction equipment, to Yabazhuang to take care of construction. For more than two years, the construction progressed according to design and with quality workmanship. On the eve of the Spring Festival in 1981, on behalf of the oilfield's Party Committee, I went to visit the construction site. Commander Li personally received me. We both acknowledged that the construction of the school was a joint effort of the Army and people.

About this 190,000 m^2 school, there are three things that I remember clearly to this day:

First: with a total area of 13,000 m^2 and rising to seven stories, the school building was built of prefabricated concrete and steel. It had 96 classrooms and many labs and offices, and it could accommodate 5,000 students at any given time. This was at that time the largest and tallest building in the Renqiu and Cangzhou area, so it towered like a crane among chickens. To oversee construction, I would take two hours every week climbing the stairs and walking among the reinforced concrete slabs. Actually, I knew nothing about constructing a building—I was just eager to see its completion. Thirty years later, when I climbed this building again, it was like counting my own family treasures—I could tell how many rooms this building had, to the surprise of those who accompanied me there.

Second: The cafeteria building, which had a student cafeteria to accommodate 3,000 and a staff cafeteria to accommodate 800, was built with an advanced technique for the time: beamless cable-stayed steel mesh construction, which eliminated low, heavy ceilings. Both spaces were roomy, bright, and grand.

Third: There were ten dormitory buildings, each with 5,000 m^2 of space and five stories high. At first the oilfield had only allowed for the construction of two buildings, and said they would discuss the remaining eight the following year, so the construction crew initially built the foundations for two buildings. The associate principal responsible for construction asked me what should be done. I told him, "Make the foundations for all ten buildings. If there is trouble, I'll take responsibility." Afterward, I talked it over with relevant oilfield division managers. Even though they were reluctant, they deferred to my superiority as a deputy commander of the oilfield and allowed the project to go forward. The following spring, the whole oilfield tightened expenditures. All school projects that had not yet started were delayed. School colleagues commented that fortunately the foundations of all ten buildings had been built, otherwise the project would have been delayed for years.

Another task was the acquisition of a new Daqing I oil rig. In the winter of 1980, when the Ministry's Drilling Engineering Meeting was held in Huabei Oilfield, Minister Song Zhenming and Deputy Minister Jiao visited the tent school. Song asked me what I needed. I asked for a new Daqing I rig for students to practice on. At the time, this kind of rig cost 500,000 yuan and was in short supply. Many oilfields needed them for oil production. Probably Minister Song did not want to turn me down and agreed, against his own will. To this day, the rig still towers over the spacious school, providing an ideal practice site for students of drilling.

From its beginnings as a tent school, after three years of construction, Huabei Petroleum Technical School could now accommodate 5,000 students. All the other technical schools in the oilfield were closed and merged with this one. Overall, this school signified that Huabei Oilfield had become a leading center for petroleum vocational training. It broke the long petroleum industry tradition of recruiting young peasants and army veterans to fill front-line technical positions. It improved the quality of employees and contributed to Huabei Oilfield's production and construction.

At the beginning of the new millennium, the school's name was changed to Bohai Petroleum Vocational College. I had not visited it in years but still cherished it in my mind. I finally revisited it in the middle of the summer in 2010. Because it was during summer vacation, the 300-acre school was green and quiet, its differently colored buildings with more than 200,000 m^2 of space nestled among trees and lawns. The school president, Wang Yining 王义宁, guided me to visit the instruction building, the library, and the newly-built electric and chemical labs. As we walked and looked around, he explained that after adjustments, the school had become a three-year technical college, with 4,500 full-time students accepted from around the nation. In addition to various petroleum-related majors, it also offers majors such as auto operation and maintenance, medicine and health care, tailoring, and cooking. Another 10,000

FIGURE 20.3 Huabei Oilfield Petroleum Technology College in 1982

students attended different short-term training courses there annually. These part-time students mainly came from Huabei Oilfield and other oilfields in the nation. It was a very prosperous school and could assume sole responsibility for its profits and losses. It had received many awards from Hebei Province and China National Petroleum Corporation (CNPC). In 1991, it also received the National Vocational Education Leader award jointly issued by the Ministry of Education, Ministry of Labor and Personnel, Ministry of Finance, and Federation of Trade Unions.

I was very pleased with this visit. It was due to Huabei Oilfield's financial strength and Deputy Minister Jiao's full support that the school was built. I also made my meager contribution. Like those reports and articles I wrote in Daqing Oilfield in the sixties, this school embodies a small trace of the work I did for improving the nation's petroleum vocational education.

I was transferred from Huabei Oilfield back to Beijing at the end of 1982. How I left the oilfield remains fresh in my memory. In 1982, Huabei Oilfield continued to be depressed, like a house rattled by a storm. It suffered for a couple of reasons. One was that in order to investigate the Liu Boping 刘伯平 case (to be discussed in Part 5, Chapter 24 below), officials from Hebei

Province's Discipline Inspection Commission came to speak with people and gather evidence almost every day. The investigation went on for more than two years, leaving the oilfield's leadership in fear, so that basically their hearts were not set on production. The second contributing factor to the oilfields' plight was that it had been developed in "plunder mode" in its early days. After 1980, its crude oil production started dropping off, and within four or five years, it had turned from a "golden baby" that everyone praised to an "orphan" that everyone shunned. It was like a night-and-day difference!

One evening at the end of 1982, four old friends had dinner with me to see me off. An older bureau director brought an expensive bottle of Maotai especially for the occasion. With ten eyes looking at each other and spirits very low, we only drank one-third of the bottle of liquor. For forty minutes, we just sat there, speechless as to the state of the oilfield. Some even had tears in their eyes. I got up, shook their hands, and said good bye to them. I got into the jeep, which soon disappeared in the night fog in the depth of the winter. Eight years before, I had come to campaign there, full of pride and enthusiasm. That day, I left the stage, dejected and depressed. During the car ride back to Beijing, overwhelmed by feelings of all sorts, I wrote the following lyric (*ci* 词) poem to commemorate my departure.

Goodbye to Huabei Oilfield
(to the tune of "Fragrance Filling the Courtyard" 满庭芳)

Clear is the water of Baiyang Lake,
Far the Taihang Mountains—
And on the plains, rigs are roaring.
Halt the flying car for now,
Let's have a parting toast.
So many old oilfield events!
Vain it is to turn back:
They pass before our eyes like mists and clouds.
In the evening haze,
Ten thousand flowery lanterns twinkle,
Apt metaphors for brotherly love.
Stop! Stop! It's all in the past,
We have poured out our sorrows,
Should not have many complaints.
Bosom friends are hard to find,
But you know my heart.
After I leave, we shall meet again—

Don't let hot tears stain your clothes!
In closing,
May each of you take care.
I'll spend the rest of my life in the sea of books.

January 22, 1982, Beijing

CHAPTER 21

Ten Years of Endeavor, Leading Industry with High Speed

1 Ten Years of Endeavor, Five Changes

On June 4, 1975, after Deng Xiaoping took the reins of the CCP Central Committee for the second time, it issued "CCP Central Committee Document No. 13," which included this comment: "It is especially pleasing that the oil industry has been in the lead, developing at a high speed." So, how did the oil industry take the lead in high-speed development during the ten-year Cultural Revolution? What did it achieve? This can be summarized by five changes that took place.

(1) Change in production. Crude oil production had increased continuously, reaching 105 million tons in 1978. At that time, China ranked as the eighth largest oil-producing country in the world. In these ten years, the average annual crude oil increase was 7.4 million tons, with the annual rate of increase being 18.6%.

(2) Change in oil distribution. In 1965, there were only 11 oil and gas fields in the country's 31 provinces and autonomous regions (not including Taiwan). By 1975, the number had increased to 21, thus altering the overall layout of the oil industry and laying the foundation for new developments in that industry.

(3) A big increase in natural gas production. Before 1966, there was only one gas field in Sichuan with an annual production of around 1 billion m^3. However, there was major development of natural gas from 1966 to 1976, first in Sichuan, and then in other oilfields as more and more attention was placed on natural gas exploration and utilization. Thus, by 1976, annual production of natural gas had reached around 3 billion m^3.

(4) Modes of oil and gas transport. Before 1966, oil transport relied mainly on railroads, rarely on trucks, and there was only one pipeline from Karamay to Dushanzi. On August 3, 1971, the construction of the 8/3 Northeast Pipeline began, and this led the way, within a few short years, for the construction of the Northeast Pipeline, the construction of the Huabei and Huadong Pipelines, and then the linkage between the latter and the Northeast Pipeline. There were other pipelines, too, like the natural gas pipeline in Sichuan, the oil pipeline in Xinjiang, and the natural gas pipeline in Changqing Oilfield. Pipelines became the main mode of transportation for oil and gas.

 | DOI:10.1163/9789004402737_022

(5) The beginning of offshore oil exploration, so that oil could be found both on land and offshore. In the ten or so years after 1966, after a difficult exploration process, oil workers made breakthroughs in both the Bohai and South China Sea, and they built oil bases in Tanggu 塘沽, Tianjin and in Zhanjiang 湛江, Guangdong. They also began to import some special ships and equipment, which paved the way for large-scale offshore oil exploration.

2 Ten Years of Struggle, Three Major Contributions

First, the oil industry satisfied the needs of the national economy, national defense, and the population in general. It created wealth for the country. In 1965, before the Cultural Revolution, oil and natural gas only accounted for 9.4% of total national energy production. By 1975, it had increased to 25%. In terms of capital accumulation for the country, in 1965 the oil industry accounted for 3.2% of the national industrial output. In 1975, this increased to 5.6%. As for revenue, the oil industry's revenue accounted for 4% of the total national revenue in 1965, but by 1975, it had increased to 10.3%.

Second, the oil industry did much to preserve much-needed social stability by meeting the country's energy needs. Due to long-term turmoil during the Cultural Revolution, coal production was sporadic and the rail transportation of coal was unreliable. As a result, many big cities in the country burned oil in lieu of coal. In 1966, the country burned 2.34 million tons of oil; in 1970, it burned 10.75 million tons; and in 1975, it burned 30 million tons, i.e., more than a third of the country's annual crude oil production. There was no alternative. This emergency measure greatly eased the energy shortage, thus helping to stabilize society and meet the basic needs of the population.

Third, oil exports earned foreign currency for the country. In the 1950s, China relied mainly on imported oil, but in doing so, it spent a lot of valuable foreign currency. Due to the development of the Daqing Oilfield in the 1960s, it became fully self-sufficient in oil and oil products and began to export oil in 1965. In 1975, oil exports earned US$1.877 billion and accounted for 7.1% of China's total export income.

3 The Cultural Revolution Seriously Damaged the Oil Industry

The Cultural Revolution severely damaged the oil industry, as it did other areas of life. Although the oil industry did achieve many things, as described above, and even though the oilfields kept drilling and producing oil, all other aspects of the industry were halted or set aside. It would simply take up too much space to list all the damage the Cultural Revolution inflicted on the individual oil workers, the industry's management, and its long-term development.

PART 5

Shengli Era (1980–1989): When the Gods Fight, Mortals Suffer: Rolling a Boulder Uphill

Greek mythology has it that Sisyphus had offended Zeus, so Zeus punished him by making him roll a huge boulder up a steep hill. Just before Sisyphus got the rock to the top, the boulder would roll back down the hill. Zeus made him do the same useless action repeatedly for eternity.

Due to changes in the political and economical situation, the oil industry during the 1980s was like Sisyphus rolling the boulder uphill—it moved forward, but with much difficulty. Old oilfields stabilized and increased production, and new oilfields were discovered, which enabled oil production during these ten years to increase by 3.8 million tons per year. In 1985, oil exports reached their maximum, 37.75 million tons, which accounted for 23% of China's total exports.

However, various internal and external factors meant that during the 1990s the Chinese petroleum industry would face enormous challenges.

CHAPTER 22

Hua Guofeng Endorses Kang Shi'en's Proposal of a Dozen Daqings

1 The Fast-Developing Oil Industry

In October 1976, under the leadership of Hua Guofeng 华国锋, Ye Jianying 叶剑英, and others, the Gang of Four were arrested and publicly denounced, thus ending the ten years of chaos brought about by the Cultural Revolution. The whole country was jubilant. Growing the national economy as soon as possible was the common wish of all the people. On building the economy, Hua Guofeng proposed to "liberate thinking a bit more, be a bit bolder, use a few more methods, and take faster steps."

In December 1976, at Minzu Hotel in Beijing, the Ministry of Petroleum held its first managerial conference of various oilfields since the debacle of the Gang of Four. Managers from the front lines of production were very happy and felt energized. I attended this conference as a representative from Huabei Oilfield, and after the conference, there was a surge of activity in each oilfield.

First, it was Daqing Oilfield. In 1976, Daqing Oilfield's oil production reached 50.3 million tons. To stabilize production at 50 million tons, the first Ten-year Production Stabilization Plan was proposed, according to which crude oil production in Daqing would increase from 50.3 million tons in 1976 to 51.501 million tons by 1980.

Second, at Shengli Oilfield, crude oil production increased from 17.5 million tons in 1976 to 19.46 million tons in 1978.

Then Huabei Oilfield, just discovered and developed that year, was already performing well. After only one year of development, by the end of 1976, annual production capacity had reached 10 million tons. In one year, Huabei Oilfield had been proven, developed, and built, and it had also repaid the country's investment.

Among the group of fast-developing oilfields in the oil industry, there were also the new oilfields of Liaohe and Zhongyuan in Bohai Bay Basin, Jilin Oilfield in Songliao Basin, Karamay Oilfield, Changqing Oilfield, Yumen Oilfield, Qinghai Oilfield, and Yanchang Oilfield in the northwest, as well as the Jianghan Oilfield, Henan Oilfield, and Sichuan's natural gas. All made considerable contributions.

 | DOI:10.1163/9789004402737_023

FIGURE 22.1 "Building a Dozen Daqings" (1977)

2 Kang Shi'en's Proposal

From April 20 to May 14, 1977, the CCP Central Committee held the National "In Industry, Learn from Daqing" Conference in Daqing, which brought together more than 7,000 managers and enterprise representatives from different central government ministries, provinces, and autonomous regions. This was the first major conference for industrial, transportation, financial, and trade enterprises after the downfall of the Gang of Four. Hua Guofeng, the Chairman of the CCP Central Committee and Premier of the State Council, presided over the meeting. The meeting played an active role both in spreading the Daqing Experience and repairing enterprises severely damaged by the Cultural Revolution. It was also a mobilizing event where the Central Committee continued to rely on the oil industry, using it as a bellwether for realizing rapid national economic growth.

Yu Qiuli, Vice Premier of the State Council in charge of national economy and Director of the State Planning Commission, gave the keynote speech at the conference. In it he said, "Premier Zhou, per Chairman Mao's instructions, in his governmental work report for both the Third and Fourth People's

Congresses announced the two-step hypothesis for developing our national economy. The first step to be carried out by 1980 is to build an independent and relatively complete industrial system and national economical system. The second step, to be carried out by the end of the century, is to fully modernize agriculture, industry, national defense, science and technology, in order to place our national economy at the forefront of the world."

Kang Shi'en, Vice Premier of the State Council in charge of industrial and transportation enterprises, Director of the State Economic Commission, and Deputy Director of the State Planning Commission, spoke at the conference and proposed the idea that "the oil industry will build a dozen Daqings by the end of the twentieth century." This had been the target that Kang had set in 1970 for the Fourth Five-year Plan for the Chinese oil industry, after he had studied the American oil industry. It meant that China's oil industry would try to produce 400 to 500 million tons of oil per year by the end of the twentieth century. This was a dream that all Chinese oil workers shared. Kang's proposal was also fully endorsed and promoted by Hua Guofeng.[1]

1 Kang Shi'en zhuan bianxie zu, ed., *Kang Shi'en zhuan*, 332.

CHAPTER 23

Hua Guofeng Resigns, and the Oil Industry Faces a Storm

1 Attacked Like Rats Crossing the Street

From 1980 to early 1981, change swept China's political arena. Hua Guofeng resigned from his positions as Chairman of the CCP Central Committee, Premier of the State Council, and Chairman of the Central Military Commission. Deng Xiaoping became Chairman of the Central Military Commission. The title of Chairman of the Central Committee was changed into General Secretary, and Hu Yaobang assumed that role. Zhao Ziyang 赵紫阳 became the Premier of the State Council. Meanwhile, as Chinese economic work underwent a three-year adjustment starting in 1979, Chen Yun, who had been Vice Chairman of the CCP Central Committee before the Cultural Revolution and who was the main state leader in charge of economic work, proposed "curb demand and stabilize prices; abandon development to secure stability; slow down reform and emphasize adjustment; concentrate the big and disperse the small." Soon afterward, criticism of "ratio imbalance" in economic work and "overheating the economy" became popular throughout the country.

Because of the oil industry's past performance and the fact that it had been heavily relied on and praised, it bore the brunt of the criticism, becoming the proverbial rat crossing the street—at which everyone shouts, "Smash it!" All of a sudden, criticizing the oil industry became popular. In fact, the criticisms of the oil industry at this time were attacks by proxy on Hua Guofeng, who had resigned.

2 The Concept of "a Dozen Daqings"—Not a Big Mistake

In 1977, while working my hardest in Huabei Oilfield looking for oil, I was very happy to hear the news of a dozen Daqings. I felt so proud to hear that our country's oil production could reach 400 to 500 million tons annually. I had felt the same way when I heard Premier Zhou Enlai's call to realize the Four Modernizations.[1] But two years later, the political weather had suddenly

1 The Four Modernizations (in agriculture, industry, national defense, and science and technology) were goals initially proposed by Premier Zhou Enlai in 1963.

 | DOI:10.1163/9789004402737_024

changed. The national media, scholars, experts, and economists criticized in unison that "a dozen Daqings" was the model for "the false, the big, and the empty," the sign of a "ratio imbalance" and of "overheating the economy." This surprised and puzzled me. After careful consideration, I felt that as an oilman's dream, there was nothing wrong with a dozen Daqings. If there was anything wrong, it was that of being out of sync with the financial and material situation of our country. It was unfair to criticize it as a model for "the false, the big, and the empty" and as an expression of economic "ratio imbalance" and "overheating the economy."

When criticism of the oil industry reached its height, one expert argued that one of the mistakes of the dozen Daqings was that it "enticed many units to change coal-burning furnaces into oil-burning furnaces, and then they had to change them back ... Such going back-and-forth was very wasteful." The switch from coal-burning furnaces to oil-burning furnaces had occurred because during the Cultural Revolution many of the coal mines stopped production and railroads stopped running. There was a great shortage of coal in many cities, including Shanghai. Premier Zhou had to order the Ministry of Petroleum to supply oil. During one of those years, more than 30 million tons of oil was burnt! That was a record at the time. After the fall of the Gang of Four, however, coal mines resumed production and railroads were back and running. For this reason, oil-burning furnaces were changed back into coal-burning furnaces. It had nothing to do with the dozen Daqings.

3 Risk in the Oil Business and "the False, the Big, and the Empty"

I do not think it appropriate to lightly accuse the oil industry of being false, big, and empty. The oil industry is an extremely risky business. Oil is buried deep underground, where it is all but invisible and intangible. At that stage, you can rely only on indirect means to evaluate it. The industry is both high-tech and high-risk. Due to different geological conditions and people's understanding of them, people will argue for years, even decades, over whether such-and-such area has oil, how much oil it has, how big the investment should be, and how big the plan should be. In the past, as we have seen, many people had indeed fallen into the trap of making plans that were false, big, and empty. We should have steered clear of that, but at the same time, if we were too conservative and did not think big and forward-looking, then the oil industry's hands would have been tied, and nothing would have been achieved.

In the first half of 1980, I attended an enterprise management workshop in Beijing. Nationally famous scholars and experts in the field were invited to speak. One day, an economics scholar from the Chinese Academy of Social

Sciences spoke at length on economic “ratio imbalance,” “overheating the economy,” and so on. He cited the fledgling Zhongyuan Oilfield as an example. Zhongyuan Oilfield, which was put into production on July 1, 1979, had an annual capacity of 300,000 tons. In that initial year it produced 226,000 tons of crude oil. However, a pipeline 32.7 cm in diameter was constructed from Puyang 濮阳, Henan to Shandong. Since the oil production was small at the time, in the beginning, the oil had to be pumped through the pipeline in batches, so that any batch required three stages of pumping in order to reach Shandong. This was called “kicking the ball.” In the short term, this no doubt was wasteful. Therefore, the scholar used this to criticize the Ministry of Petroleum's projects as false, big, and empty and a “big waste,” in order to show that he himself was aboard the “criticize the oil industry” train. Little did he know that oil production at Zhongyuan Oilfield would increase drastically. In two years, Zhongyuan Oilfield had to build a much bigger pipeline right next to the original pipeline. Under the circumstances of the time, China had few economists with genuine knowledge and talent.

4 A Philosophical Question: Production First or Livelihood First?

Some criticized the oil industry's slogan “Production Before Livelihood (先生产后生活)” as uncaring for employees' well-being. In the beginning of the Daqing Oil Campaign, that slogan had played an important role in promoting production and development. Tens of thousands of people in Daqing had accepted the slogan; they interpreted it as another way of saying “building an enterprise through arduous effort.” Following this slogan, people in Daqing and in the oil industry dramatically increased production and improved their livelihood. Isn't that a good thing? Who would have guessed that many years later someone claiming to be “a defender of oil workers' rights” argued that “Production Before Livelihood” showed indifference towards the employees? Copiously citing authoritative works, he even argued that the slogan defied economic laws.

I would like to talk about how this slogan came to be, about its background and premise. As I have just mentioned, the oil industry is very risky. Suppose that you design things for a large-scale oilfield, but in the end, the oilfield turns out to be smaller than anticipated. In the early 1950s, when the Soviet Union and China jointly designed and developed Dushanzi Oilfield in Xinjiang, living quarters for employees were built first, according to Soviet construction procedure. After people settled in, then oilfield development began. But contrary to expectations, Dushanzi Oilfield turned out to be very small. Most of the employee living quarters were not used, resulting in unnecessary waste. Wasn't this a hard lesson on "building tall buildings first, then looking for oil"? It was on the basis of this experience that people in Daqing Oilfield proposed the slogan "Production Before Livelihood."

Of course, "Production Before Livelihood" was a slogan proposed under the historical conditions of the time. It was an expedient measure, not a timeless one. In all fairness, whether it was "Production Before Livelihood" or "Livelihood Before Production" or "Grasp Production and Livelihood Together," for us it was like the old "Which comes first, the chicken or the egg?" question, which is difficult even for philosophers. Someone had to go and find bones in an egg and use them to nitpick at the oil industry. What could we say?

5 What is the Right Way?

When Daqing Oilfield and the oil industry were being criticized, there was an argument that "Daqing's ways of doing things, and that of the oil industry for the past twenty years, are products of Mao Zedong's extreme leftist line." This was a very popular and frightening label. People in the oil industry know from plain, practical experience that Daqing and the oil industry had made many achievements production-wise, and that is a fact. If this fact were due to "Mao Zedong's extreme leftist line," then we would be admitting that the line was correct. Many critics of the oil industry also thought highly of Deng Xiaoping's famous "cat theory": "It doesn't matter whether a cat is white or black, as long as it catches mice." If one were to examine the oil industry's past decades according to the cat theory, the result would be like this: it doesn't matter whether the line is extreme leftist or extreme rightist—the line that improves oil production is the correct one.

Of course, needless to say, in the past, for reasons understood by everyone, some of the work of the oil industry and Daqing Oilfield was negatively impacted by the extreme leftist line.

CHAPTER 24

The Oil Industry Reels from Accidents, Scandals, and Vendettas

1 The Handling of the Bohai No. 2 Accident Harms the Oil Industry

When it rains, it pours. When criticism of the oil industry and Daqing was in full swing, an accident occurred which shocked the whole nation. In the predawn hours of November 25, 1979, the Bohai No. 2 jack-up drilling ship of the Ministry of Petroleum's Offshore Oil Exploration Bureau capsized in a big storm while it was being towed to its new well location. This resulted in the death of 72 of the 74 crew members and great loss of national property, and at the time it was China's worst-ever offshore oil accident. On November 26, Kang Shi'en, the Vice Premier in charge of the oil industry learned of the accident and immediately reported it to Li Xiannian, the Vice Premier in charge of the economy. According to Li Ye, Kang's former secretary, at a relevant Central Committee meeting, Kang Shi'en wailed, blaming himself and begging for punishment.

On August 25, 1980, the State Council made a formal decision, pointing out, "The capsize of Bohai No. 2 was due to the fact that the leadership of the Ministry of Petroleum did not do things according to objective law, did not respect science, and did not attach importance to production safety, the opinions of employees, and historical lessons." It decided to relieve Minister of Petroleum Song Zhenming of his position and recorded a major demerit for Kang Shi'en. Then, in Tianjin, four bureau-level managers of the Offshore Oil Exploration Bureau were each sentenced to a few years in prison. Such severe punishments for one production accident, disciplining even the Vice Premier, was unprecedented since the founding of the new China.

There were probably deeper reasons for Song's punishment. The following two stories will shed some light on this. The first story was told to me in 1984 by Duan Zhigao 段志高, who was a senior bureau chief in the Ministry of Petroleum and who had been (from 1979 to 1982) the Vice President of the Central School of the Communist Youth League of China. He told me: "In early 1980, Feng Wenbin 冯文彬, the Director of the General Office of the CCP Central Committee, had a talk with me, and he asked me to relay two messages to the Ministry from the soon-to-be General Secretary Hu Yaobang.

 | DOI:10.1163/9789004402737_025

One was that the Ministry should not have treated the Bohai No. 2 funeral as it would a wedding (the Offshore Oil Exploration Bureau had lauded the good deeds of the dead oil workers at the funeral); instead it should have been investigating the accident's causes and who was responsible. The other message was that back in 1977, after Daqing had the big 'In Industry, Learn from Daqing' Conference and Hua Guofeng went there to see the sites, it erected steles to commemorate the occasion. Hu Yaobang said to take these 'earth god shrines' down as soon as possible, and 'If you have money, you should buy more text books for the children.'" Duan also told me that he had conveyed these messages to Deputy Minister Jiao, who immediately relayed them by phone to Minister Song in Daqing. But Song's response was, "It's no big deal—ignore it."

Li Ye told me the second story. In the first half of the 1980s, when criticism of the oil industry and Daqing Oilfield was peaking, the CCP Central Committee held a meeting at Jingxi Hotel in Beijing. One day, Petroleum Minister Song Zhenming was dining at the same table as Cai Chang 蔡畅, the Chair of the All-China Women's Federation, and Kang Keqing 康克清, the widow of Zhu De, the Chairman of the Standing Committee of the National People's Congress from 1959 to 1976. The two women asked Song about the dozen Daqings. For some reason, the two women felt slighted by Song's impolite demeanor and answers. They complained about him to Chen Yun. Chen Yun was infuriated and told them that he would fire this minister.

The sacking of Song Zhenming was warranted. He was promoted from chief of Daqing Oilfield to Minister of Petroleum in 1978, right when the popularity of "In Industry, Learn from Daqing" was at its height (1977–1979), so he moved to Beijing to take up his new position. Two things happened during his administration. The winter he moved to Beijing, he transferred workers from Daqing to dig a vegetable cellar in the spacious courtyard house that had been given to him as a Minister. This "country bumpkin" was about to store cabbage there just as he had in Daqing! But the excavation destroyed part of the Beijing municipal government's complex underground communication network, so the local government bureaus were furious with him.

The second incident involved his chauffeur. Song took with him from Daqing his personal chauffeur, Li Bendong 李本东, to Beijing. While Song was out of Beijing on a business trip, the chauffeur drove his fancy car to Beijing Railway Station and lured a girl (one of the "sent-down educated youth [下乡知识青年]") who was transferring trains in Beijing on her way from the northeast countryside back home to Shanghai. He raped and killed the girl, then dumped the body in the wilderness in Pinggu County. He was soon found, arrested, and sentenced to death. The incident shocked all levels of society.

These incidents shed a different light on the petroleum industry from the 1977–1979 period, when everyone was enthusiastically trying to learn from Daqing. However, the law of dialectics is merciless, and when things reach their height, things can soon reach their opposite if not handled well. Although none of them relate directly to Bohai No. 2, it is difficult to say whether they had no bearing on Minister Song Zhenming's removal.

As for what went on behind the scenes with Kang Shi'en, in 2013, my good friend Liu Xisan 刘锡三, who at the time of the calamity was the deputy Party secretary of the Offshore Oil Exploration Bureau and took part in dealing with the accident, said to me, "Bo Yibo insisted on handling it this way." Bo Yibo at the time was the deputy director of the Party's Central Advisory Commission. This remark jogged my memory—of that time in the summer of 1967, when Bo Yibo was escorted to the Ministry of Petroleum to be denounced at a mass meeting, and Kang Shi'en publicly exposed what Bo Yibo had said about Daqing: that the oil campaign was like a big melee, and that Chairman Mao's favoritism towards Daqing was like that of the Tang Emperor Xuanzong's favoring Yang Yuhuan over his 3,000 other concubines (see Part 3, Chapter 14, Section 9). Bo Yibo meant basically that Chairman Mao favored the oil industry above the others—but Kang Shi'en accused him as "being against the red banner of Daqing" and "being against Chairman Mao." This, too, does not relate at all to the Bohai No. 2 accident, but it does relate to Bo Yibo's insistence on punishing Kang Shi'en. This sort of coldly calculating "eye-for-an-eye" behavior may be considered a lingering aftereffect of the Cultural Revolution.

Many years after the accident, the retired Hua Guofeng said to the visiting Li Ye, then the Lieutenant-governor of Shandong, "The handling of the Bohai No. 2 accident impaired the oil industry's vitality!"

2 The Liu Boping Case: Hebei Province Harasses Huabei Oilfield

Again, when it rains, it pours—in buckets. Around the same time of the Bohai No. 2 accident, the Liu Boping 刘伯平 case occurred, which also shocked the whole country. Remember that between 1976 and 1977, Huabei Oilfield had become the country's third largest oilfield after Daqing and Shengli Oilfields, contributing significantly to China's annual oil output of more than 100 million tons. Perhaps its contributions and fame made it a natural target for all sorts of criticisms. As the saying goes, "People should fear fame as pigs fear getting fat." A small example: because the newly-built hotel in Huabei Oilfield had used some slip-preventing mosaic tiles in its bathrooms (of a kind no longer used), no less than the *People's Daily* criticized the oilfield as going in for grandiose, foreign products and incurring great expense.

In 1979, Liu Boping, the deputy manager of Huabei Oilfield's Office in Shijiazhuang, the capital of Hebei Province, purchased a few dozen smuggled Japanese color TVs and cassette recorders through a friend in Guangzhou, ostensibly for "educational purposes" in the oilfield, and then resold them to various directors and managers in Hebei Province and the Huabei Oilfield. This was undoubtedly wrong. Afterward, using the same method, he purchased the same number of color TVs and cassette recorders for the Educational Division of Huabei Oilfield. For this, the Hebei Provincial Discipline Inspection Commission spent more than two years thoroughly investigating Liu. Although they found no criminal evidence of bribery or corruption, the Commission just would not let it go. They spent a great amount of manpower and financial resources and carried on internal and external investigations, sparing no effort to find the slightest infraction. Just before the 1983 Spring Festival, they arrested and imprisoned Liu Boping. Meanwhile, the story had spread beyond the oilfield and gone national, so that *Hebei Daily* (Shijiazhuang), *People's Daily*, and *Workers' Daily* (Beijing) were constantly dinning criticisms of him into people's ears. It was as if the oil industry had produced a "big shark." Liu Boping was sentenced to eight years of prison. Years later, Liu's sentence was overturned through the intervention of Qiao Shi 乔石, the Secretary of the Central Commission for Discipline Inspection. Liu Boping, who had been languishing in jail for six years, could once more see the light of day. The reason I mention this case is to show how Hebei Province used it to harass Huabei Oilfield.

Why did the Hebei Discipline Inspection Commission dare to investigate so vigorously and sentence Liu Boping? Given the macro-climate of the time, it had become common throughout the nation to make the oil industry, particularly Daqing, the butt of criticism. Naturally, Hebei Province did not want to appear out of step. More specifically, however, the Commission was being backed by Gao Yang 高扬, the First Secretary of the Communist Party Committee of Hebei Province. Gao Yang had been a senior industrial manager. In the 1950s, he was a Deputy Minister of Metallurgy and was later promoted to Minister of Chemical Industry, but he was overthrown when the Cultural Revolution began. In 1967, the State Council (through arrangements made by Yu Qiuli and Kang Shi'en) appointed Xu Jinqiang, then Deputy Minister of Petroleum, to be the Minister of Chemical Industry. When Xu arrived at the Ministry of Chemical Industry, he supported the faction that overthrew Gao Yang, and Gao never forgot it. After the Cultural Revolution, on Deng Xiaoping's recommendation, Gao Yang was appointed the First Secretary of the Communist Party

Committee of Hebei Province. In view of this, the framing of Liu Boping was unavoidable. From this, one can see the how the "nest-fighting" and other ill effects of the Cultural Revolution continued to linger years after it had officially ended.

3 The 1983 Shake-Up and the Collapse of Huabei Oilfield

The Liu Boping affair should have ended in early 1983, after Liu's arrest, but Petroleum Minister Tang Ke instructed First Deputy Minister Zhao Zongnai to lead a working group to Huabei Oilfield just to shake things up. They stayed there more than six months doing so.

Zhao's group disregarded the past achievements of Huabei Oilfield. As soon as they stepped out of their cars, they determined that severe problems existed among the oilfield's Party Committee and several leading managers, so using measures even more severe than those used in the Cultural Revolution, they took aim at Huabei Oilfield's bureau- and division-level managers:

- They held meetings of upper-level managers that lasted for weeks at a time, day and night. Zhao Zongnai, coupling threats with promises, would order these managers to confess on their own or expose one another, and tell them that the meetings would not be dismissed for as long as clarity was not brought to the situation. He also ordered each of the small groups within these meetings to print bulletins daily. From the small group meetings for investigating cases alone, more than 120 bulletins were produced.
- People were dispatched to secretly "monitor" key bureau- and division-level managers. People were assigned to have "heart-to-heart-talks" with these managers, in order to trick them and force confessions.
- They set up "mass dictatorship teams" that detained and investigated division managers illegally for months, leaving their families and friends in the dark.
- They carried out illegal searches of the homes of bureau- and division-level managers and illegal confiscations of their property. The investigations on the whole were outside of any law or court.
- Without the approval of any authority and without any concrete evidence, they organized exhibitions that publicized the "mistakes" and "crimes" of these managers. Zhao Zongnai ordered the newspaper *Huabei Petroleum* (*Huabei shiyoubao* 《华北石油报》; Renqiu) to name and criticize the managers, but the newspaper's president, who had some legal knowledge, refused to do so.

- The Ministry working group (whose members had come from other oilfields) sent subordinate working groups to all divisions in the oilfield, holding "classes" and meetings for managers, ordering them to confess or expose one another's deeds. They also set up several special case groups to investigate "key" division managers.
- They completely changed the oilfield's leadership, removing or demoting more than one hundred bureau- and division-level managers. Most of the bureau managers were demoted and changed—and these were replaced with managers from other oilfields.
- For more than six months, the Deputy Petroleum Minister focused on the shake-up but seldom asked about production.

I can list all of these shake-up techniques with little difficulty, because Zhao had applied so many. He inflicted massive chaos on the Huabei Oilfield. During the shake-up he orchestrated, people in the oilfield were worried about their own safety, not knowing when their houses could be searched and their property confiscated, or when they might be detained. From this catastrophic upheaval, this once lively and robust oilfield never recovered. Its oil production plummeted, and the downward spiral could not be checked.

4 Oil Production Drastically Declines Nationwide

Things in the oil industry were so strange! Under the kinds of circumstances described above, from 1980 on, China's oil production started to drastically decrease, and it could not maintain annual production at 100 million tons.

First, it was Shengli Oilfield, which had played an important role in stabilizing and increasing oil production. It reached 19.46 million tons in 1978. Due to a sharp decrease in investment and labor, in 1979 and 1980, its production dropped to 18.65 million tons and 17.58 million tons respectively.

In October 1980, Shengli Oilfield set the following production goals in its Five-Year Production Plan from 1981 to 1985: 1981, 15 million tons; 1982, 14.03 million tons; 1983, 13.09 million tons; 1984, 12.22 million tons; 1985, 11.31 million tons.

Huabei Oilfield was next. Its oil production reached its peak of 17.28 million tons in 1978. In 1982, it was 13.3 million tons; 1983, 10.5 million tons; 1987, 8 million tons; after 1992, it was below 4.8 million tons.

On January 30, 1981, at a meeting of Communist Party Group of the Ministry of Petroleum, Wang Naiju 王乃举, the Deputy Chief of the Development Bureau, who was in charge of crude oil production, pointed out, "The Development

Bureau has reported an extraction rate and a five-year trend. Extractable reserves are 2.2–2.3 billion tons. The oil production plan for the next five years, in sequence, is 100 million tons for 1981, 95.35 million tons for 1982, 87.29 million tons for 1983, 72.99 million tons for 1984, and 66.57 million tons for 1985."

How to deal with the continuous decrease in crude oil production? According to Kang Shi'en's biography:

> At the time (the winter of 1980), facing one difficulty after another, Deputy Minister of Petroleum Jiao Liren had no choice but to write a letter to the State Council recommending that the oil industry receive increased essential investment and that the oil production plan for 1981 be adjusted to 100 million tons. Otherwise, it would be difficult to maintain annual production at 100 million tons. The leadership of the State Council realized the gravity of the problem. Two days after receipt of the letter, in a meeting discussing the plan for 1981, Jiao Liren made his report.
>
> Premier of the State Council Zhao Ziyang said, "Oil production cannot drop, otherwise, the [plans for] national economy will need to be readjusted." At the meeting, they basically accepted his recommendation and increased by 500 million yuan the investment for the oil industry.[1]

1 Kang Shi'en zhuan bianxie zu, ed., *Kang Shi'en zhuan*, 359.

CHAPTER 25

Rolling a Boulder Uphill

1 Kang Shi'en Receives an Order

Faced with the news that annual oil production could not be maintained at 100 million tons, Premier of the State Council Zhao Ziyang could not sit still. He had no choice but to turn once more to the 65-year-old Kang Shi'en, who was hospitalized due to bladder cancer.

One afternoon in December 1980, Zhao Ziyang visited Kang Shi'en in the hospital, seeking advice on how to maintain national annual oil production at 100 million tons. Based on his conversations with Yu Qiuli and others, Kang recommended the idea of a 100 million tons annual guarantee. Then, Zhao inquired if Kang would be willing to be Minister of Petroleum again and promote the oil industry. At the time, since Kang Shi'en was most concerned about how to prevent the 100 million ton figure from slipping and felt obligated to help the nation overcome this difficulty, he happily agreed to Zhao Ziyang's offer. In the morning of January 19, 1981, at a meeting of the State Energy Commission, Yu Qiuli, the new Director of the Commission, announced that the CCP Central Committee had decided that Vice Premier Kang Shi'en would double as Minister of Petroleum.

2 Vigorous Efforts to Turn the Tide: the 100 Million Ton Guarantee

Starting in 1979, the nation began a three-year economic adjustment period. Decreasing investment in capital construction was the main objective, so during this period the country drastically reduced investment in the oil industry. The planned investment for 1981 was only half that of 1980—only 1.7 billion yuan.[1] Under these circumstances, people paid close attention as to how the oil industry could keep producing 100 million tons annually.

In the fall of 1980, Yu Qiuli, the Vice Premier and Director of the State Energy Commission, proposed to the Ministry of Petroleum, "If the country asks for 100 million tons annually, you should produce 100 million tons. Let's be reasonable ... Now the country has some problems. You should have a revolutionary spirit. You should dare to think, dare to speak, and dare to act, liberate your

1 Kang Shi'en zhuan bianxie zu, ed., *Kang Shi'en zhuan*, 358.

 | DOI:10.1163/9789004402737_026

thinking, and accomplish it within three years!" Li Renjun 李人俊, the Deputy Director of the State Planning Commission and long-time Deputy Minister of Petroleum, also said, "The Ministry of Petroleum should think of ways to produce and save an extra 1 million tons of crude. We could consider giving it all to you." While still in the hospital, Kang Shi'en met with relevant people to discuss how to get the oil industry out of this difficult situation of reduced funding. They had similar ideas, and they formed the plan for the oil industry's guarantee of 100 million tons per year.

The plan, in a nutshell, was that the Ministry of Petroleum would guarantee production of 100 million tons of crude oil annually, but the state would allow the Ministry to use any oil produced in excess of that 100 million tons for export. Proceeds from the exports would then be used to supplement the state's investment in the oil industry. This was a break from the past, when the central government controlled all revenues and expenditures, when "money for buying soy sauce could not be used for buying vinegar." Under these new circumstances, the Ministry of Petroleum now had some financial autonomy and could direct some of the revenues as it saw fit.

In December 1980, when Zhao Ziyang visited Kang Shi'en in the hospital, he agreed to this plan. On April 22, 1981, the State Energy Commission, State Planning Commission, Ministry of Finance, and Ministry of Petroleum collectively wrote a letter to the State Council titled "Report on the Need to Increase Drilling and Oilfield Development This Year in Order to Guarantee Crude Oil Production of 100 Million Tons Next Year (关于保证明年原油产量稳住一亿吨，需要增加今年钻井和油田建设工作量的报告)." Four days later, Zhao Ziyang instructed State Energy Commission Director Yu Qiuli and Yao Yilin 姚依林, the Director of State Planning Commission, to carry it out and suggested that the Ministry of Petroleum should have a "three-year contract" for the 100 million ton guarantee. On May 14 of the same year, the State Energy Commission invited eleven government agencies including the State Planning Commission, the Ministry of Foreign Trade, the Ministry of Finance, the Ministry of Petroleum, and the Bank of China to discuss and decide on the export of the Ministry of Petroleum's surplus crude oil. By this time, the plan to encourage the oil industry to take initiative and develop "the 100 million ton guarantee" was formally unveiled. It played an important role in the oil industry's continued development.

3 Daqing Oilfield—a Model for Chinese Industries

Out in society, the wind of criticism directed at Daqing Oilfield and the oil industry in general was still blowing hard. In September 1980, First Secretary of Heilongjiang Provincial CCP Committee Li Li'an 李力安 wrote a letter to the CCP Central Committee suggesting that the latter should issue a definite, formal statement on the issue of "In Industry, Learn from Daqing," in order to unify people's understanding. Hu Yaobang wrote comments on the letter, requesting the Secretariat of the Central Committee research and discuss the questions raised in the letter. With Hu Yaobang personally presiding, on December 18, 1980, the Central Committee issued "CCP Central Committee Document No. 47 (中共中央文件 47 号)," entitled "Notice of the CCP Central Committee Forwarding the Report of the Party Group of the State Economic Commission on the Issue of 'In Industry, Learn from Daqing' (中共中央转发国家经委党组关于工业学大庆问题的报告的通知)."

The notice gave high praise to the model of Daqing:

> Under the leadership of our Party, armed with Mao Zedong thought, Daqing oil workers in the early sixties, in the face of Soviet hegemonism and under extremely difficult circumstances, were determined to win honor for the country and strive on the people's behalf. With the will to make the country strong, relying on Chairman Mao's two essays, "On Practice" and "On Contradiction" for guidance, they built Daqing Oilfield, one of the most advanced in the world, ending China's dependence on foreign oil. In the early seventies, while the country was severely sabotaged by the Lin Biao and Jiang Qing anti-revolutionary cliques, the workers stood strong and brave. They shunned all interference and greatly increased crude oil production, playing an important role in the development of the national economy during the period of great turmoil. After the destruction of the Gang of Four, when oil production in old oil-producing areas had been decreasing for years, they, in a thousand and one ways, through hard and skillful work, stabilized high production, kept annual crude oil production at 50 million tons and continued contributing greatly to the country. Through self-reliance and arduous enterprising practice, the Daqing oil-working masses created Daqing Oilfield's history of continuous forward movement and development. The whole set of experiences in Chinese self-established and self-managed modern enterprise, which the Daqing cadres and masses created, demonstrates that Daqing has proven itself to be an advanced model for our industries, and that Daqing workers have proven themselves to be an advanced part of our nation's workforce.

The issuance of this "Notice" provided great encouragement to the oil industry as it strove to reach the 100 million ton goal.

4 Hu Yaobang Mobilizes

Faced with the sharp drop in oil production, Zhao Ziyang urged Kang Shi'en to return to the Ministry of Petroleum, and he ratified the 100 million ton plan. As General Secretary of the CCP Central Committee, Hu spared no effort to visit various oilfields and offer encouragement and support wherever he went.

On November 12, 1981, while Hu Yaobang was visiting the northeast, he said to Deputy Minister of Petroleum Sun Xiaofeng 孙晓风, who accompanied him on the trip, "If you can produce 120 million tons of oil, our economy will be better, and the atmosphere among the people throughout the country will be different. People in the whole country will be grateful to you."

In May 1982, Hu Yaobang visited Dagang Oilfield.

In August 1982, Hu Yaobang visited Daqing Oilfield and praised it as "a lustrous pearl of our nation's socialist modernizations," and he showed Yu Qiuli and Kang Shi'en "due respect."

In October 1982, when Hu Yaobang made a detour to visit the new exploratory area of Zhongyuan Oilfield, he said, "If you are sure of the resources, then don't delay. You have to develop it as soon as possible."

In May 1983, Hu Yaobang visited Karamay Oilfield in Xinjiang. He wrote in calligraphy: "Energy is the lifeblood of our nation's four modernizations. It is the greatest honor for comrades to try to explore and find more oil. Salute all comrades on the oil front who strive and surmount every difficulty."

In August 1983, Hu Yaobang visited Daqing Oilfield for the sixth time.

In February of 1984, Hu Yaobang visited Shengli Oilfield and wrote in calligraphy: "One History of Hard Pioneering Work, One Million Earth-Shaking Men."

In March 1984, Hu Yaobang visited Huabei Oilfield.

In May 1984, while passing by Liaohe Oilfield, Hu Yaobang encouraged them, "Strive to be the third largest oilfield."

In October 1984, while visiting Bohai Petroleum Company, Hu Yaobang expressed the hope of "creating an offshore Daqing and a new generation of oil people."

In short, during the four years from 1981 to 1984, Hu Yaobang frequently visited oilfields to boost morale and spur the oil "horse" to run faster. Someone jokingly said afterward, "When the oil production is about to decrease, the General Secretary ushers in the east wind." Since oil workers' will and enthusiasm to produce more oil for the country have always "grown with a little sunshine and blossomed with a little rain and dew," with Hu Yaobang's encouragement and urging, their will and enthusiasm soon turned into actual deeds that increased oil production.

The CCP Central Committee's attitude toward the oil industry had taken a 180-degree turn. While power struggles among the state leadership had impacted the oil industry, Hu Yaobang's efforts to vindicate the industry were touching. Had they known that the earlier criticism would have such a deleterious effect, they probably would have refrained from it.

5 Two Areas of Focus: Daqing and Shengli Oilfields

To fulfill the 100 million ton guarantee, the primary focus was Daqing Oilfield. Ever since its discovery and development in the early 1960s, Daqing Oilfield had been the pillar of the oil industry. In 1976, it had reached the annual production scale of 50 million tons. Faced with the target of 100 million tons, Kang Shi'en invited Daqing executives Chen Liemin and Wang Sumin 王苏民 and some experts to a 13-day conference in Beijing, discussing if it was possible for Daqing Oilfield to top 50 million tons by increasing crude oil production another 1.5 million tons. With no new reserves, Daqing could barely manage producing 50 million tons. To increase production by that much was both difficult and risky. It would be disastrous if Daqing Oilfield overreached and collapsed. In addition, the 100,000 Daqing workers who had made great contributions to the country were still living in the adobe houses from the beginning of the Daqing Oil Campaign. In other words, they lived in poor conditions. Much needed to be done in terms of equipment, oilfield construction, and living facilities. However, managers and workers in Daqing had long had a strong tradition of solidarity. If the other oilfields ran into problems, Daqing Oilfield, as the oil industry leader, would assume the task of producing 1.5 million tons more crude oil without hesitation.

The second focus was Shengli Oilfield. As the second largest oilfield, oil workers there worked hard to reach a production level 19.46 million tons in 1978, following the fall of the Gang of Four. Afterward, due to insufficient funds, oil exploration diminished and crude oil production decreased year after year. In the winter of 1980, Deputy Ministers of Petroleum Jiao Liren and Huang Kai 黄凯 led a working group to Shengli Oilfield, thinking that they needed to bring about an upswing in Shengli Oilfield production and immediately put a halt to the decline. They decided to transfer Li Ye, who had been in charge of Zhongyuan Oilfield, back to Shengli Oilfield.

6 The Resurgence of Shengli Oilfield

First, Shengli Oilfield relied on its geologists. Soon after the Spring Festival in 1981, on February 23, Shengli Oilfield held a geological conference attended by more than 100 experts and scholars. It was both a geological "diagnostic

FIGURE 25.1 Shengli Oilfield's Gudong Oil Campaign in 1986

meeting" and "planning meeting" for Shengli Oilfield, extremely important for short-term oilfield development. Those who attended the conference, which lasted more than ten days, were invigorated and talked freely. Thirty-four geologists spoke and gave comprehensive reports, which incorporated copious amounts of data and information reflecting Shengli Oilfield's complexity and vast richness. During the conference, some of the experts gathered and talked through the night, calculating, weighing evidence, arguing—and gradually the arrangements and plans for further oil exploration were hammered out. At a meeting near the close of the conference, Li Ye, the Secretary of the Party Committee of Shengli Oilfield, pointed to the gigantic geological map and said: "Summing up the various experts' opinions, the arrangements for the short-term exploration of Shengli Oilfield can be carried out according to the 'three kinds and thirty chunks' proposed by Comrade Liu Xingcai and others. We might find ten or so chunks of the first kind with reserves under one million tons, ten or so chunks of the second kind with reserves under 10 million tons, and ten or so chunks of the third kind with reserves under 100 million tons. But these three kinds just point the direction—there's no great certainty." As it turned out, though, oil production later on surpassed this prediction.

In early 1981, Liu Xingcai, who was the Deputy Chief of the Geological Exploration Division of Shengli Oilfield, gave each member of the oilfield management team a copy of his lengthy geological investigation report. This well-documented report was the result of his twenty years of research into Shengli Oilfield. Backed by strong arguments and valid reasons, he showed that Shengli Oilfield had oil, and lots of it. In the past, he had lectured and submitted numerous reports to that effect, but he was always given the cold shoulder because "the time is not right," or he was deprecated as "Big Braggart Liu." He could not do anything about it. This time, however, the Shengli Oilfield's leadership appreciated his report, and soon afterward at the February 23 geological conference, he performed superbly. Before long, even though Liu was not even a member of the Communist Party, Shengli Oilfield's Party Committee promoted him first to Deputy Chief Geologist, and later to Deputy Commander and Chief Geologist of Shengli Oilfield. He had both position and power. From then on, many achievements were made due to the good advice given by Liu Xingcai and his colleagues. Shengli Oilfield made good use of many talented people like Liu Xingcai.

Second, Shengli Oilfield relied on advanced oil exploration and development technology, especially on advanced drilling technology. Yao Fulin 姚福林, the Deputy Commander who was in charge of drilling, had repeated discussions with Chief Drilling Engineer Yu Wanxiang 于万祥 and Drilling Headquarters Commander Zhang Zhaoli 张兆礼; they proposed concrete measures to "liberate drilling speed, liberate mud, and liberate oil beds." He organized special task forces to tackle key technologies. In 1986, they finished the whole year's drilling of 3.5 million m fifty days ahead of schedule. In December 1986, in a celebration meeting held at Drilling Headquarters, Kang Shi'en, who was then over seventy years old, invited to the podium those drilling crew captains and drilling engineers who had performed with such distinction. As they stood before an audience of thousands, Kang respectfully bowed to them three times. In 1987, they pressed on without letting up and drilled 1,959 wells, a total of 4.34 million m.

Third, Shengli Oilfield relied on workers' loyalty and cared about workers' well-being. At the time, only 3,000 workers had their wives and families with them—a matter that had been decided before the Cultural Revolution. After 1970, however, the number of oilfield employees already exceeded 100,000. Even though some workers' wives had been organized a few years earlier to take part in the agricultural work of the oilfield, many of them still held IDs of rural areas and could not accompany their husbands to the oilfields—and even if they did, they were still technically unauthorized to be there. They were "illegal residents," so they were not provided food or housing, not allowed to take part in agricultural or sideline production, and their children could not go to school. In 1974, Duan Qitong 段其通, a worker, stopped Li Ye's car, threw himself on the ground, and said, “My wife and two children are illegal residents. I truly can't support them with my monthly salary of a couple of dozen yuan. Please let my wife take part in agricultural work to make a little money to live.” In 1975, three women took their complaints to Li Ye. One cried out, “We have been living illegally in the oilfield for more than ten years. We have been going to the free market secretly to buy food at high prices, so we can only afford corn and dried yams. During New Year celebrations, we dare not let our children visit our neighbors' homes. Other people eat white steamed buns (*mantou* 馒头). Our children have never even seen steamed buns. When one was hospitalized, we could not afford to buy him any fruit. We need to eat and live. Please let us work and make some money.” Another worker in his forties or fifties told Li Ye, “I have been separated from my wife and children for more than twenty years. They are helpless in the country. You let us know when our families could join us. I will wait, even until I am eighty years old. At least, I have something to hope for.”

These problems, due to many reasons, had never been solved and had become the number one problem for Shengli Oilfield. In 1981, Shandong Province gave Shengli Oilfield a few farms and hundreds of thousands of acres of saline-alkaline land. The leadership of Shengli Oilfield decided to appoint Hou Qingsheng 侯庆生, the Deputy Commander in charge of logistics and employee families, to oversee the work. At last, the oilfield completely solved the problem of oilfield workers separated from their families. They announced immediately that as long as they were willing, the wives of oilfield workers could move to the oilfield and reclaim wasteland in order to farm. Accepting as many wives as possible to come to the oilfield was a first in the oil industry. What unexpected good news! The workers were happily reunited with their families. The whole oilfield was with one heart and one mind—battling for oil!

As they say in the military, “It is better to win the hearts of the people than to attack a city.”[2] This measure adopted by Shengli Oilfield did increase production. Oilfield leaders have a responsibility to care for the employees and improve their livelihood.

---

2 The statement is attributed to Ma Liang 马良 (187–222) in *Xiangyang ji* 《襄阳记》 [*Record of Xiangyang*], quoted in the Pei Songzhi 裴松之 (372–451) commentary to Chen Shou 陈寿, *Sanguo zhi* 《三国志》 [*History of the Three Kingdoms*] (Beijing: Zhonghua shuju, 1960), 983.

7 Oil Production Stabilized at 100 Million Tons, and Growth Restored

From 1981 to 1985, led by the two key oilfields of Daqing and Shengli, oil workers in the oil industry steadily increased crude oil production.

Shengli Oilfield took the lead. In 1981, while the annual plan called for 15.9 million tons, Shengli actually produced 16.11 million tons. In 1985, while the annual plan called for 11.31 million tons, it actually produced 27.05 million tons. Whereas it was originally planned to produce a total of 65.65 million tons of oil in five years, Shengli actually produced 100.86 million tons. Its newly increased production over that five-year period amounted to nearly half that of the newly increased national output. In five years, its actual oil production exceeded planned production by 35.21 million tons. Based on international oil prices of the time, that amounted to US$6 billion, or 50 billion yuan.

Liaohe Oilfield meanwhile became China's third largest oilfield. Situated in the northeast of Bohai Bay Basin, this oilfield was discovered in 1970. After many years of endeavor, production and development continued. Especially worth noting is the discovery near Shenyang of the small but abundant Damintun 大民屯 Oilfield in the early 1980s. In 1986, Liaohe Oilfield produced 10 million tons of oil, truly becoming the third largest oilfield.

Zhongyuan Oilfield took off around the same time that Shengli Oilfield's surged anew. Its crude oil production rose steadily: 3.03 million tons in 1983, 4.01 million tons in 1984, and 5.5 million tons in 1985. It became the fourth largest oilfield in the country.

In the northwest, after decades of effort, there were new developments in Xinjiang's Karamay Oilfield. Its production reached 4.994 million tons in 1985. There were also obvious contributions from the new Changqing Oilfield as well as the older Yanchang Oilfield, Yumen Oilfield, and Qinghai Oilfield.

Others such as Dagang Oilfield in Tianjin, Jilin Oilfield in Jilin, Nanyang Oilfield in Henan, Jianghan Oilfield in Hubei, and Jiangsu Oilfield all had new achievements during this period.

As for Daqing, China's number one oilfield, it still ranked first. In 1985, besides producing the required 50 million tons, it increased production by 5.289 million tons. In 1990, it produced 55.6 million tons of crude oil. It had stable yields for 15 years.

Summing up oil productions from all oilfields, one could see that from 1981 to 1985, the Chinese oil industry once again started to grow.

Within five years, the downward spiral of crude oil production had ceased. In 1981, the national crude oil production reached 101.21 million tons, and in 1985, it was 125 million tons. China became the world's sixth largest oil producing nation.

China's oil exports—37.75 million tons—reached their peak in 1985 and earned US$6.96 billion. This exceeded the annual target of oil export revenue of $5 billion set by Deng Xiaoping and accounted for 23% of China's total exports that year.

Within five years, the oil industry self-funded 12.6 billion yuan for oil exploration. This investment increased at an average rate of 19.5% a year, so that self-generated funds now surpassed that of state investment in the industry. Thus a negative situation of decreased investment, reserves, and production was transformed into a positive situation wherein the increase in reserves exceeded the increase in production.

Finally, the state's "100 million ton guarantee" policy allowed the oil industry to export any oil in excess of 100 million tons, and because the oil industry could then keep 15% of the revenues from the sale of these exports as bonuses and benefits for oil employees, the oil workers' incomes, housing, standards of living, and medical conditions were all improved.

CHAPTER 26

The Debate on Shengli Oilfield's Second Daqing Plan

1 The First Round: Yu Qiuli Proposes Making Shengli Oilfield "the Second Daqing"

On February 11, 1984, CCP General Secretary Hu Yaobang, accompanied by PLA General Political Department Director Yu Qiuli, visited Shengli Oilfield. The following is a transcript of a conversation recorded in the morning of February 12, 1984, during Hu Yaobang's visit, when Li Ye, the Secretary of the Party Committee of Shengli Oilfield, made his report to Hu Yaobang.

Yu Qiuli: "As an advisor, permit me to make this report to the General Secretary ... Our country has an area of more than 9.6 million km^2, and in more than 4 million km^2—a very large area—oil may possibly be found. You've been working in Shengli Oilfield for twenty years, so you should have enough experience [to know that] with an oilfield 700 km^2 in area and more than 2 billion tons of reserves, it should be developed as another Daqing, because Daqing has only developed some 800 km^2, with 2.4 billion tons of reserves ..."

Hu Yaobang: "If our country could have a second Daqing, then the eighties for us will have great potential. This is an important factor! The thinking behind the proposal to build a dozen Daqings was, we can think this way—why can't we think this way? It was just that at the time [in 1977 when it was first proposed], the estimate was overly optimistic."

(Then Hu and Yu start doing the math of a dozen Daqings, using their fingers.)

Hu Yaobang: "Two in the northeast, one in Shengli, ... find the fifth, the sixth, the seventh in the nineties ... If our country has more oil, say, reaching 300 million tons or even more than that in this century, then there will be stronger material conditions for quadrupling [the national economy]. There are many factors for such material conditions, but if we could get 300 million tons of oil, that would be the most important factor ..."

Hu Yaobang, turning his head, and asking Li Ye: "Are you thinking to reach 20 million tons this year?"

Li Ye, replying: "We're thinking of reaching more than 20.5 million tons this year." [In the end, Shengli's actual annual production that year was 23.01 million tons.]

 | DOI:10.1163/9789004402737_027

Hu Yaobang: "When will you get 30 million tons? When will you get 40 million tons? Once you have 40 million tons, you'll have one Daqing."

Li Ye, replying: "[We] originally planned to get 3 billion tons of reserves and 30 million tons' production by 1990."

Hu Yaobang: "[You] need to set a plan for making a second Daqing" [i.e., annual production of 40 million tons].

After the meeting, Hu Yaobang dedicated a work of calligraphy to Shengli Oilfield: "Build a second Daqing, dedicate it to the country's fortieth anniversary."

On the third day, at a meeting attended by a thousand cadres, Hu Yaobang spoke at some length, and then he and Yu Qiuli sang "The Song of the Military and Political University of Resistance Against Japan." All of a sudden, Hu Yaobang changed the lyrics: "On the banks of the Yellow River, gather the fine sons of the Chinese people. To achieve the Four Modernizations and find energy sources, it is up to you to take charge ..."

The last thing Hu Yaobang said at the meeting was: "I will cheer you on. Let's meet again on the balcony of Tiananmen in 1989"—by which he was essentially repeating the idea of "Build a second Daqing, dedicate it to the country's fortieth anniversary."

2 The Second Round: Zhao Ziyang Implies That the Shengli Oil Reserve Figures Are Unrealistic

On April 8, 1984, Premier of the State Council Zhao Ziyang, accompanied by Minister of Petroleum Tang Ke, visited Shengli Oilfield. That afternoon, Li Ye presented a report entitled, "Building The Second Daqing, or Realizing the Idea of Two Doublings (建成第二个大庆或实现两个翻番的设想)," and said, "We are determined to complete building the second Daqing in the eighties and to achieve two doublings by the end of the century, that is, reach an annual production of 80 million tons."

On April 10, when Zhao Ziyang spoke at a big meeting of Shengli Oilfield's cadres, he said:

> For our oil industry, we are faced with the problem of achieving greater development of oil on-land, and making our greater development stand on a scientific, reliable foundation. Today, insufficient geological reserves are still an important issue for us ... Strengthening geological exploration and increasing geological reserves should still be the focus of our work. This point Comrade Kang Shi'en has told me repeatedly. We should never

> forget that. For instance, at a meeting in Karamay, when asked about [the plan for] Karamay Oilfield, Comrade Kang Shi'en said, "We will consider it only when 2 billion tons of reserves have been found in Karamay. As for the problem with traffic and other problems, we should not consider them before securing 2 billion tons of reserves. This is the big picture. We can consider how to lay out the construction only after reserves of a certain scale and greater reliability have been found."

On April 11, Zhao Ziyang and his party travelled from Shengli Oilfield in Dongying, Shandong Province to Zhongyuan Oilfield in Puyang, Henan Province. Zhao clearly told the leadership of Zhongyuan Oilfield that their oil reserves were realistic, unlike those of Shengli Oilfields.

3 The Third Round: Wang Zhen Says, "Beware of Those Who Blow Cold Air"

He Jianming 何建明 tells a story he may have heard from Li Ye:

> On March 28 [1985], Yu Qiuli, on behalf of the Politburo of the CCP Central Committee, invited Shengli Oilfield's leadership to dinner in the Great Hall of the People. Wang Zhen, Yao Yilin, Chen Muhua [陈慕华], Hao Jianxiu [郝建秀], Kang Shi'en, Wang Bingqian [王炳乾], Song Ping [宋平], and others attended. Petroleum Minister Tang Ke also attended. At the dinner, General Yu Qiuli said, "The General Secretary Hu Yaobang asked me to continue being in charge of petroleum on behalf of the Politburo." Wang Zhen, who had proven himself a seasoned general of the battlefield, pointed out, "On the issue of getting a second Daqing, beware of someone blowing cold air. Like in this room, as soon as the cold air blows, old comrades like us will catch cold and be hospitalized, maybe lose our old lives ..."[1]

Among those present at the dinner—leaders from Shengli Oilfield, Politburo members, vice premiers, ministers—General Wang Zhen was the most senior and highest ranking. The rest were all his subordinates. When the PLA marched into Yumen Oilfield in September 1949, it was he who recommended

1 He Jianming 何建明, *Buzhang yu guojia* 《部长与国家》 [*Minister and Country*], chapter 10 (3), http://www.laiyangdianshang.com/2007/2007-01-12/47651.html (accessed 2015). The online version of this text is more extensive than the printed version.

Kang Shi'en to General Peng Dehuai to be the chief military representative of Yumen Oilfield. When Wang Zhen talked about "someone blowing cold air," insiders all knew that he was criticizing Petroleum Minister Tang Ke and Premier Zhao Ziyang.

4 The Fourth Round: Hu Yaobang and Zhao Ziyang Respond Differently to the Shengli Plan

On January 14, 1986, the Ministry of Petroleum formally submitted Shengli Oilfield's "Plan for the Second Daqing (第二个大庆规划)" to the CCP Central Committee and the State Council. This plan was carefully compiled and formulated over a period of nearly two years under the personal supervision of Kang Shi'en, who was then a member of the State Council. The plan was attached to a letter addressed to Hu Yaobang and Zhao Ziyang and written by Kang Shi'en.

On January 16, Hu Yaobang wrote his responding comments and instructions under the "Yaobang" section provided in Kang Shi'en's letter: "Please ask Ziyang to read and comment (it seems the plan is rather carefully thought out and can be ratified)."

On the same day, Zhao Ziyang wrote his response under the "Ziyang" section of the same letter: "I agree in principle. Please turn it over to the State Planning

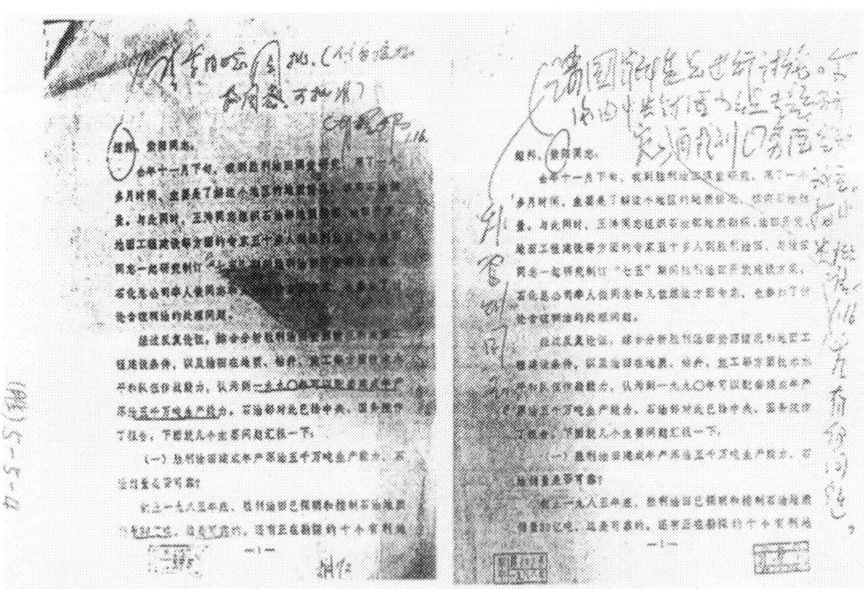

FIGURE 26.1 Hu Yaobang and Zhao Ziyang's handwritten instructions concerning Shengli Oilfield's Plan for a Second Daqing

Commission for discussion first, to see if it has any problems. Afterward, have the Central Committee's Leading Group for Financial and Economic Affairs to study it, and then have the State Council to agree on it and the CCP Central Committee to ratify it." From the copy of the actual letter, one can see that Zhao inserted the words, "I agree in principle," and, "to see if it has any problems," and circled them, after much consideration, after writing his original comment.

After this, the matter got shelved.

5 The Fifth Round: an Anonymous Letter and Zhao Ziyang's Instructions

In March or April of 1986, two or three months after Hu Yaobang and Zhao Ziyang wrote their instructions on the Plan, someone in Shengli Oilfield wrote a letter under the pseudonym Shi Weimin 史为民, complaining about serious problems with Shengli Oilfield's Commander Li Ye. The letter soon landed on the desk of Zhao Ziyang. Zhao immediately gave his instructions.

On April 26, Kang Shi'en spoke in Shengli Oilfield:

> The purpose of our trip to Shengli Oilfield this time is to carry out Premier Zhao's orders and investigate Shengli Oilfield's problem of undertaking construction projects without authorization, not according to the procedures for infrastructure-building; to investigate the oilfield's problem of failing to carry out the country's repeated orders to shrink its scale of capital construction; and to investigate how an organization such as Shengli Oilfield, which is under contract and has achieved success, disregards the country's regulation and goes its own way. These three layers of meaning, which are very clear in Premier Zhao's instructions, are the content of our investigation.

6 The Sixth Round: a Six-Hour Argument between Zhao Ziyang and Kang Shi'en

Between 1999 and 2004, Li Ye, the target of the pseudonymous Shi Weimin's letter, told me many times about a six-hour discussion that took place after Kang Shi'en saw the letter and Zhao Ziyang's instructions.

In brief, Kang discussed with Yu Qiuli whether he should let Li Ye see it. Yu said, "Let him see it. What the fuck are we afraid of?!"

A few days later, Zhao asked Kang over to talk. Kang took the complaint letter, with Zhao's instructions written on it, to see him. Zhao grabbed the letter and his instructions from Kang's hand and shouted: "How did this letter end up in your hands? I instructed that it not be passed down."

According to Kang Shi'en, this talk lasted from nine in the morning until three in the afternoon, with heated arguments.

The following is part of their conversation:

Zhao: "This Second Daqing Plan includes too many projects—the plan is too big."

Kang: "The Ministry of Petroleum has always been like this. Back then, Daqing, Shengli, Sichuan, and the Xinjiang Karamay campaigns were each separate, individual projects. The Chairman [Mao] and Premier [Zhou] immediately ratified them."

Zhao: "Some projects started without prior approval, against procedures for infrastructure-building."

Kang: "This plan has been under discussion for two years. Some of the projects have been reported to you—you never evaluate them, so I am always illegal. Hu Yaobang named the airport; Peng Zhen named the seaport." [At the time, Peng Zhen was Chairman of the Standing Committee of the National People's Congress.]

Zhao: "I don't care who named what. If I say no, it's no!"

Zhao: "This plan spends too much money, invests randomly ..."

Kang: "This plan does not ask a penny from the country. It's all self-funded. Some projects seem to be for local government, actually they're auxiliary projects for the development of the oilfield."

Zhao: "Old Kang, you take a few people to Shengli Oilfield to investigate."

Kang: "It is inappropriate for me to go and implement your instructions ..."

Zhao: "It is still better for you to go, I don't want to blow this matter up ..."

7 The Seventh Round: Li Ye Admits His Mistakes

On April 26, 1986, Kang Shi'en led the investigation group to Shengli Oilfield. During the managerial meeting held on April 27, the newly appointed Petroleum Minister Wang Tao read Zhao Ziyang's instructions. Afterwards, Kang Shi'en spoke:

> We must fully understand the spirit of Premier Zhao's instructions, seriously deal with the mistakes pointed out by Premier Zhao's instructions, be strict with ourselves, carefully investigate, and willingly investigate.

> As for the errors in the three projects of Dongying Airport, Guangbei Reservoir, and the Shaying-Dawang Highway, as they are with regard to Shengli Oilfield, the chief responsibility lies with Comrade Li Ye, so he should be investigated in earnest. The Ministry of Petroleum also has responsibility and should be investigated. As an investigator, I should also investigate myself, because I have either seen or heard of these projects. However, I did not stop them, so I need to do self-criticism.

On May 11, 1986, the investigative group wrote a formal report to Zhao Ziyang, with Li Ye's self-criticism, dated May 3, attached. Besides admitting "serious mistakes" and expressing the desire to learn "hard lessons," Li sincerely examined the cause of these mistakes:

> After the crude oil production contract policy was implemented, exploration and development happened relatively fast in Shengli Oilfield. Oil reserves were greatly increased, and crude oil production increased year by year. In addition, we were facing the task of new developments. Under such circumstances, a "many bones, little meat" mismatch between production and construction that had existed for many years in Shengli Oilfield became more pronounced, especially where water sources are insufficient, where water injection cannot keep up, where there is big underground voidage, where production in old oilfields is slipping, and where there are difficulties in developing new oilfields; road construction lags lead to constant traffic jams, causing difficulties in transporting materials and equipment; various construction projects in the oilfield are incomplete, and social service facilities inadequate. After the establishment of Dongying Municipality, the economic foundation of the three districts and three counties has been weak, with people living in poverty. Confronted with these urgent issues, I myself was concerned and worried, and I became impatient. In addition, crude oil production has increased a lot in recent years. I lost my cool and clear-headedness, on the one hand, by wanting to build the oilfield into another Daqing, and on the other hand, by wanting to develop the Yellow River Delta. I was eager to speed up the oilfield's development and eager to move people out of poverty and into wealth. Such urgent desire resulted in the deviation from the guiding ideology of my work....

A few who attended the meeting told me that Li Ye was sobbing when he delivered his self-criticism.

After the investigation group submitted its report on the Shengli Oilfield investigation to Zhao Ziyang, Zhao Ziyang gave the following comments and instructions: "I agree with the investigation report. You should carry on according to the suggestions given in the report. I also agree with the speech of Comrade Wang Tao and the self-criticism of Comrade Li Ye of the Ministry of Petroleum. Please distribute copies these three documents to comrades of the State Council and Secretariat, for perusal. May 12, 1986."

8 The Eighth Round: a Change of Shengli Oilfield Leadership

In January 1987, Hu Yaobang resigned. Yu Qiuli and Kang Shi'en both completely exited the scene of the oil industry. Late in the summer of 1988, Petroleum Minister Wang Tao ordered a change in leadership at Shengli Oilfield. First, Li Ye was relieved of all duties in the oilfield. In June 1989, Zhao Ziyang was removed from office because of the June 4th Incident. At the end of July 1989, the Ministry of Petroleum transferred Shengli's First Deputy Commander Zhang Wenyan 张文彦 to Henan Oilfield; those managers who had implemented the Second Daqing Plan were all relieved of duties, transferred, or given limited employment. A new group of leaders for Shengli Oilfield was organized. When Shi Xunzhi 史训之, the Director of the Cadre Bureau of the Ministry of Petroleum sought out Zhang Wenyan for a talk, he bluntly told him, "We're transferring you people because you're not on the same page as the Ministry."

More than ten years later, when Zhang Wenyan in his seventies recalled this episode, he was still angry. On May 11, 2000, in Henan Oilfield, he told me, "We were working together for a whole eight years, from 1981, when Shengli Oilfield was facing reduction in output, to 1988, and even though there were shortcomings and blunders here and there, these three achievements cannot be taken away from us: first, Shengli Oilfield's reserves doubled from 1.26 billion tons in 1981 to 2.6 billion tons in 1988; second, Shengli Oilfield's production increased from 17.58 million tons in 1980 to 33.3 million tons in 1988—a production increase in eight years of 15.72 million tons, or an average annual increase of around 3 million tons—including two years with increases of more than 4 million tons, which was an oil industry miracle; third, we truly cared about the workers' lives, solved at one stroke the problem of long-term separation between workers and their families, thus reducing their worries at home and stabilizing the troops. Why did those in power pay no heed to or ask about these things, but instead remove or transfer us all at once on the pretext that we weren't on the same page? Where would we go to argue?"

9 The Ninth Round: the Second Daqing Plan is Completely Discarded

The volume devoted to the history of petroleum industry (i.e., the Shengli Oilfield) in the *History of Shandong* series (*Shandong sheng zhi* 《山东省志》) drily reports:

> On July 30, 1989, CNPC decided to change the name of the Shengli Oilfield Campaign Headquarters to Shengli Petroleum Administration Bureau ... On November 3 to 7, 1989, [the new leadership (Deputy Petroleum Minister Zhou Yongkang, who doubled as Director and Secretary of Party Committee of Shengli Petroleum Administration Bureau)] held an expanded meeting of the Shengli Petroleum Bureau Party Committee's Standing Committee. At this meeting, they summarized the main achievements and experiences of the 25 years of the Shengli Oil Campaign, concretely analyzed main dilemmas in the oilfield's production and development, and made the important decision to transfer the work focus to oilfield development.[2]

That is to say, Zhou Yongkang announced at the meeting that the future focus at Shengli Oilfield would be "oilfield development," meaning they would abandon the search for more oil reserves. With this, the Second Daqing Plan, which had then been under way for more than five years, would be entirely discarded.

In 2001, Li Ye told me the following story. On the second day after Zhou Yongkang assumed his position as the chief of Shengli Oilfield, he went to visit Li Ye at his house. The two men had nothing to say to each other, and after a few minutes, Zhou stood up to leave. As soon as he left the house, Li blurted out, "Things won't end well for this fellow." Who would have thought that more than a decade later, Li Ye's prediction would come true?[3]

2 Shandong sheng difang shizhi bianchuan weiyuanhui 山东省地方史志编纂委员会, eds. *Shandong sheng zhi: shiyou gongye zhi* 《山东省志·石油工业志》 [*History of Shandong Province: History of the Petroleum Industry*] (Jinan: Shandong renmin chubanshe, 1996), 486, 722.

3 On June 11, 2015, Zhou Yongkang was sentenced to life in prison for bribery, abuse of official authority, and deliberate disclosure of state secrets. Once one of the members of the Politburo Standing Committee of the Party Central Committee, he was the highest official since the founding of the PRC to have been given a prison sentence for corruption.

10 The Tenth Round: Conclusion

Thanks to the Second Daqing Plan, Shengli Oilfield's annual production in 1993 reached its peak of 33.5516 million tons. The next year, it began a permanent decline.

As an observer, after years of pondering, I believe that this debate originated from the question of whether the reserves were proven. This is normal. No matter who is right or who is wrong, it should be an open and on-going debate. Work should be done even while the debate is going on, and adjustments should be made along the way. However, the people in power, like Yu Qiuli, would not listen to differing opinions and skirted the key issue of oil reserves in the debate, so that it became simply a high-level power struggle of who should be in charge of the oil industry. This harmed Shengli Oilfield—a lamentable outcome.

In addition, Zhao Ziyang's instructions only criticized Shengli Oilfield for not having followed procedures when carrying out three infrastructure-building projects, but he did not completely veto the plan. After Zhao's removal in June 1989, the newly appointed Petroleum Minister Wang Tao and First Deputy Petroleum Minister Zhou Yongkang used Zhao's instructions as an excuse to eliminate this plan. They then made a decisive turn westward and marched on a large scale into Tarim Basin to try to find the Second Daqing over there. This was a baffling move.

CHAPTER 27

Seven Situations and Severe Challenges

1 "Galloping the Horse on Little Hay"

Reviewing the history of the speedy development of Chinese petroleum, one can see that each of the industry's achievements has been tied to an increase in the country's investment and support. In the early 1980s, the annual national production plan of 100 million tons could not be secured. Starting in 1979, production at the pillar of the industry, Shengli Oilfield, began slipping annually at a rate of 1 to 2 million tons per year. As a result, the State Council took Kang Shi'en's advice and adopted the "100 million ton guarantee," which increased investment in the oil industry. It stabilized national annual production at 100 million tons and brought renewed growth. Shengli Oilfield saw an uptick in growth thanks to the policy.

This guarantee policy was sharply slashed in the mid-eighties. According to Liang Hua and Liu Jinwen:

> The state portion of the total national investment in the oil industry slipped from 41.9% in 1981 to 18.3% in 1987. Although the oil industry's self-accumulated funds through the guarantee [policy] rose from 58.1% in 1981 to 81% in 1987, even though that rate increased, the absolute amount of funding drastically decreased. In 1988, for example, according to the content and conditions of the guarantee, the oil industry could use the 137 million tons of oil produced that year to raise 12.6 billion yuan in exploration funds on its own, but in fact it raised only 4.78 billion yuan, a 62.1% decrease. This shows that the total investment in the oil industry was greatly reduced. Furthermore, the country through various means gradually increased taxes on the oil industry. In those years, cost of raw materials, electricity, and land requisition rose sharply, which further burdened the oil industry. For instance, in 1988, the industry used 16 billion kWh of electricity, but the country provided it an electricity quota of only 10.3 billion kWh, which meant that the industry had to self-fund the remaining 5.7 billion kWh. Then there was the cost of land. In 1988, land cost 2,496 yuan per *mu* [6.6667 metric acres], but three years later

 | DOI:10.1163/9789004402737_028

> it was 10,301 yuan per *mu*. Likewise, prices for other raw materials also increased.[1]

Thus, while investment decreased, payments to the government and expenses sharply increased. Under the planned economy system, this state policy of "galloping the horse on little hay" greatly harmed the oil industry.

2 "In Industry, Learn from Daqing" Becomes Past Tense

Due to the drastic reduction of state investment in the oil industry as well as the aftermath of the Cultural Revolution, the oil industry by the eighties already showed signs of fatigue: even though production increased in old oilfields, it was difficult to curb sharp production decreases in some; even though new oilfields were occasionally discovered, there were no major breakthroughs. Under these circumstances, the oil industry could not live up to the task of creating big production increases year after year. Thus the underlying message of "In Industry, Learn from Daqing," which was, "The national economy depends on the oil industry," inevitably became a thing of the past. The debate on the Second Daqing Plan was a sign of this trend, but Yu Qiuli and Kang Shi'en at the time were unaware of it.

3 The Ministry of Petroleum Fantasizes about Oil Campaigns

The Ministry of Petroleum specialized in "waging campaigns." From my personal experience, I know that all the decision-makers in the oil sector fantasized about successful oil campaigns. They could not get out of their heads the exceptionally grand success of the 1979 Huabei Oil Campaign, in which they dug up a "big golden baby" (that year in Huabei Oilfield, they established an annual production capacity of 10 million tons and produced 5.97 million tons). In the eighties, there were still quite a number of oilfields where they were keen on doing oil campaigns and digging up "big golden babies." A good example of this would be the massive march into Tarim Basin in the mid-eighties. Little did they know that within the context of the nascent market economy, a great deal of malfeasance would occur if you continued using the same oil

1 Liang Hua 梁华 and Liu Jinwen 刘金文, *Zhongguo shiyou tongshi, juan si* 《中国石油通史，卷四（一九七八-二000）》 [*A Comprehensive History of Chinese Petroleum, Vol. 4 (1948–2000)*] (Beijing: Zhongguo shihua chubanshe, 2003), 99, 137–39.

campaign model that had been used in a planned economy, or that greater effort would result in much less production.

4 Good News from Tarim Basin

In the eighties, finally good news arrived from Tarim Basin, after long-term efforts in oil and gas exploration there by the Ministry of Geology and the Ministry of Petroleum. The good news consisted of three main items. First, in 1983, the Ministry of Petroleum captured the gigantic 60,000 to 80,000 km^2 structure of Tazhong-1 塔中一号, located 5,000 m below the earth's surface in the depths of the desert. Second, in 1984, on the Yakela 雅克拉 Structure in the north part of the basin, the Ministry of Geology drilled a high-yield well that produced 1,000 m^3 of oil and 2 million m^3 of natural gas per day. Third, in 1988, on the Lunnan 轮南 Structure near Yakela Structure, the Ministry of Petroleum drilled the high-yield Lunnan-2 well. These three achievements brought excitement to the offices of the Ministry of Petroleum. All attention was on Tarim Basin.

5 Foreign Companies Take a Strong Interest in Tarim

In the latter part of 1985, Kang Shi'en led a delegation of a dozen oil experts to visit the United States. This coast-to-coast trip lasted a few weeks. Bosses from big American oil companies went after them like a flock of ducks, all wanting to discuss with Kang their desire to participate in the exploration and development of the Tarim Basin. Wu Xunduo 吴训铎, the Ministry of Petroleum's representative in the United States at the time, accompanied Kang through the whole trip as his interpreter. He told me, "These were all bosses from the world's top oil companies, like Exxon, Chevron, Texaco, and Amoco. Kang Shi'en warmly met with them, telling them that Tarim Basin was so big that it would be difficult for any single company to handle alone, but if they were willing to get involved, it would be better if a few of the companies were to team up. He suggested that they should come to Beijing to discuss the details ... He showed sincere willingness to bring in foreign investment."

6 Disagreements over the "Stabilizing the East and Developing the West" Policy

Given the circumstances discussed above, heated debates on the oil development strategy of the nineties began up and down throughout the Ministry of

Petroleum. Numerous oil geologists gave their opinions and proposed the policy of "stabilizing the East and developing the West (稳定东部，发展西部)."

As for how to implement this policy, there were two sets of completely different opinions and plans on two key issues. One was whether the focus of the nineties should be immediately transferred west to the Tarim Basin or kept in the east; the other was whether oil exploration in Tarim Basin should include foreign investment or be entirely domestically financed. In hindsight, the debates over these different plans were essentially manifestations of the planned economy versus market economy debate.

Yan Dunshi, the former Chief Geologist and Deputy Petroleum Minister, told me on October 13, 1999, of the plans of the seasoned oilman Kang Shi'en, who at the time was a member of the State Council: "Around 1985, Kang Shi'en held conferences attended by leadership and geologists in the oil industry to discuss oil development strategies for the nineties. He vividly summed up his thoughts. First we must first rely on our own strength, build up the oilfields in the East, and continue to stabilize and increase oil production in that area. Second, using foreign investment, we should wage a two-pronged attack on offshore oil and Tarim. If we fight these three battles successfully, by the end of the twentieth century we can achieve an annual production of 200 to 300 million tons." Kang Shi'en stated very clearly that in the nineties, the oil Industry's focus should still be on the old oilfields in the East, whereas foreign investments should be directed toward Tarim Basin. Using foreign investments would have to be carried out under the conditions of a market economy.

7 Kang Shi'en Recommends Wang Tao as Petroleum Minister

In 1985, Minister of Petroleum Tang Ke was still serving even though he was past retirement age, and the highest echelons of the leadership had been discussing for a long time who should succeed him. A close friend of mine, Li Bocheng 李伯诚, who was Secretary of the Discipline Inspection Commission at the Research Institute of Petroleum Exploration and Development, told me the following story back in 1986:

One day, Kang Shi'en was at home when he said, "The CCP Central Committee was discussing who should succeed Tang Ke for a long time and could not make a decision. At the time, Tang Ke recommended Zhao Zongnai, but I didn't agree. I recommended Li Ye, but Tang Ke didn't agree. Premier Zhao Ziyang said, 'In that case, we won't consider either. Old Kang, you should recommend a few others.' After some consideration, I recommended ten people to Premier Zhao, the fourth was Wang Tao. I said, 'Wang Tao studied geology in the Soviet Union. In 1965, he became Deputy Commander and Chief Geologist of Dagang Oilfield, later he was transferred to Liaohe Oilfield. In 1983, he became the General Manager of China Offshore Oil Nanhai East Corporation. This company was the first in China to carry out reform on a large scale.' At this point, Zhao interrupted my talk and said it would be him. 'You send someone to Guangzhou to investigate, and we'll settle it fast.'

Afterward, the Ministry of Petroleum sent Wei Xushun 魏绪顺, the Deputy Director of the Cadre Bureau, to Guangzhou to investigate. Another good friend of mine, Zou Jiazhi 邹家智, who was then the Secretary of the Party Committee of Nanhai East Corporation, told me, "We were all shocked when we heard the news! We reported to Wei Xushun all kinds of bad conduct on the part of Wang, but to no avail. Soon, his appointment was announced." I said, "Wei's investigation was only for show, for the common people—you shouldn't have taken it seriously." Zou was dumb-founded.

In the summer of 1985, the Ministry of Petroleum held a managerial meeting attended by more than a thousand people to introduce the newly-appointed Petroleum Minister Wang Tao. Kang Shi'en, who was over seventy years old, gave a long talk listing Wang Tao's accomplishments and placing great hope on Wang Tao. Sitting beside me was Zheng Hao 郑浩, the Director of the Supply Bureau. He ventured a comment to me: "Who knows if this fellow is good enough?"

As for Wang Tao's own response to these events, we can read about it in his voluminous memoir, the first paragraph of which leads off with the immodest statement, "The CPC Central Committee promoted me directly from a bureau director-level cadre to a formal, ministerial leadership position, and I became the Minister of Petroleum."[2] It seems that Wang Tao attaches the utmost importance to this promotion up two official levels.

The seven situations described above were real. Let us see how the decision-makers' decisions affected the rise or fall of the oil industry.

2 Wang Tao 王涛, *Zhengzhan siwang zhi hai: Talimu shiyou huizhan* 《征战死亡之海—塔里木石油会战》 [*Battling the Sea of Death: The Tarim Oil Campaign*] (Beijing: Zhonggong dangshi chubanshe, 2013), 2.

PART 6

The Tarim Era (1990–2000): Bad Decisions Lead to Zero Increase of On-Land Oil Production

In the fall of 1999 and summer of 2001, I visited Tarim Basin, each time for about a week. I visited Tazhong Oilfield, Lunnan Oilfield, and the Desert Highway, met and talked with dozens of workers, geologists, and managers, and gathered dozens of items of textual material. After returning to Beijing, I also consulted with relevant experts in order to gain a more comprehensive understanding of the whole oil and gas exploration process in Tarim Basin, especially during the 1986–1997 Tarim Oil Campaign. In accordance with Confucian teaching, I "transmit but do not create" (*shu er bu zuo* 述而不作), i.e., I pass down what is already there without producing anything new. Every story has its origin.

In short, because of faulty decision-making in the Tarim Oil Campaign, Chinese on-land oil production in the 1990s increased a meager 2 million tons, from 138 million in 1990 to 140 million in 2000. Hence, during these ten years, there was basically zero growth in on-land production, so that the Chinese oil industry, which had grown so quickly in the previous thirty years, was brought to an embarrassing halt.

CHAPTER 28

Origin of the Tarim Oil Campaign

1 The Allure and Risk of Oil and Gas Exploration in the Tarim Basin

The Tarim Basin, situated in the southern part of Xinjiang Uyghur Autonomous Region, has an area of 560,000 km². It is the largest basin in China and the second-largest in the world. Its total area makes up one-sixth of the nation's sedimentary rock area. According to two resource assessments by an authoritative national agency, its oil and gas resources total 19.15 billion tons. With such rich oil and gas reserves, this basin has long attracted the world's attention. In 1942, geologists led by Huang Jiqing explored the basin, found oil seepage in the northern Kuqa area, and predicted the possibility of finding an oilfield of a similar magnitude to those of Baku of the former Soviet Union and Venezuela's Maracaibo Basin. Tarim Basin inevitably became an important base for energy resources in western China. This spurred the drive to seek oil and gas there.

However, to find a high-yield oil and gas field, you need to take great risks. In the Tarim Basin, there were two major risks: one was the dangerous natural conditions above the ground, another was its complex underground geological conditions.

As for the above-ground conditions, the center of the Tarim Basin is dominated by the Taklamakan Desert. Taklamakan means "place of no return," and it has also been called the "Sea of Death (死亡之海)." With a total area of 330,000 km², it is also the world's second largest moving desert area, covered with constantly drifting yellow sand. During sandstorms, anything beyond two meters away disappears. It receives less than 100 mm per year of precipitation, but the equivalent of 2,500 to 3,400 cm of water vapor per year evaporates. Summer temperatures reach 38°C, while winter temperatures drop to −20°C. To circle the perimeter of this desert you must travel three or four thousand km. To cross it from north to south, you must travel five or six hundred km. The whole basin is sparsely populated, but along its edges are settlements with residents from many different ethnic groups. It has always been economically backwards. Since it is far from China's traditional domain to the east and far even from Wulumuqi, the capital of Xinjiang Uyghur Autonomous Region, roads were bad and communication was poor. All of these posed tremendous difficulties and risks for oil exploration here, compared with the old oilfields in the east.

 | DOI:10.1163/9789004402737_029

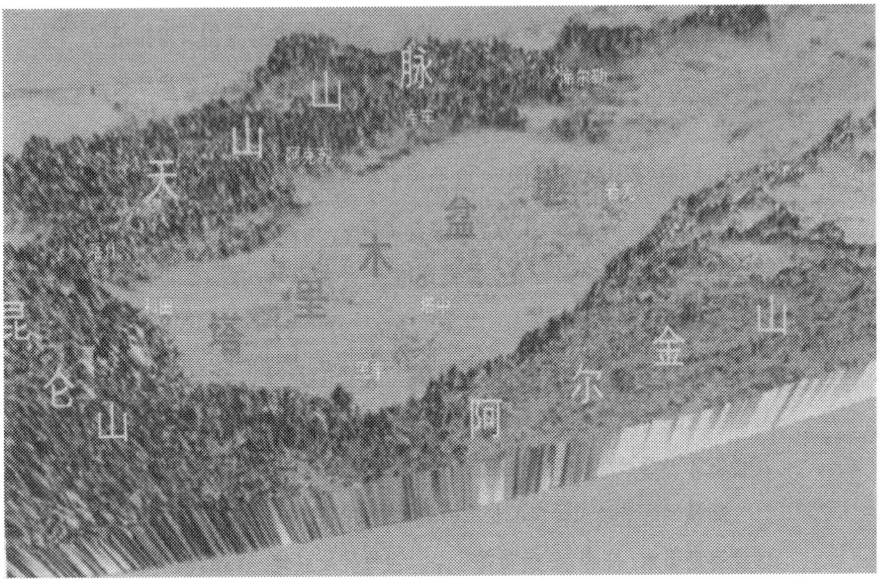

FIGURE 28.1 Tarim Basin, 560,000 square kilometers

As for conditions underground, one must first keep in mind that its vastness encompasses great geological variety. Due to tectonic movements, there are many geological faults and intermixed marine and continental strata, with some strata missing or eroded, and some lacking good cap formation. In addition, the cap throughout the sedimentary basin is very thick (between more than 5,000 m and 16,000 m). Moreover, little work had been done here in the past. In other words, finding oil and gas fields within this 560,000 km^2 area, at more than 5,000 m underground, is like "finding a needle in a hay stack." Easier said than done! Such is the risk!

2 The Hard Work of the Ministries of Geology and Petroleum

The Ministry of Geology started working in this "treasure bowl" in 1955. Headed by the first Geology Minister Li Siguang and Chief Geologist Huang Jiqing, the Ministry of Geology spent decades working in this basin. However, it was after 1977 in particular that a group of experts headed by geologist Kang Yuzhu 康玉柱 had some success. They stepped up exploration in the Kuqa area, in the northern part of the basin. On August 12, 1983, drilling began at Shacan-2 沙参2井 on the Yakela structure. On August 23, 1984 at 5,391.2 m, the well had a strong blowout, and high-yield oil and gas flow was achieved.

The work of the Ministry of Petroleum in the region can be traced back to 1952, when Sino-Soviet cooperation started. By 1986, over a span of 34 years and under the leadership of Xinjiang Petroleum Bureau, they had made six exploration attempts in the basin and entered and exited the Sea of Death nine times, with no major breakthroughs. On July 3, 2001, former Deputy Petroleum Minister Li Jing 李敬 told me, "I was appointed chief of Tarim oil exploration in 1978, and I was already the fifth chief since the 1950s. For decades, thousands of oil workers deep in the desert, braving untold dangers, quietly contributed. The number of people who died just because of production accidents, getting lost in the desert, or being killed by wolves was close to a hundred ..."

On August 18, 1958, the six members of the 113 Geological Team of the Xinjiang Petroleum Bureau were conducting geological investigations at various locations in the northern part of the basin when they were trapped and engulfed by a sudden, rare torrential rain, the magnitude of which had not been seen in a hundred years. When their bodies were found in the valley that the flash floods had raged through, some still had geological records clutched in their hands. The team leader, Dai Jian 代健, was a talented female graduate of the Dept. of Geology at Xibei University (Xi'an), and the names of the other members, all young, were Li Yueren 李月人, Chen Jieping 陈介平, Li Naijun, 李乃君 (female), Yang Xiulong 杨秀龙, and Zhou Zhengtu 周正涂. Soon, an oil rig was erected at the place where they had died in the line of duty. The same year, the first oilfield in Tarim Basin, Yiqikelike Oilfield (依奇克里克油田), was established. Afterwards, oil workers named this valley Jian Ren Valley (健人沟), after Dai Jian and Li Yueren, and erected a stele to commemorate their fallen colleagues.

3 Outcomes of the Ministry of Petroleum's Exploration of Tarim Basin (1952–1988)

During this period, the following things were accomplished. Ministry geologists first succeeded in acquiring an initial understanding of Tarim Basin by gathering large amounts of geological data, which established the basis for further geological surveys. For instance, between 1952 and 1954, Xinjiang Petroleum Bureau dispatched 13 geological crews to carry out surveys in Kuqa and Kashgar, as well as four drilling crews to dig seven exploratory wells. Between 1958 and 1963, it engaged 16 drilling crews and 21 geological crews. Between 1964 and 1969, nine seismic crews finished 2,395 km of seismic line, and ten drilling crews drilled 32 exploratory wells. In 1985, the Petroleum Geophysical Exploration Bureau alone sent out nine seismic crews and finished 4,989 km of seismic line and 3,099 km of gravity coverage. Even though vast human, material, and financial resources had been invested in this basin over a more than thirty-year period, the main focuses had been in the Kuqa area along the basin's northern edge and in the Kashgar area, on the basin's southwestern border.

Secondly, the Ministry explored and developed Yiqikelike Oilfield. Xinjiang oil workers had been exploring in Yiqikelike, the area in front of the mountains in Kuqa, starting in 1952. It was in this area that the six survey crew members lost their lives in 1958. On October 9, 1958, a huge oil blowout occurred at 468 m at Yiqikelike-1. From its discovery in 1958 until termination of drilling in 1987, the oilfield produced 957,900 tons of crude oil, contributing to the economy and frontier development of southern Xinjiang.

Third, Kekeya Oil and Gas Field 柯克亚油气田 was also explored and developed. In 1975, after long-term cooperation with the Ministry of Geology, the Ministry of Petroleum decided to strengthen its work in Tarim Basin, focusing on the Kashgar area in the basin's southwest. In May of 1976, the drilling of Kekeya-1 in Yecheng County began, and on May 17, 1977, a strong blowout took place at 3,783.1 m. In the beginning, it produced 1,300 m^3 of crude oil and 260 m^3 of natural gas per day, and it sustained the pressure and amount of production for one month. This type of high-yield, high-pressure well was rare in Xinjiang at the time. This was the initial well of Kekeya Oil and Gas Field. In later years, Xinjiang Petroleum Bureau explored and developed this high-yield, high-pressure oil and gas field, completing its construction in 1986. By the end of the twentieth century, Kekeya Field had cumulatively produced 3.2 million tons of crude oil, 7 billion m^3 of natural gas, and 20,000 tons of liquified petroleum gas, greatly contributing to the development of southern Xinjiang. This was especially important for the economic development of the remote and impoverished region of Kashgar.

Fourth, Zepu Petrochemical Plant (泽普石化总厂) was built. In August 1983, after presiding at the Western China Petroleum Exploration Meeting in Karamay, Kang Shi'en, the State Councilor in charge of the petroleum industry, accompanied by Song Hanliang 宋汉良, the Vice Chairman of Xinjiang Uyghur Autonomous Region, visited southern Xinjiang and Kekeya Oil and Gas Field, then still in development. After his visit, Kang proposed the following to the CCP Central Committee and the State Council: (1) developing Kekeya Oil and Gas Field into a field with annual oil production of 150,000 tons, and building an oil refinery that could process that much oil annually, in order to meet the oil needs of Kashgar, Hotan, and Tibet's Ngari; (2) building a plant for producing 60,000 tons of synthetic ammonia per year, since the oil and gas field produced 200 million m^3 of natural gas per year, and the plant would help to alleviate the poverty in rural Kashgar; (3) building a plant producing 10,000 tons of LPG per year, using the oilfield's light crude to partially satisfy the need for fuel in a few cities in southern Xinjiang. The CCP Central Committee and the State Council granted Kang Shi'en's proposal, and Zepu Petrochemical Plant was completed in September 1990. At the end of the twentieth century, with its annual production reaching more than 2 million tons, this petrochemical plant had become "a golden flower" among the industries in the southwestern part of Tarim.

Fifth, a base for future oil exploration was also built in Tarim. After more than 30 years of effort in the area, oil workers had developed a broad, solid foundation to help them achieve their goal of finding a big oilfield in Tarim Basin. For example, they had formed broad friendships with different ethnic groups in southern Xinjiang; they had gained a deeper understanding of the local terrain and landforms and accumulated geological information and valuable experience in finding oil and gas in the basin; they had trained skilled petroleum exploration troops. In strategic areas they had created necessary infrastructure, such as roads, electricity, communication facilities, and houses. Sizable oil bases had been established in Kuqa County on the basin's northern border and Yecheng County on its southern border. All these accomplishments would play important roles in the later oil campaign.

4 Three Achievements Set Off Heated Debate

During the 1952–1988 period, the Ministry of Geology and the Ministry of Petroleum achieved three goals of special significance.

First, in accordance with the oil exploration plan jointly laid out in December 1981 by experts of the Ministry of Petroleum and the Ministry of Geology, in January 1982, the Ministry of Petroleum decided to import advanced foreign technology and cooperate with the Geophysical Service Inc. (GSI) of the United States; it bought American equipment and hired American technical personnel. It then formed three desert seismic crews, two of which had American personnel, with people who came primarily from the Third Division of the ministry's Petroleum Geophysical Exploration Bureau. In June 1983, they marched into the hinterland of Taklamakan Desert to conduct large-scale systematic seismic explorations. By the end of 1984, 20 seismic crews were engaged. They crossed the desert from south to north 13 times and from east to west nine times. They put down 11,310.83 km of seismic line and 28,038.6 km of gravity line, finished 19 large seismic cross sections totaling 5,782.2 km, conducted gravity surveys along the lines, and completed 1:200,000 aeromagnetic surveys. They captured the gigantic Tazhong-1 Structure, which lies 5,000 m below the surface and has an area of 6,000 to 8,000 km^2.

Second, in January 1978, the Ministry of Geology dispatched crews to Tarim Basin. One crew of experts, headed by geologist Kang Yuzhu, after a few years of careful research and full preparation, had a strong blowout at Shacan-2 on the Yakela Structure in Kuqa on August 23, 1983 (see Section 2 above). The well was producing more than 1,000 m^3 of crude oil and 2 million m^3 of natural gas per day, but a big fire of unprecedented strength and duration soon broke out. It was finally extinguished on November 6, 1987, three years and 45 days from the date it began. Once the fire was put out, the blowout was under control. This was a momentous event in Tarim Basin oil exploration.

Third, in February 1986, the Ministry of Petroleum held the Tarim Basin Petroleum Exploration Deployment Meeting, with Chief Geologist Yan Dunshi presiding. Based on the findings of the Petroleum Exploration Bureau and the good news about the Ministry of Geology's Shacan-2 well, the Ministry decided to drill exploratory wells at Lunnan Structure, in the southern part of Luntai County and not far from Yakela Structure. Of the five exploratory wells drilled, three obtained commercial oil flow or showed oil and gas. This was especially true of Lunnan-2, which generated high-yield oil flow on November 25, 1988.

FIGURE 28.2 Oil exploration in the "Sea of Death"

FIGURE 28.3 Geology crew leader Kang Yuzhu and Shacan-2 well (1985)

These three achievements, though exciting, were no more than a few drops in the ocean in terms of oil exploration in the immense Tarim Basin. To find actual oil and gas fields with commercial value is not an easy matter! Heated debates broke out on two issues within the Ministry of Petroleum and the higher echelons of the government. The first issue centered on whether to shift focus immediately from the east to Tarim. The second issue was whether to get foreign help or to carry out the oil campaign on our own.

In May 1979, the American oil tycoon Armand Hammer visited China. At the Great Hall of the People, while meeting with Deng Xiaoping, he expressed his interest in having his Occidental Petroleum work in Tarim. Deng gladly accepted. Kang Shi'en attended this meeting.

On July 24, 1985, when CCP Central Committee General Secretary Hu Yaobang visited Karamay Oilfield in Xinjiang for the third time, he spoke at a meeting:

> We have consistently held that petroleum on land should be opened up to foreign companies as well. Xinjiang is so big. We need to be resolute in getting foreign investment for petroleum exploration. Comrades Xiaoping, Chen Yun, Ziyang, and I all hold this view .... Capital has been accumulated overseas for a few hundred years, and it has been accumulated in the United States for 200 years. We have had only 30 some years—not to mention the mistakes made during this time. You cannot solve the funding shortage problem in two or three years, and the Ministry of Petroleum cannot solve it, either. The only thing we can do is to try solving it together and coming up with different approaches, including that of getting foreign investment.

Kang Shi'en also held the same opinion and laid out his ideas to the Party Central Committee in a proposal entitled "The Petroleum Industry Development Strategy for the Nineties (九十年代石油工业发展战略)," the content of which was summed up by Kang's biographers:

> First, what should we do with our country's eastern region? Kang Shi'en believes that old oilfields in the east, including Daqing, Liaohe, Shengli, Huabei, Dagang, and Zhongyuan, are the base of the petroleum industry and the focus of exploration and development. Its crude oil production is 70% of the whole nation's production. A stable production policy should be adopted. Secondly, what should we do with the western region? Kang Shi'en says that the western area is vast, rich in resources, and has good prospects for exploration in the Tarim Basin. At the same time, [however,] this area is a desert, conditions above ground are bad, and conditions underground complex, so that the difficulties and risks are great. It will inevitably take a relatively long time, large amounts of money, and advanced technology and equipment. Based on the experiences and outcomes of foreign cooperation in offshore petroleum exploration, Kang Shi'en proposes we should open up to foreign companies ... attract foreign investment to do it ...[1]

1 Kang Shi'en zhuan bianxie zu, ed., *Kang Shi'en zhuan*, 533.

CHAPTER 29

Wang Tao Decides: Go All Out in Tarim without Foreign Investment

1 Shifting Focus to the West

Meanwhile, there were other, diametrically opposing opinions. In the fall of 1985, Wang Tao, the newly-appointed Minister of Petroleum, was of the opinion that "the focus should be shifted to the west." Early in 1986, at a Ministry of Petroleum managerial meeting of petroleum enterprises, he argued, "The development of oilfields in the east has passed its prime—we must turn our line of vision to the west."[1]

2 The Ministry of Petroleum's Report on Foreign Investment

In Ministry of Petroleum Document No. 15, dated January 7, 1988, Wang Tao formally submitted to the State Council "A Report on the Question of Opening Xinjiang Tarim Basin to Foreign Investment" (关于新疆塔里木盆地石油对外开放问题的报告), which began:

> Per the instructions of Acting Premier Li Peng, we have carefully analyzed and researched the issue of opening petroleum exploration in Xinjiang Tarim Basin to foreign investment. We believe that the work of petroleum exploration and development of this region is very different from offshore petroleum exploration. In the beginning, we had neither the money nor the technology for offshore, whereas we have already done many years of foundational work in the northwest, possess considerable capability in seismic exploration and drilling in the desert, and can self-fund the exploration. Once we find oil and gas, we can still borrow foreign money as needed for development. Therefore, we should do the exploration and development of Tarim Basin ourselves.

The report first talked about the "significant and good" prospects of finding a big oilfield in Tarim Basin. Then, it gave an account of 22 foreign petroleum

1 Bainian shiyou bianxie zu 百年石油编写组, ed., *Bainian shiyou* 《百年石油》 [*One Hundred Years of Petroleum*] (Beijing: Shiyou gongye chubanshe, 2002), 356.

 DOI:10.1163/9789004402737_030

companies from seven countries (Kang Shi'en had made contact with some of these companies during his US visit), all of which had expressed deep interest in cooperating with China on the Tarim Basin. Based on a detailed analysis of the proposals from some of these world-famous petroleum companies, Wang Tao remarked, "The purpose of their cooperation with us is to find in Xinjiang the opportunity to participate on large, high-yield oilfields, so as to make up for the production decline in these corporations' current oilfields." He then drew these three conclusions:

(1) Foreign companies' key objective in proposing cooperation is to occupy territory and control resources.

(2) According to the foreign companies' cooperation proposals, it will be very difficult to increase production in the near future.

(3) The costs of cooperation with foreign companies on exploration and development are very high. Even if oil is found, we have little to gain.

In the end, the report clearly states, "We urge the State Council to agree with the opinion of our Ministry on the issue of foreign cooperation in Tarim Basin, make a decision, and through an appropriate medium announce it internationally." Subsequent developments in the oil industry followed the recommendations set forth in this report. Oil and gas exploration in Tarim Basin was entirely "done by ourselves."

3 Skirting Kang Shi'en and Scrapping a Foreign Investment Contract

Beginning in 1986, many rounds of negotiations had already taken place between the Ministry of Petroleum and a certain foreign oil company. On August 19, 1988, in the Diaoyutai State Guest House, the parties agreed in to sign the first foreign venture capital contract to do geophysical exploration in Tarim. Kang Shi'en, as State Councilor, was to preside over the ceremony. While Wang Tao and Kang Shi'en were riding in the same car to Diaoyutai, Wang Tao took out a document and showed it to Kang Shi'en to read. It was the aforementioned Ministry of Petroleum Report to the State Council that recommended doing the Tarim Basin oil exploration ourselves without foreign participation. The report had been ratified by Premier Li Peng, with his signature and instructions. Kang was shocked when he saw it and said only, "Since this is so, I don't need to go." Wang Tao then hurried to the Guest House by himself and announced that the ceremony was cancelled. In this way the first contract to use foreign investment to conduct geophysical exploration in Tarim Basin was destroyed by Wang Tao, and the Ministry of Petroleum lost credibility with foreign investors.

CHAPTER 30

The State Council Ratifies the Tarim Oil Campaign

1 Premier Li Peng's Speech

On December 18, 1988, China National Petroleum Corporation (CNPC, formerly the Ministry of Petroleum), headed by Wang Tao, submitted to the State Council its "Report on Accelerating Petroleum Exploration in Tarim Basin (关于加快塔里木盆地油气勘探的报告)." According to Wang Tao's autobiography, on January 13, 1989, when the State Council held a meeting in Zhongnanhai's 4th Conference Room, Premier Li Peng listened to Wang Tao's report and ratified it. At the end of the meeting, Premier Li Peng said, "In terms of Tarim oil exploration and development, our general principle is: rely mainly on our own troops, explore oil on our own, and develop the oilfield on our own.... In the next few years, your staff doing the work of exploration and development in Tarim Basin should follow the CNPC's report."[1]

On November 23, six months after the campaign had started, Premier Li Peng visited Tarim. After hearing Wang Tao's report, he said, "In the future, the development of Chinese oil industry mainly depends on Tarim ... The strategic transfer of oil industry from the East to the West is a major decision for the development of our national economy."

2 Tarim Oil Campaign Headquarters Established

In April 1989, Tarim Oil Campaign Headquarters was established in the city of Korla 库尔勒. Zhou Yongkang, CNPC's First Vice President, became the campaign commander and Secretary of the Party Committee. At the meeting to celebrate its establishment on April 10, Wang Tao announced, "This is an important, strategic move that will greatly increase our country's natural gas production in the near future and our annual crude oil production by the end of this century. Whether this battle can be successfully fought will decide, on a large scale, the rate of development of our nation's oil industry." With this, the self-financed oil campaign started full-scale at a few key locations in Tarim Basin.

1 Wang Tao, *Zhengzhan siwang zhi hai*, 107.

 DOI:10.1163/9789004402737_031

3 Oil Workers and Equipment Transferred to Begin the Campaign

From 1986 to 1989, the Ministry of Petroleum (or CNPC) had successively issued mobilization orders to key petroleum enterprises nationwide. Heeding that call, oil workers traveled day and night to the Tarim Basin from faraway oilfields such as Daqing, Shengli, Huabei, and Zhongyuan, as well as from Sichuan and Karamay. By 1991, the "Sea of Death" had 24,830 workers and 45 large-scale rigs, including 36 6,000 m rigs, eight 7,000 m rigs, and one 9,000 m rig. In terms of speed, dedication, the advanced level and scale of equipment used, and duration, this oil campaign surpassed all others that had taken place in China up to that point, and there were few parallels for it in world petroleum history.

CHAPTER 31

Seven Achievements of the Tarim Oil Campaign

Based on extensive research and information gathering, seven achievements of the Tarim Oil Campaign from 1989–1997 may be listed as follows:

1 Proving the Existence of 200–300 Million Tons of Oil Reserves

On December 20, 1997, Zhou Yongkang claimed, "In eight years, thanks to the efforts of all campaign employees, we have achieved splendid results … Mainly: exploration has led to the discovery of a group of oil and gas fields, and the completion of 160,000 km^2 of 2D seismic lines and more than 8,000 km^2 of 3D seismic lines; 294 exploratory wells were drilled, with a total drilling depth of 1.58 million m. Collectively we have proven and probable reserves of 550 million tons, of which oil reserves are 370 million tons, and natural gas reserves are 180 billion m^3."[1]

Usually, however, the "probable reserves" that Zhou Yongkang talked about are normally not included in the count. On March 25, 1997, Tarim Oil Campaign Chief Geologist Liang Digang 梁狄刚 said: "In eight years of campaign, Tarim cumulatively had proven oil reserves of 248 million tons and condensate reserves of 41.75 million tons, totaling 290.1 million tons."

2 Establishing an Annual Production Capacity of 4 Million Tons

A group of medium and small oilfields with an annual production of more than 4 million tons was developed. According to information provided by Tarim Oilfield Research Institute geologists in a September 2, 1999 conference:

> As of the end of 1998, Tarim had 250.91 million tons proven reserves in place, had developed eight oilfields with a cumulatively built production capacity of 4.7 million tons, and was producing 4 million tons of oil annually. The oilfields are mainly situated at Lunnan Low Uplift 轮南低凸起, Tazhong Low Uplift 塔中低凸起, and Halaha Depression 哈拉哈凹陷,

1 From the text of a Ministry document. Zhou's calculation here uses the approximation of 1,000 m^3 of natural gas for 1 ton of oil.

 | DOI:10.1163/9789004402737_032

forming three oilfield groups of Lunnan, Tazhong, and Donghetang 东河塘. Among the eight developed oilfields, the four main force oilfields have an annual capacity exceeding 400,000 tons, i.e., Lunnan Oilfield, Donghetang Oilfield, Tazhong-4 Oilfield, and Tazhong-16 Oilfield. In 1998, these eight oilfields' total oil production amounted to 3.92 million tons.[2]

3 Purchases of Advanced Equipment and Technology

The Tarim Oil Campaign purchased and made use of a large quantity of advanced equipment and technology. At the time, bank loans in US dollars were available. Each year, tens of millions of dollars were allocated for the procurement of advanced equipment for geophysical exploration. Advanced equipment and technology were also purchased for geological research, drilling, well logging, well completion, production testing, downhole operation, etc. A large number of gigantic desert transportation vehicles were purchased from abroad each year. As Confucius once said, "A worker who wants to do a good job must first sharpen his tools." These purchases greatly improved the equipment being used and played an important role in the oil and gas exploration and development in Tarim.

4 Adoption of New Technology

With its new advanced equipment, the campaign could develop and apply new technology. For instance, in the area of geophysical prospecting, the technology used to conduct seismic exploration in desert areas covered by shifting sand dunes was unique in the nation. Its image logging technology was also on the leading edge. The technology it developed to drill wells more than 6,000 m deep quickly was foremost in the nation.

5 Contractual System of Administration

Learning from the industry's offshore experience, the oil campaign, from the start, contracted out its various projects to other companies in the oil industry.

2 From a two-page handwritten summary of Tarim Basin achievements presented in a conference at the Petroleum Hotel, Korla; the conference was attended by the Tarim Oilfield Research Institute's deputy chief geologist, head geologist, and other geologists.

The campaign headquarters of Tarim Oilfield, acting as an investment and management center, did not form any construction workforce of its own. It only set up production management and a research institution, which together employed only a few more than 4,000 people. Various national oilfields would bid on respective projects. At the height of the campaign, more than 20,000 people worked under these projects. These workers worked provisionally and left Tarim upon completion of their assignment. They did away with the previous practice whereby wives and children would accompany workers to campaign sites. No doubt, this was a step forward.

6 Support of Local Development

The campaign actively supported local development, helping different ethnic groups to rise out of poverty. According to formal statistics, in nine years the campaign poured 11.66 billion yuan into Xinjiang Uyghur Autonomous Region, an amount equivalent to 49.4% of the total investment of the campaign. From this amount 400 million was given to the region gratuitously, and 4 million was donated to China Youth Development Foundation's Hope Project. It jointly built ten roads with Xinjiang, including a highway through the desert, totaling more than 1,000 km. It invested 22 million yuan in the electricity project in Dongsixiang, Luntai County and 24 million yuan in the Hotan and Aksu liquified natural gas stations. It spent 70 million yuan on railroad construction in southern Xinjiang and another 6 million yuan on earthquake relief, etc. Thus the Tarim Oil Campaign made important contributions to Xinjiang's economic and social development. For example, Bayingolin Mongol Autonomous Prefecture's revenue grew from 70 million yuan in 1988 to 560 million yuan in 1998. Korla, the capital of the prefecture, used to be rundown and backwards, ranking last in all of Xinjiang. With the help of the petroleum industry, it became one of the best cities in Xinjiang, second only to Wulumuqi, the capital. It was also named as a national model city for sanitation.

7 Proved Natural Gas Reserves

Finally, the campaign discovered extensive natural gas reserves. Most of the gas reserves were proven in 1994 and 1996. Yingmaili 英买力 Gas Field was discovered in 1992, and its reserves of 31 billion m^3 were proven in 1994. Yaha

牙哈 Gas Field was discovered in 1993, and its reserves of 35.8 billion m^3 were proven in 1996. Yangtake 羊塔克 Gas Field, discovered in 1994, had reserves of 30 billion m^3, which were proven in 1996. In addition, this long oil-finding campaign laid the foundation for discovering other natural gas fields in the new millennium.

CHAPTER 32

Falling between Two Stools

To provide balance to our investigation, we must also consider the mistakes of this campaign. As a retired oilman, I only point out these faults in the hope that such a discussion will prevent them from being repeated. I shall limit my critique to the decision-makers of the campaign. I have no words of criticism for the other levels of leadership—the geologists, engineers, technicians, or laborers, whose hard work and dedication moved me. We will never take for granted their blood and sweat. That said, looking for oil and gas is a risky business. Mistakes and blunders are hard to avoid. However, we should face reality and draw lessons.

1 Falling between Two Stools

Kang Shi'en used the Chinese idiom *liangtou luokong* (两头落空) or "falling between two stools" to describe one of the mistakes. His biography records:

> In the proposal "Petroleum Industry Development Strategy for the Nineties," Kang Shi'en made clear for the first time that due to the extremely tight budget of the oil industry at the time, if the limited funds were devoted to the exploration and development of Tarim Basin in the west, stable yield from important oilfields in the east might be negatively affected. Exploration in the Tarim area might not be successful in the short term, and there was the danger of "falling between two stools." Kang Shi'en was most concerned that this situation might occur, bringing adverse consequences for the national economy.[1]

Unfortunately, Kang Shi'en's prediction of "falling between two stools" came to pass in the Chinese oil industry in the 1990s. Concentrating on Tarim Basin did not reap immediate benefits and harmed stable oil production in the established eastern oilfields.

In 1990, China produced 138 million tons of petroleum on land (excluding oil production by the Ministry of Geology and Mineral Resources), but by 2000,

1 Kang Shi'en zhuan bianxie zu, ed., *Kang Shi'en zhuan*, 535.

 | DOI:10.1163/9789004402737_033

that figure had only increased to 140 million tons, meaning that the cumulative increase of ten years was only 2 million tons, averaging out to 200,000 tons annually. That is to say, in the last ten years of the twentieth century, there was basically zero growth in on-land oil production. The Chinese petroleum industry, which had been growing quickly for thirty years, had skidded to an embarrassing halt.

2 Failing to Find a Big Oilfield in Tarim Basin

On March 6, 1997, in his speech "Tarim Campaign: A Brief Retrospective (塔里木会战的简要回顾)," Zhou Yongkang, the general manager of CNPC said:

> In retrospect, through the eight years of the campaign, we implemented a series of large-scale strategies. When we started in 1989, we based our work on the 11 big profiles and earlier work by Xinjiang Petroleum Bureau. Building on the success of the Lun-2 well in Lunnan area, we focused on big structures. We first threw ourselves onto the Lunnan Uplift, and at the same time we prepared to tackle the Tazhong Uplift, which is a massive 8,000 km^2 site. At the time, we had big dreams … We were still successful in the first round, in the Lunnan region. Meanwhile, under very difficult circumstances and after detouring a few thousand kilometers, we entered Tazhong and drilled Tazhong-1, which yielded an abundant flow of oil. We had been excitedly focusing on big structures, anticipating a find on the scale of Daqing, and this was our first big hit. But what followed was frustration. The breakthrough at Lunnan, including the Ordovician system at Lunnan-8, occurred when we drilled into a cavity, and it produced high yields of oil. Drilling in surrounding areas yielded nothing. After Tazhong-1 produced oil, another well drilled nearby produced nothing. The second series of attempts took place in the Carboniferous system at Donghetang, where we drilled at the sandstone stratum that yielded in excess of 1,000 tons per day … As a result of this breakthrough, the breakthrough at Donghetang, we thought we found a big oilfield. But after searching around, it turned out that there was no big breakthrough after all. We ran into many such setbacks. We were getting excited over what turned out to be nothing. In the third of a series of attempts, we discovered Tazhong-4, where we obtained 108 m of oil sand, super-thick … This oil sand was full of oil, not like ordinary oil sand, so this caused quite a stir. But when testing was done, only the top 20 m or so tested oil, below that there was only water. Who would have thought that there was no oil,

> only water, under such a thick oil-bearing formation?! At the time, it was really giving us a lot of frustration.[2]

We can see from Zhou Yongkang's speech, made after Wang Tao had retired at the end of 1996, that this campaign failed to find any big oilfields in this basin.

To summarize Zhou Yongkang's speech, Tarim Oil Campaign went through four phases, with no big oilfield found in the end. The first phase: they hastily tackled the area around Lunnan-2, but a group of exploratory wells did not produce oil. The second phase: attempts in Tazhong also failed—it was like seeing the moon in a well and trying to fish it out. The third phase: efforts in Donghetang mostly failed. The fourth phase: In Lunnan, at the Paleozoic Era erathem (Ordovician system), with the exception of a few exploratory wells that showed oil and gas, most wells had none.

3 Mature Oilfields in the East Failed to Maintain Stable Yields

To amass the large amount of funds needed for the Tarim Oil Campaign, the Ministry of Petroleum either applied for loans or diverted the limited funds from the eastern oilfields' "compensated use fees on reserves (储量有偿使用费)," the "Exploration and Development Fund (勘探开发基金)," and so on; other funds that came from the profits off oil sales that were targeted for reinvestment got channeled instead into Tarim. The result was that investment for those eastern oilfields was drastically slashed, and it negatively impacted the production and development of all oilfields, including Daqing.

On October 13, 1999, Yan Dunshi, the Ministry's former chief geologist, told me, "Since the mid-1980s, there has been a severe lack of funds for the oil and gas industry on land in our country. The outside factor was that the state greatly reduced investment in it. Inside the industry, appropriating the East's limited funds and redirecting it to the West, going all out in Tarim, resulted in the cancellation of many projects in the eastern oil and gas fields. The production of the eastern oilfields dropped dramatically, and 'stabilizing the East' became just empty talk. For instance, in the East, projects to look for new oil

2 Zhou Yongkang, "Talimu huizhan de jianyao huigu 塔里木会战的简要回顾 ["The Tarim Campaign: A Retrospective"]. A five to six-page typescript of this speech, prepared from a recording, was given to me on August 30, 1999, during a conference at the Petroleum Hotel, Korla; the latter conference was attended by the Planning Department of Tarim Oilfield Company, the deputy chief geologist, the vice-director of the Petroleum Exploration and Research Institute of Tarim Oilfield Company, and six or seven other geologists, who described for me the outcomes of the Tarim campaign.

and gas fields were cancelled, and many projects aimed at expansion, increasing production, and gauging potential were halted. Even simple repair projects to maintain production were shelved due to lack of money. At the time, there were 1,050 land-based rigs nationwide. About half of them ceased production during those ten years [1990–2000]. Out of the 170 rigs in Shengli Oilfield, 90 stopped production, and out of the 70 rigs in Dagang Oilfield, 50 stopped production. Because drilling stopped, hundreds of thousands of oil workers were laid off. Tons of equipment became idle. Rigs worth close to 20 million yuan each lay idle in the fields, exposed to wind and rain. After a few years, they became nothing but piles of scrap metal. We estimated that 35,000 wells could have been drilled during these ten years. Fewer wells were drilled, so naturally oil production decreased—and these things happened during a long period when oil 'supply could not meet demand.'"

Here is an example of the consequences of diverting funds away from the eastern oilfields. The exploration and development of Henan Oilfield in Nanyang, Henan Province began in 1970. After more than ten years of effort, by the mid-eighties, it reliably and profitably produced 2.5 million tons of oil per year. This small but rich oilfield became the envy of the oil industry. After the start of the Tarim Oil Campaign, CNPC required every oilfield to raise funds on their own and send troops to Tarim to the areas contracted for exploration. Thus Henan Oilfield also contracted an area. The first exploratory well produced oil, whereupon Wang Tao urged people to follow the Henan Oilfield's example. With Wang's encouragement, Henan Oilfield over time poured more than 800 million yuan (CNPC also invested 800 million yuan) into Tarim, in an effort to find an oilfield with annual production of 1 million tons. However, things went contrary to their wishes. At the end of the twentieth century, annual production was only some 300,000 tons. On May 12, 2000, when I talked with a few retired executives of Henan Oilfield, their anger lingered, saying that if they had used the 800 million yuan in expanding their own oilfield, results would have been much better.

CHAPTER 33

Too Much Investment, Too Much Waste, Too Few Results

1 Input and Output Ratio

On December 19, 1998, after his retirement, Wang Tao spoke in Tarim, "During the ten years of the Tarim Oil Campaign, the CNPC gave us 13.5 billion yuan, and the Party Central Committee gave us a loan of 1.2 billion US dollars, which was equivalent to 10 billion yuan—an investment of roughly 23.5 billion yuan. We achieved approximately proven plus probable reserves of 800 million tons and built an annual production capability of 4.2 million tons." According to a formal Tarim Oilfield report provided to me on August 30, 1999, "Until the end of 1998, the cumulative investment in Tarim was 23.59 billion yuan." Based on the above data, Tarim Oilfield's general ledger was an investment of 23.5 billion yuan, proven reserves of 200 to 300 million tons, and built annual production capacity of 4 million tons. We need to point out that probable reserves (*kongzhi chuliang* 控制储量) historically have not been included in the balance sheet. Experts consider "probable reserves" to be "window dressing."

Now let us compare this with other initiatives:

Tahe Oilfield: This was originally managed by the Ministry of Geology and Mineral Resources in Tarim Basin and is located only twenty to thirty km away from the CNPC's Lunnan Oilfield. Conditions for both oilfields were the same. Exploration time was the same. This oilfield, towards the end of the twentieth century, had 197 million tons of proven reserves and an annual production of 2.47 million tons. It is now the largest oilfield in Tarim Basin. Its total investment was 6.42 billion yuan. By comparison, the Tarim Oil Campaign's investment was twice as much.

Tuha Oilfield: This oilfield, adjacent to Tarim Oilfield, is situated in Turpan Basin in Xinjiang. Its above-ground conditions in some cases are worse than Tarim Basin's. Oil and gas exploration started here in the 1980s. After more than ten years of hard work by people from Yumen Oilfield, an annual production of 3.5 million tons was achieved. The investment in Tuha was 12 billion yuan. By comparison, the investment in the Tarim Oil Campaign was 1.6 times that of Tuha Oilfield.

Gudong 孤东 Oilfield is an oilfield in the Yellow River estuary discovered in the 1980s by Shengli Oilfield. From focused exploration to full-scale

 | DOI:10.1163/9789004402737_034

development, it took only two or three years to bring it up into production. In 1986, it borrowed 150 million yuan from the bank for development. Two years later, it had an annual production of 5 million tons and had fully paid off the loan. The comparison is even more shocking! The investment in the Tarim Oil Campaign was 18 times that of Gudong Oilfield.

What is more, the investment figures usually cited in connection with Tarim do not include all expenditures. For instance, they do not include the CNPC's direct expenditures during the campaign on its Geophysical Prospecting Bureau or on the transportation company for the purchase of those giant imported trucks suitable for the desert; they do not include the cost of constructing the long-distance pipeline between Korla and Shanshan, the Korla Petrochemical Plant, the high-quality 522 km desert highway, and so on. Plus many other oilfields from around the country went to Tarim Basin with their own funds and their own equipment. If we add up all the accounts—above-the-table, under-the-table, highway and byway, not-fully-accounted-for and missing, as well as funds raised by other oilfields—the total investment in the Tarim Oil Campaign would be much higher than what Wang Tao declared.

2 Spending Huge Amounts of Money for Nothing in Tazhong

The name "Tazhong" means "middle of the Tarim Basin": the middle of this big desert of sand and sand dunes that cover hundreds of thousands of square kilometers.

FIGURE 33.1 A truck convoy in the "Sea of Death" (1988)

FIGURE 33.2 The movable desert airport runway in Tarim (1990)

On September 3, 1999, Zhong Jianmin 种建民, the Deputy Chief Geologist of Tarim Oilfield, told me that CNPC had gone all out in Tarim Basin, with its main target Tazhong, this "land of treasure." This assessment can be verified in a CNPC bulletin dated November 26, 1988, entitled, "Tarim Basin Exploration Achieves Important Breakthrough (塔里木盆地勘探又获重大突破)":

> The Corporation intends to transfer forces from throughout the nation next year to appraise and drill in the Lunnan area where an important discovery has been made, to gain control of the oil-bearing areas and reserves. We will focus on the 6,700 km² Tazhong-1 Gigantic Structure in the hinterland of Taklamakan Desert in order to find large oil and gas fields. We will speed up drilling in potential structures and venture to make major discoveries over the next two years, so as to bring about a new surge in reserves in the oil industry.

The so-called "Tazhong-1 Gigantic Structure" is that 6,000 to 8,000 km² "big fellow" jointly discovered in cooperation with the US Geophysical Service Inc. in 1982.

Starting in 1986, while the Ministry of Petroleum was going all out in Lunnan Area, it was also preparing to take on Tazhong. Toward this end, it spent huge amounts of money, buying many large desert trucks in the European, American, Japanese, and Russian markets as well as two large mobile airplane runways from the PLA Air Force. Each runway was 40 m wide and 800 m long. The news that China had found a big oilfield in the hinterland desert of Tarim Basin in Xinjiang shocked the world.

After preparation, in June 1988, six desert bulldozers and 14 large desert trucks spent 24 days clearing a transport route through the desert. At the same time, high-quality mesh steel boards were used to build a mobile desert airport. Then in August 1988, a rig capable of drilling 7,000 m was moved, as was its 20 m x 20 m x 118 m reinforced concrete rig substructure. Dozens of desert trucks, carrying a thousand tons of equipment and using a temporarily-built pontoon bridge, crossed Tarim River. After driving hundreds of km, they reached Tazhong-1 well in the middle of the desert. Sparing no money, wasting no time, they moved into position.

As it turned out, even though the first well, Tazhong-1, had a high yield in the beginning, there was a series of accidents, and in the end it only produced water instead of oil. They continued drilling other wells. Due to the lack of adequate seismic data or comprehensive geological studies at the time, no precise well positions could be provided, so at one go, they drilled a whole series of exploratory wells such as Tazhong-2, Tazhong-9, Tazhong-5, Tazhong-3, Tazhong-8, Tazhong-17, Tazhong-18, Tazhong-4, and Tazhong-16. Only Tazhong-4 and Tazhong-16 produced oil and gas. Each of these 5,000 m to 7,000 m wells cost close to 100 million yuan. The drilling cost alone for all these wells ran into the billions of yuan.

On October 13, 1999, Yan Dunshi, the former chief geologist of CNPC told me, "The cost of drilling a super-deep well, one that is more than 5,000 m deep, in the middle of the desert in Tarim, is extremely high. Just to make a big rig substructure in the desert would cost around 15 million yuan, almost a hundred times more than it would cost in the oilfields of the east. Vegetables had to be frequently flown into the desert. One *jin* (half kg) of tofu cost 40 yuan."

Most of the super-deep exploratory wells in Tazhong-1 Structure yielded nothing. The reason for this could be attributed to the fact that the seismic survey lines jointly done with the American company on this 6,000–8,000 km^2 "big fellow" were too sparse. In addition, the target stratum was broken up and fragmented. This big structure, like "the bright moon in the water," was not workable. Under such circumstances, they should have used geophysical methods, laying more seismic survey lines to solve the problem. Instead, they directly drilled, thereby increasing the cost dozens of times over. At last, the 40 km^2 Tazhong Oilfield was built, based on the Tazhong-4 and Tazhong-16 wells; it had 80 million tons of reserves and an annual production of 1.5–1.8 million tons. With its annual production at less than 2 million tons, the cost for the exploration and development of this small oilfield, the key oilfield in Tarim Basin, was the highest among similar oil and gas fields in China.

I need to clarify that all of these investment figures are approximate, because precise statistics and accounting were lacking at the time. However, when I list a number, I tend to list conservatively. I would rather list a few hundred million yuan less than list a penny more. Otherwise, I lose the foundation of my thesis.

3 Tazhong's Abandoned Airport and Fleet of Imported Trucks

On August 30, 1999, we rode a Japanese SUV a very long distance straight to Tazhong Oilfield in the middle of the desert. As we approached the oilfield, the first thing we saw was a desert field where large desert trucks were parked. There were around 70 or 80 of them—German, Japanese, and Russian. There was an 800 m x 40 m mobile runway as well as a big pile of high-quality mesh steel boards intended for another runway—all abandoned.

It was very quiet there. It had once been the location of boisterous activity. One could imagine the hubbub: a lot of traffic, the air filled with sand, dust, and truck exhaust. A few years later, the costly desert highway was built. But the climax of Tazhong Oilfield exploration and development had already passed. Maintaining the oilfield's normal operation needed only 40 or 50 people. Ms. Hu, the manager who was in charge of importing these desert trucks, sighed, "Now we have the desert highway. Those desert trucks I imported that year are all idle over there, badly damaged by the sun and wind. What a pity!" A desert truck driver we met on site told us that the giant German Benz tow truck that he drove was valued at 3.8 million yuan. It cost 7 yuan per ton per km to use, whereas a regular truck only cost 0.3 yuan per ton per km. They had been rarely used in recent years, only two or three times in a month. All year round, the trucks just remained parked there. They would become piles of rusted iron in a few years. What a waste!

4 The Two Gains and Three Losses from Building the Desert Highway

CNPC's leaders Wang Tao and Zhou Yongkang insisted on building a high-quality 522 km highway through the Taklamakan Desert. They did so with complete confidence of finding a large oilfield, despite the lack of any in-depth appraisal or proof that would give some certainty of success. Obviously this

FIGURE 33.3 The 522-km Desert Highway (1995)

met with strong opposition. The road started in Luntai County at the northern border of the desert, crossed Tazhong-1 Structure in the middle of the desert, and stretched all the way to Minfeng County at the southern border. Since there were no precedents for building a highway in a desert in China, Wang and Zhou first asked for a short study to be conducted. Before the design was fully tested, however, they were already building an extra-thick highway foundation through the dunes at great cost, as well as sand fortification nets more than ten m wide along both sides of the foundation. These nets were built of rhombus-shaped reed bundles, each approximately two m^2 in size, which were sunk 40 to 50 cm in the sand, with 20 to 30 cm remaining above. Countless reed bundles were used. To build this highway, construction workers lived and ate on-site in holes dug in the sand, braving the extreme heat of the summers (38°C) and extreme cold of the winters (–20°C)—for a number of years. The road was completed in 1994. Its high cost, large scale, complexity, and the hardships endured by the construction workers were rare in the history of road construction in China.

I rode on the highway twice, once in the fall of 1999 and then in the summer of 2001. The asphalt road was of high quality, meandering like a giant black dragon in the boundless sea of yellow sand dunes. Occasionally, a strong wind would blow, leaving patches of yellow sand on the road. Every three or five minutes, a car would pass by from the other direction. Fewer than two hundred vehicles a day would travel on the road, and among these, only three or five vehicles were servicing Tazhong Oilfield. According to our young driver, there used to be a toll station, but it was abandoned because the toll collected did not even cover the salary of the toll collectors. Since the road was situated amidst drifting sand, it cost a lot of money to maintain annually. Even so, there was no guarantee that the road would not be buried or cut off in the near future by moving sand.

I think that there were three losses and two gains from building this road. The three losses were: high investment, zero benefit, heavy burden. The two gains were: the world's longest stretch of highway through drifting sand and a modern, paved one at that.

5 The Torch of Tarim Oilfield Burned Night and Day

In September 1999, we stayed overnight at the employee dormitory of Tazhong Oilfield. Some 200 m away, a giant flare, approximately 50 m high and 10 m wide, was burning furiously. Its crackling could be heard more than ten km away.

These flares were there because during the oil-gathering process, the campaign's other auxiliary engineering projects were lagging behind, and the natural gas and light oil produced at these wells could not be transported—they could only be burned off! Thirty years before this, the Ministry of Petroleum had strictly forbidden the wasteful behavior of burning off natural gas and light oil in newly developed oilfields, but this wasteful practice went on in both Lunnan Oilfield and Donghetang Oilfield.

In the summer of 2001, I was informed formally by the deputy manager of the Tarim Oilfield Production Office that every day 2 million m^3 of natural gas and light oil were being burnt off by Tarim Oilfield. An expert told me that with that a fertilizer plant could produce 450,000 tons of synthetic ammonia and 600,000 tons of urea annually.

6 Construction of a 500 km Pipeline

On October 20, 1999, Zhang Jinquan 张金泉, the head of the Geology Department of the Research Institute of Petroleum Exploration and Development told me, "In July 1994, the Tarim Oil Pipeline Hearing was held at Shengli

Hotel, in Beijing's northern suburbs. Deputy Director Ye Qing 叶青 of the State Planning Commission presided over the meeting, which was attended by experts in the field. I also attended. Zhou Yongkang from CNPC also brought a group of his people, and he insisted on building the pipeline. Experts from the Commission disagreed, saying that Tarim had not found a sufficiently large oilfield to justify building a pipeline, so that this project should be delayed. One senior expert asked, 'In the beginning, we told you not to build a desert highway, but you said that wouldn't do; if the road were not built, it would affect finding oilfields. Now that the highway has been built, how many new oilfields have you found?' At the end of the meeting, Deputy Director Ye of the State Planning Commission asked Yan Dunshi, the Corporation's Chief Geologist, to speak. Yan Dunshi said that judging from Tarim's current exploration results, future prospects, and comparisons of existing railway transport with pipeline transport, the pipeline should not be built at the present time. Deputy Director Ye said, 'A few of that railway's branch lines are being built especially for transporting oil. Since they are near completion, I agree with Comrade Yan Dunshi's opinion. This pipeline should not be built.' This hearing foiled the Corporation's plan."

However, the Corporation still launched the pipeline's construction in 1996, and after three years of construction, the Korla-Shanshan long distance pipeline, with a capacity of 5 million tons per year, was completed. When it went into operation, it basically caused the already completed railway branch lines to go idle.

7 The Waste of Searching for Oil at Lunnan

The search for oil in the Paleozoic erathem (the Ordovician carbonate rock) in Lunnan was a prime example of disregarding financial consequences and being too wasteful. Once the Lunnan-1 well produced oil from the Paleozoic erathem in May 1988, Wang Tao advocated going all out, but some geologists believed that there was insufficient seismic data to warrant this. If one drilled into the deeper Paleozoic erathem too hastily, the losses might be huge, given that drilling a single well deeper than 6,000 m could cost as much as 100 million yuan. They urged caution. But Wang Tao would not listen. His only basis was that "all wells in the world with a daily output of 10,000 tons come from carbonate rock oilfields." According to the second and third chapters of his memoir, in 1989, 22 wells were drilled over the course of three cycles, and even though there were strong signs of oil and gas, only three wells produced oil, and that soon rapidly decreased, so they failed to "pass the test satisfactorily."

Thus when General Secretary Jiang Zemin 江泽民 visited Tarim on August 23, 1990, he dedicated a work of calligraphy, which read: "Hope you work hard, turn in your test with a passing grade as soon as possible." This was one of the instances described as a setback. Six or seven years later, another exploratory well produced oil at the Lunnan Paleozoic erathem, so another six exploratory wells were drilled in 1998, producing "less than ideal results." However, based on the decision-makers' "new geological understanding," between 2000 and 2008, another 35 exploratory wells were drilled.[1]

Wang's memoir also records that by 2004, an oilfield with 60 million tons of reserves and an annual production of 258,000 tons was found at the Lunnan Paleozoic erathem.[2] Finally, after 16 years, the expenditure of billions of yuan, and the drilling of 50 to 60 deep wells, only this pocket-sized oilfield was discovered. Lunnan's oil production is only one-twentieth that of Tarim's current production of 4 million tons. Was it worth it? What has been the return on investment?

8 The Flight of the Tazhong-1 Core

Page 350 of the 2001 proof copy of the book, *One Hundred Years of Petroleum* (*Bainian shiyou* 《百年石油》) included the following passage:

> At 00:00 [midnight] on September 26 [1989], a core dripping with oil was obtained from the Tazhong-1 well. It was perhaps a coincidence that exactly thirty years to the day earlier, Songji-3 well had started to produce oil, which marked the discovery of Daqing Oilfield. That very night, the oil-bearing core was flown to Daqing on a chartered plane, for on September 26, the 30th Anniversary Celebration of Daqing Oilfield was being held there. General Manager Wang Tao, holding the core, announced to the meeting: "In the hinterland of the Taklamakan Desert, in Tarim Basin in the northwest, we have obtained an oil-bearing core from Tazhong-1 well." The gathering burst into prolonged applause.

Wang Tao was the chief consultant of the group that wrote the book.

Now in order for General Manager Wang Tao to hold that core and win the "prolonged applause" from the audience, he spared no expense. He chartered a special plane to fly a core weighing less than a kilogram thousands of

1 Wang Tao, *Zhengzhan siwang zhi hai*, 125–139, 246–311.

2 Ibid., 314.

kilometers from the northwest all the way to Daqing! Was this appropriate? However, for some unknown reason, when the book was published in 2002, this splendid account of how "the oil-bearing core was flown to Daqing on a chartered plane" and about "General Manager Wang Tao holding the core" was deleted.

As the saying goes, "Whatever one is doing, God is watching." God, in this case, is the tens of thousands of oil workers who took part in the Tarim Oil Campaign.

CHAPTER 34

Overstating Achievements and Experiences

1 Every Year a New Slogan, Every Year an Exaggerated Target

Between August 31 and September 2, 1999, I talked at length with two senior geologists at Tarim who shared with me their recollections of the past. They said that after the campaign started, Minister Wang Tao would always be sure to preside in person over the annual geological exploration forum. At each of these meetings, he was always passionate in his speech, whether he was making lengthy reports or freely chiming in with comments on others' speeches, and it was always about how a big oilfield would definitely be found in Tarim. From 1988 onward, he would produce a new slogan every year, as shown below:

1988: "Transfer 40,000 people, quickly install 100 rigs, struggle a few years, obtain 1 billion tons of reserves, establish 20 million tons' annual production" (the details were spelled out in a Ministry of Petroleum report dated September 1, 1988).

1989: "Struggle a few years, pull a second Daqing in Tarim" (with annual production of 50 million tons).

1990: "During the eighth Five-Year Plan Period [1991–1995], build an annual production capacity of 20 million tons" (see the Ministry of Petroleum Report dated July 9, 1990).

1991: "Target '855'" (that is, 800 million tons of reserves, 5,000 people, and 5 million tons of annual production).

1992 and 1993: "With one hand get the big picture, with the other build an oilfield producing 5 million tons per year." By "the big picture," he meant securing reserves of 1 billion tons and an annual production of 20 million tons.

1994: "Work hard for three years, make the big picture a reality."

After 1995, he proposed that Tarim currently was on "the eve of the big picture," that the "the dawn light of the big picture" was already visible, that we were already "approaching the big picture," "closing in on the big picture," "realizing the big picture" …

In 1996, before Wang Tao retired, he still proposed achieving an annual production of 8 million tons. Someone suggested that if that included both oil and gas, it was possible, but Wang Tao still insisted that natural gas should not be included.

At the end of 1996, CNPC's slogan was: "In the next decade, the general guiding principle for Tarim is to use three years to clarify the oil and gas big picture."

 | DOI:10.1163/9789004402737_035

Such seriously unrealistic slogans went on unchanged for more than ten years. Mr. Zhong, the Deputy Chief Geologist of Tarim Oilfield, told me, "In those years, year after year for more than a decade, at the beginning of each year, Minister Wang would encourage us by saying, 'Our Tarim Oil Campaign is now in the 'darkness before the dawn,' and at the end of each year, Minister Wang would say we had to 'listen where it is silent for the sudden clap of thunder.' But we have never seen the first light of dawn, nor have we heard any sudden clap of thunder!"

2 An Exaggerated Report about Oil in Lunnan

The Corporation's report, "On the Development of Oil Exploration and Future Deployment in Tarim Basin (关于塔里木盆地石油勘探进展及今后部署的报告)," dated July 9, 1990, listed three achievements. The first of these was, "We have proven that there are three main oil-bearing series in the Lunnan region of Tarim Basin. We have obtained approximately 1,000 km² of oil-bearing area with either high-yield oil and gas streams or good hydrocarbon shows." Wang Tao's report was submitted to the relevant department of the State Council. The State Planning Commission was deeply concerned about it and dispatched Du Yonglin 杜永林, a petroleum geologist and the Division Chief of its Chemical Engineering Division, to investigate in Tarim. On March 6, 2003, Du Yonglin told me, "When I went to Tarim, I read each well's data one by one and had many conferences with geological personnel. After more than one month's work, I thought that the only ones that truly showed oil and natural gas were those two 'reference wells' on the Lunnan structure—Lunnan-1 and Lunnan-2. But since they are so far apart and isolated, they cannot be considered as being in one area." Therefore, this division chief reported the truth: at that time, no 1,000 km² oil-bearing area had been discovered in Tarim. Li Ye told me that afterward, Zhou Yongkang, the Corporation's Vice President, went to see Ye Qing, the Deputy Director of the State Planning Commission, demanding that Du Yonglin be punished. Ye Qing refused.

3 Another Exaggerated Report

On October 13, 1999, Yan Dunshi told me that in August 1990, the CCP Central Committee General Secretary Jiang Zemin, accompanied by the Party Secretary of Xinjiang Uyghur Autonomous Region Song Hanliang, came to inspect the Tarim Oil Campaign for the first time. Wang Tao also came. On August 23, they held a conference at the Tarim Oil Campaign headquarters in Korla, and Song Hanliang invited a manager from the Ministry of Geology and Mineral Resources to attend. During this conference, Wang Tao reported, "In terms of our battle in Lunnan, we have secured a whole oilfield with 100 million tons of reserves ..."

That manager from the Ministry of Geology and Mineral Resources immediately interrupted him: "We have been working here for more than ten years and are more familiar with the underground situation. As far as we know, up until now, a whole oilfield with 100 million tons of reserves has not been found in Tarim Basin ..."

Afterward, on the train ride back to Wulumuqi, when Jiang Zemin asked Wang Tao about the matter of the oilfield with 100 million tons of reserves at Lunnan, Wang Tao answered ambiguously, "The underground situation is relatively complex, and the situation does change ..."

Song Hanliang had been an oilman himself—he worked in Xinjiang for decades as a petroleum geologist. As an old acquaintance of Yan Dunshi, he told Yan Dunshi this story.

4 Showing Off Tazhong-1 to the Whole World

The 2002 edition of *One Hundred Years of Petroleum* records the following:

> On November 2 [1989], five days before the Fifth Plenary Session of the 13th CCP Central Committee, CNPC held a news conference in Beijing, announcing to the whole world: "While drilling the first exploratory well, Tazhong-1, on Tazhong Structure in the hinterland of the desert in Tarim Basin, an oil-bearing stratum was found contiguously for 117 m. A test midway [in drilling] produced 576 m^3 of crude oil and 360,000 m^3 of natural gas per day." This news shocked the whole country, and it also shocked the international oil industry.[1]

As for the "good news" of Tazhong-1's 576 m^3 per day crude oil output, Wang Heng 王恒, head of the Production Management Division of the Tarim Oil Campaign headquarters, told me on September 1, 1999, "On the eve of National Day, Tazhong-1 well, which was in the process of being drilled, showed oil and gas. Wang Tao immediately ordered that this exploratory well, originally meant

1 Bainian shiyou bianxie zu, ed., *Bainian shiyou* (Beijing: Dangdai Zhongguo chubanshe, 2002), 362.

FIGURE 34.1 Tazhong-1 gushes oil (1989)

to test for oil, be put into production, so that the good news could be in time for the report to the CCP Central Committee. At first, Wang Tao complained the output was too little and kept demanding that the choke be opened up. The dispatcher on duty at the Production Management Division was on the phone day and night relaying Minister Wang's orders. At last, on November 2, they reported the good news of more than 500 m^3 daily output. However, this 500-plus m^3 daily output not only contained a lot of water, it was not the actual output per single setting." Another senior person at Tarim told me in August 2001, "At the time, the oil-gathering only took two or three hours, which was used as the basis for estimating the daily output."

Soon after the good news was reported, the borehole walls of Tazhong-1 collapsed and water gushed from the bottom of the well. The well would not produce another drop of oil. Later, it took more than one year and a cost of 100 million yuan to repair the well and deal with the accident. The well did not finish its geological mission, nor did it produce much oil, so the well to this day still sits abandoned in the middle of the desert—and several surrounding exploratory wells that were drilled at the time all turned out to be dry holes as well.

Was it appropriate for Wang Tao, as a decision-maker, to use an estimated daily output figure, based on only two to three hours of oil-gathering, to amaze the world with "good news"? Was it appropriate to announce this well as "good news" when it soon became worthless? And was it appropriate to continue doing so a dozen years later, by including it, without inhibition, in his cherry-picked history *One Hundred Years of Petroleum*?

5 Exaggerations about the "New System" and the "New Technology"

In his memoir *Battling the Sea of Death: The Tarim Oil Campaign* (*Zhengzhan siwang zhi hai: Talimu shiyou huizhan*), Wang Tao writes at length describing the "new system," i.e., the system of contracting oilfields and companies within the Corporation for specific projects (甲乙方项目管理), as the main achievement and most valuable lesson of the Tarim Oil Campaign. I disagree that it is. I believe that any talk of the advantages of the so-called "new system" without the yardsticks of "market" and "benefit" is empty bragging. At best, this "new system" may be treated as a relatively good lesson, similar to Daqing's "job responsibility system." Overemphasizing its importance can only arouse skepticism. Furthermore, if this "new system" were already so perfect, that would make the systems at China National Offshore Oil Corporation and the Ministry of Geology and Mineral Resources' Tahe Oilfield near Lunnan Oilfield only second-rate.

Here are some comparisons. Tarim Oilfield, which Wang Tao had so bragged about, had an annual production of 4 million tons, engaged more than 4,000 employees, and had its base on location in Korla. China National Offshore Oil Corporation's Nanhai Gas Field produced 3.5 billion m^3 of gas, the equivalent of 3.5 million tons of oil; it had around 80 to 90 employees, and was based in Zhanjiang, Guangdong. The Ministry of Geology and Mineral Resources' Tahe Oilfield's had an annual production of 2.5 million tons, employed 40 to 50 people, and had its base in Wulumuqi, the capital of Xinjiang.

We should also discuss the "new technology" of direction or slant drilling, which has undergone very rapid development in this century, with the United States taking the lead. Around 2000, China in this regard was still at the beginning stages. In 1999 and 2001, when I visited the Tarim Oil Campaign showroom, there was a directional well exhibit. The well was not deep, and the horizontal section of the well was not long—it was totally experimental. However, *Battling the Sea of Death* presents this as having reached "advanced international standards." Anyway, if one brags behind closed doors like this, none of those who are locked in can compare the claims against the outside world!

6 Bragging and Unrealistically High Targets Plague the Oil Industry

In its record of the Tarim Oil Campaign, Wang Tao's *Battling the Sea of Death* uses the word "big" or "large" (*da* 大), and its related compounds (*weida* 伟大, *zhongda* 重大, etc.) throughout—"the big Tarim Oil Campaign," "big new oil campaign," "the big picture," "great scientific practice," "important strategic significance," "especially large gas field," "big green oil and gas field," and so on. It makes constant use of phrases that emphasize importance, so that the campaign becomes:

> "A heroic undertaking (*yingxiong zhuangju* 英雄壮举)"
> "A glorious chapter (*huihuang pianzhang* 辉煌篇章)"
> "An historic breakthrough (*lishixing tupo* 历史性突破)"
> "A strategic breakthrough (*zhanlüexing tupo* 战略性突破)"
> "An historic step (*lishixing de yibu* 历史性的一步)"
> "A new page turned in history (*lishi xiekai le xin de yiye* 历史掀开了新的一页)"
> "A heroic deed on the order of creating the universe (*kaitian qidi de zhuangju* 开天辟地的壮举)"
> "A sea of oil and gas (*youqi zhi hai* 油气之海)"

When I read all of this, though, I really felt a bit of deja-vu. I recall that in the beginning of the Daqing Oil Campaign in 1960, the word "big" or "large" (*da*) was also used everywhere: "the big campaign," "large study," "big commendation," "big appraisal," "big basic training," "the big tour of underground palaces" (studying geology). In October 1963, in preparation for the CCP Central Committee Northeast Bureau Meeting held in Daqing, Kang Shi'en and Wu Xingfeng believed that the word "big" should be used as sparingly as possible, and even more so for words such as "great (*weida* 伟大)" and "glorious (*huihuang* 辉煌)" for describing work accomplished. I remember Wu Xingfeng, in particular, asking me to take down the wooden sign I had written, "Political Department of the Big Songliao Oil Campaign," and change the wording to "Political Department of the Songliao Oil Campaign." At the Daqing Exhibition in Beijing in early 1966, Kang Shi'en and Wu Xingfeng strictly forbade the use of their photos. This shows that they were much more cautious in those days. I guarantee that words such as "big campaign," "great," and "glorious" were no longer used in the major documents coming out of Daqing.

Looking for oil is a very risky business. I remember that before 1977, none of the oil campaigns in the country had specific targets when they were starting out. However, as "In Industry, Learn from Daqing" spread, in 1977, at the National "In Industry, Learn from Daqing" Conference, Kang Shi'en came up with the concept of building "a dozen Daqings by the end of the twentieth century," i.e., reach annual production of 500 million tons. All the oilfields followed suit; each wanted to get one or two Daqings, that is, become an oilfield with an annual output of 50 million tons. A precedent was set. No wonder that Tarim

Oil Campaign had a new slogan and new target every year! To be fair, boasting and setting high targets was not unique to the Tarim Oil Campaign. It was a common practice in the industry, and the only area in which Wang Tao outdid Kang Shi'en. Furthermore, the oil industry's problems in this regard were also a byproduct of the state-planned economy, because by talking big and setting high targets, they could then apply for state approval for their projects, as well as obtain more state investment.

CHAPTER 35

Other Issues Relating to the Tarim Campaign

1 The Big Gas Field at Kuqa Did Not Result from This Campaign

In his memoir, Wang Tao insists that the big gas field found at the beginning of the new century in Kuqa Area was the result of this campaign. This is not true. On September 1, 1999, geologists of the Exploration and Development Research Institute of Tarim Oilfield Company told me in Tarim that the previous year (1998), they had discovered high-yield natural gas at the Kela-2 Structure in Kuqa and that they were working hard, hoping to achieve greater results. The construction of the West-East natural gas pipeline from Tarim to Shanghai began in 2002, after the big gas field in Kuqa was secured. The pipeline was finished and put in use at the same time as the big gas field at the end of 2004. These facts show that these two engineering projects were all directed and implemented by the Tarim Oilfield Company, after the headquarters of Tarim Oil Campaign had ceased to operate. They were achievements of that company, not the Tarim Oil Campaign!

If the Campaign had indeed discovered the gas field, then Xinjiang Petroleum Bureau, the Ministry of Geology and Mineral Resources (renamed the Ministry of Land and Resources in 1998), as well as Sino-Soviet Petroleum Company would all be claiming it as their achievement. The Sino-Soviet Petroleum Company, which was established in 1952, once found and developed a small gas field at Kuqa, the same area where the big gas field was discovered in the new millennium.

2 Refusal of Foreign Investment Based on Subjective Views

I have described how, on January 7, 1988, the Ministry of Petroleum submitted a report to the State Council recommending the refusal of foreign investment. The Ministry insisted on taking on Tarim by itself and cited three reasons for refusing foreign investment: (1) Foreign companies took the initiative to offer cooperation with us, because they wanted to "occupy territory and control resources." In other words, we Chinese should not let them realize their nefarious schemes but instead protect our national rights and interests. What is puzzling is that in 1993, the Ministry decided to take the initiative to *attract* foreign

 | DOI:10.1163/9789004402737_036

investment in Tarim.[1] How is it that only six years later, foreign companies would put away their "wolflike ambitions" and turn over a new leaf? (2) The report claimed that if foreign companies were allowed to work in Tarim, we would probably not see any results before 2000, which would impact the energy supply in our country. If we were to do it ourselves, we could see "a much greater increase." Obviously, the Ministry's leaders were full of confidence. But what was the result? By 2000, they managed to get only 4.2 million tons, or three percent of China's total on-land oil production, and their campaign dragged on-land production down to a zero-growth predicament. (3) The report claimed that if we admitted foreign companies to Tarim, China would have little gain. Again one might ask why, after only six years, the Ministry took the initiative to bring foreign investors in? Does that mean that foreign companies were all of a sudden showing mercy and giving us more benefits? The report also stated that if we did the campaign ourselves, then once we found the oilfield, we could still borrow money from abroad to develop it. We do not know if they actually did borrow, and if so, how much. Probably not much—otherwise, why does Wang Tao write, "Obtaining foreign loans is very difficult"? If we thus examine these issues, it becomes apparent that each of the reasons listed in the report for refusing foreign investment were concocted by Wang, the decision-maker, on the basis of his subjective ideas.

3 The Tarim Campaign's Function and Impact on Overall Oil Production

Wang Tao said that this campaign "was a campaign with great strategic significance and far-reaching impact for our country's new era of reform and opening up," that it "affected the future of China's oil industry." Well said! To clarify what that impact was, let us review Wang Tao's own figures. In an official speech that he gave on December 19, 1998, he said, "During the ten years of the campaign, we built an [annual] production capacity of 4.2 million tons."

Now, we can obtain a grasp of the overall "function" of the campaign by remembering that in 2000, national on-land oil production was at 140 million tons. If Tarim was producing 4.2 million tons, then it would have amounted to only approximately 3% of the total.

A few simple statistics on China's oil production over four decade-long periods bring into relief the Tarim campaign's impact:

1 Wang Tao, *Zheng zhan siwang zhi hai*, 276–287.

1960–1969: China's oil production increased by 2.8 million tons per year on average; the Daqing Oil Campaign started in 1960.

1970–1979: oil production increased by 7.4 million tons per year on average.

1980–1989: oil production increased by 3.8 million tons per year on average.

1989–2000: oil production increased by 200,000 tons per year on average; the Tarim Oil Campaign started in 1986 and ended in 1998.

4 Was "Dismounting" a Rotten Idea?

In his *Battling the Sea of Death*, Wang Tao denounces those who suggested discontinuing the campaign when it ran into problems in 1993 as a manifestation of "dismounting trend," and he brings out Mao Zedong's famous rhetorical question, "How long will the red flag fly?" to criticize and intimidate his critics.[2] This frames it as a political issue and implies that those who recommended discontinuing the campaign were not following a revolutionary path. After pondering over the issue for a long time, I came to the conclusion that "dismounting" may not have been such a rotten idea. This reminds me of what occurred during the Chuanzhong Oil Campaign in 1958.

In the spring of 1958, Yu Qiuli the newly appointed Minister of Petroleum, had just assumed his position when oil started gushing from three exploratory wells at three different structures, including Longnüsi, in the Nanchong region. Yu Quili was ecstatic. He immediately reported the news to the CCP Central Committee and to the nation. He sent the deputy minister in charge of drilling, Kang Shi'en, to the site to direct operations. Troops and equipment were transferred from Yumen Oilfield, Qinghai Oilfield, and Xinjiang Oilfield to the Chuanzhong Oil Campaign. After a year of hard work, however, the new wells in the surrounding area produced little. As for those three initial oil-producing wells, they stopped producing. They tried everything to get things going, but

2 Wang Tao, *Zhengzhan siwang zhi hai*, 270. A few words on the history of these phrases is in order. In 1959, after the Great Leap Forward disrupted the economy, the State scaled down planning and started consolidation. Many enterprises either stopped or reduced production. At that time the term "dismounting (*xiama* 下马)"—as in dismounting from a horse—was used to describe this trend. There was no negative connotation attached to the term, but Wang Tao's use of the term, strictly speaking, was inappropriate. As for "How long will the red flag fly?" Mao Zedong first said it in 1928–1929, when the Red Army was holed up in Jinggangshan, Jiangxi Province, and in a very difficult situation. Morale was low. Lin Biao also echoed the sentiment—and decades later, after Lin Biao's plane crashed, Lin's critics lambasted him for it. Wang Tao's usage of this rhetorical question thus likens recommendations to abandon the Tarim oil campaign to defeatist sentiments 60 years earlier. It is an instance of referencing Mao to silence opposition.

it was like getting blood out of a turnip. At the time, people realized that "one oil-producing well does not equal the discovery of an oilfield; temporary high yield does not equal long-term high yield."

In March 1959, Yu and Kang promptly decided to beat the gongs and withdraw the troops. They apologized to the CCP Central Committee and the nation for having "fired blanks." Even so, the failure of the Chuanzhong oil campaign became an asset to the oil industry, serving as a mirror for the Daqing Oil Campaign two years later. Failure is the mother of success. The lessons learned from the failed Chuanzhong Oil Campaign were transformed into the success of Daqing. Such are the dialectics of history.

How similar the Tarim Oil Campaign of 1986 was to the Chuanzhong Oil Campaign of 1958! History is truly treacherous and unfathomable—it played a joke on us in the oil industry. We marched forward for thirty years only to go back to the original spot. Would it not have been much better if decision-makers like Wang Tao had some knowledge of the experiences of Chuangzhong and Daqing? But history does not wait to see "ifs," it only recognizes results.

5 Zhou Yongkang Tries to Stifle a Different Opinion

In 2002, Gao Shubo 高树柏, the editor-in-chief of the CNPC's *Petroleum News* (*Shiyou xiaoxi bao* 《石油消息报》, Beijing), told me that on October 28, 1992, the paper had published on its front page a short article (600-plus characters) entitled "Energy Research Institute of the State Planning Commission on 'Suggestions on Tarim Petroleum Exploration' (Guojia jiwei nengyuan yanjiusuo guanyu 'Talimu pendi youqi ziyuan kantan de yijian' 国家计委能源研究所关于《塔里木盆地油气资源勘探的意见》." Its main points were as follows:

(1) Tarim Basin, rich in petroleum resources, is one of our strategic back-up oil and gas industrial bases. It will play a decisive role in our national oil supply after 2000.

(2) Tarim Basin's geological conditions are complex, making exploration difficult. One should not be too optimistic.

(3) While concentrating on exploration, experiments of a certain scale may be conducted in production and development.

(4) Since oil and gas resources are buried, surface conditions poor, and transportation distances long, the economic benefits from developing Tarim's petroleum resources will not only be inferior to those from the old oilfields in the east, they will be even lower than imported oil for a certain period of time.

(5) Exploring Tarim Basin requires a huge amount of capital, but borrowing foreign capital to explore it on our own would be risky. To lower risks and speed up exploration, certain areas for international bids may be designated; then have the foreign oil companies explore them at their own risk, and have Chinese and foreign companies jointly develop the oilfields.

The "Suggestions" also made the observation that in the east, only 50% of the on-land oil and gas resources were proven, but they were much better than Tarim in terms of the exploration difficulty and economic benefits. The article therefore recommended that, before 2000, the focus should be on developing oil and gas resources in the east.

The article reflected a different opinion, and the wise course for decision-makers should be to listen to both sides. As the saying goes, "Listen to both sides, and you will be enlightened; heed but one side, and you will be benighted (*jian ting ze ming, pian xin ze an* 兼听则明，偏信则暗)." However, CNPC's Vice President Zhou Yongkang became livid at seeing the article and immediately ordered halting the distribution of this issue of *Petroleum News*. Unfortunately for him, the newspapers had already been sent out by the post office. If the company were going to notify each oilfield, it would only look bad and draw even more attention to the article. So they let it go. Afterward, the Director of the CNPC's Science and Technology Bureau scolded the editor Gao Shubo to his face, "You're on the Corporation's payroll—how dare you sing a different tune from the Corporation!"

6 The Attitude of the CCP Central Committee

In his memoir, Wang Tao writes concretely, vividly, eloquently, and at great length about how the CCP Central Committee's leadership cared for, affirmed, granted, inspected, visited, commended, and encouraged this campaign. He devotes more space to that than to the subject of finding oil and gas. These accounts give credit for all achievements to the Party and to the country's leadership, as is typical of Chinese official writing. However, in placing so much emphasis on this, Wang Tao seems to be "draping himself in the big flag to boost his own courage." Actually, whether the Tarim Oil Campaign was correct or not had nothing to do with the attitude of the heads of state. As Francis Bacon once said, "Truth is rightly named the daughter of time, not of authority."

7 Comments from Four Insiders

The following are what four employees of the Tarim Oil Campaign had to say about the campaign. These four plain, honest, and ordinary oil workers felt no rancor toward anyone. They were not my acquaintances. I chatted with them on separate occasions in the fall of 1999. I ask the reader's indulgence for not disclosing their names, but I was moved by the truth of what they said.

One manager who worked at the campaign headquarters said, "In spending money hand over fist like this in Tarim, the key point is that it isn't their own money. It's public money, so why not spend it? No one cares."

One young truck driver said, "Whoever becomes the new Minister, he'll use the oil industry to build up his own track record. If he succeeds, he can be promoted higher and become even more famous. If he doesn't, no one will investigate him. No one will dare to investigate."

One geologist who had worked in Tarim for more than twenty years said without hesitation, "Even though foreign capitalists are wealthy, they're very careful about how they spend it. If they were allowed to come in and tackle Tarim, they might take a few more years, but they will do it with the care and precision of a sculptor. They won't spend a penny unless it is necessary. They won't loose the eagle unless they've seen the hare."

One geologist who graduated from college in 1991 told me, "Just before graduation, two stories were circulating in our college: one was that water could turn into oil, and the second was that today's Tarim would soon become the next Kuwait. As for the former, at the time we thought it was a hoax. The latter turned out to be a hoax as well."

8 Summarizing Historical Lessons

Chairman Mao taught us, "Historical lessons are worth heeding." From a historical point of view, Tarim Oil Campaign abandoned what was the best in the Daqing Model (high revolutionary spirit and strict scientific attitude) but inherited and gave full rein to its most negative aspects—talking "big," setting excessively high targets, using a dated management style, calculating only political gains instead of economical gains, and exaggerating achievements. If the Daqing oil campaign represents the Daqing Model's zenith, then the Tarim Oil Campaign thirty years later represents its nadir. The Tarim Oil Campaign caused such harm to the Chinese oil industry that it entered the present century wounded, like a warhorse that has just survived a rough battle.

9 Kang Shi'en Makes a Deathbed Confession, Wang Tao Wails over His Coffin

Li Ye told me that around 1993, when the gravely ill Kang Shi'en saw the failure of Tarim Oil Campaign, Kang wrote to the CCP Central Committee, suggesting that Wang Tao be removed as petroleum minister (another person told me that Wang Tao also had a copy of the letter). When the Central Committee discussed the issue, Premier Li Peng said that his government would not consider replacing him. Kang Shi'en, in the hospital, refused visits from Wang Tao several times, and he admitted, "The greatest disservice I ever did to the oil industry was recommending Wang Tao."

At Kang Shi'en's funeral in the spring of 1995, Wang Tao wailed and even caressed the coffin. His tears were perhaps real. What was really going on in his heart and mind? Kang Shi'en had turned his back on him, yet had been responsible for his rise. Certainly Wang Tao must have harbored mixed feelings toward Kang.

One can't help but bemoan the falling-out that occurred between these two men.

To be fair, the failure of the Tarim Oil Campaign did not rest entirely with the decision-makers, for they were also to a large extent shaped and influenced by the turbulence of their times—a situation that gave the campaign's failure an air of inevitability. But all this is in the past. As Yang Shen 杨慎 (1488–1559) put it, "Right or wrong, success or failure, all vanish in an instant ... Matters ancient or modern, great or small, all become fodder for conversation and are shaken off with a laugh."[3]

3 From a lyric to the melody of "Lin jiang xian" (临江仙) in Act III of Qin-Han section of Yang Shen's *tanci* (弹词; a narrative alternating between verse and song) entitled *The 21 Histories* 《二十一史》 (*Ershiyi shi*). The Chinese characters: 是非成败转头空…古今多少事， 都付笑谈中.

PART 7

The Offshore Era (1953–2000): Leading the Reform, Another Daqing Realized

The development of China's offshore petroleum may be divided roughly into two stages: the first, pioneering stage, lasting thirty years from 1953 to 1982, and the second stage from 1983 to 2000, coinciding with China's reform and opening up period, a time when annual oil production increased, on average, by 1 million tons per year. In 2010, when China produced more than 50 million tons, another Daqing was realized offshore. This happened thanks to the support and close attention of the whole nation, as well as the policies of reform and opening up.

Offshore petroleum is a beacon of hope for the Chinese petroleum industry.

CHAPTER 36

A Dry Duck Trying to Swim: the Difficulties of Offshore Petroleum Exploration

1 A Blue Territory, Rich in Oil and Gas

China has one of the longest coastlines in the world: more than 18,000 km. From the north to the south, there are the Bohai (Bo Sea), Yellow Sea, East China Sea, and the South China Sea, which includes the area around the mouth of the Pearl River, the Yinggehai Sea, Beibu Gulf, and the sea around the Dongsha (Pratas), Xisha (Paracel), Zhongsha (Macclesfield Bank), and Nansha (Spratly) Islands—all parts of the Pacific Ocean. These four interconnected seas have a total area of 4.73 million km^2—nearly half the size of the country's land area. The climate ranges from temperate to tropical, natural conditions are diverse. The underwater topography consists of a vast continental shelf extending out to the continental slope and deep sea basins. The approximately 1.3 million km^2 of continental shelf area has average depths of around 200 m near the shoreline. These waters are rich in marine, mineral, and especially petroleum and natural gas resources, which are very important for national economic development and prosperity.

Geologically, the land and waters of China are at the eastern edge of the Eurasian Plate. The Pacific Plate is located to the east, and the Indo-Australian Plate lies to the southwest. Beginning in the Mesozoic Era, the movements of these gigantic plates have formed a series of sedimental basins. In China's present-day waters they total an area of more than 1.3 million km^2. From the north to the south, these sedimental basins are: Bohai Basin, North Yellow Sea Basin, South Yellow Sea Basin, East China Sea Basin, Okinawa Trough Basin (including Diaoyu Islands and surrounding waters), West Taiwan Basin, Southwest Taiwan Basin, East Taiwan Basin, Pearl River Mouth Basin, Beibu Gulf Basin, Yinggehai and Southeast Qiong Basin, and the South China Sea Basins (including Dongsha, Xisha, Nasha Islands and surrounding waters). These oil- and gas-rich sedimental basins constitute the principle oil and gas containing areas of the western Pacific Ocean.

Due to many factors and conditions, China's continental shelf oil and gas exploration currently concentrate on the Bohai, Yellow Sea, East China Sea, Pearl River Mouth Basin and the continental shelf in the northern part of the South China Sea. The projected oil reserves in these basins are 27.53 billion tons, and

 | DOI:10.1163/9789004402737_037

the gas reserves amount to between 8.4 to 14 trillion m^3. Some experts hold that the projections are greatly underestimated.

2 Yinggehai Sea: the Birthplace of China's Offshore Petroleum

In the 1930s, Chinese geologists pointed out the likelihood of oil and gas on the offshore sedimental basins. On March 1, 1953, the noted geologist Li Siguang wrote a report entitled, "Our Nation's Prospects for Petroleum Exploration from the Standpoint of Continental Structure (从大地构造看我国石油勘探远景)," for the Petroleum Administration Bureau of the Ministry of Fuel Industries. This report for the first time included Bohai together with the Songliao Basin and Huabei Plain as areas for prospective petroleum exploration in the eastern part of China.

One spring day in 2000, I visited the 88-year-old Beijing University-trained geologist Zeng Dingqian 曾鼎乾 at his China Offshore Oil Nanhai West Corporation residence in Zhanjiang, Guangdong. He told me that Yinggehai Sea, by the southwestern corner of Hainan Island, is our country's offshore oil birthplace. Why does that part of the ocean have the beautiful name of Yinggehai ("Sea of Singing Orioles")? It turns out that petroleum geologists named it after a small fishing village, Yinggehai, on the southwest corner of Hainan.

Zeng Dingqian recounted that before 1956, the villagers frequently reported to the local government about several spots of oil seepage from the bottom of the sea not far from the village. Whenever the sun shone, one could see a dazzling multitude of colors on the sea surface. Tao Zhu 陶铸, the Secretary of the Guangdong Provincial Party Committee, and Guangdong Governor Chen Yu 陈郁 arranged for a gunboat from the PLA's Nanhai Fleet to take them there for an inspection. The trip further strengthened their resolve to find offshore oil. Subsequently they repeatedly requested the Ministry of Petroleum to explore Yinggehai. In 1957, the Petroleum Bureau of Guangdong Province and the Maoming Oil Shale Company (茂名油页岩公司) of the Ministry of Petroleum drafted a 1:200,000 map showing the 39 known locations of hydrocarbon seepage in this area, and they suggested that this oil and gas seepage could originate from an oil-bearing depression not far from the seashore. This oil and gas survey was the prelude to our nation's offshore petroleum exploration.

From then on, Guangdong Province and the Ministry of Petroleum started preliminary work on offshore petroleum exploration. In 1960, the Nanhai drilling crew of the Petroleum Bureau of Guangdong Province, using an anvil type percussion drill installed on a barge, drilled two wells 15 m below the surface, at a point 1.5 km from Yinggehai. The first one had a depth of 26 m and the other had a depth of 22 m. They yielded 150 kg and 10 kg of oil respectively. From 1960 to 1965, at the suggestion of petroleum geophysicist Weng Wenbo, the Ministry of Petroleum organized offshore geological research crews. Using two boats, they conducted seismic tests and reconnaissance in the shallow sea of the Yinggehai area and drew up 2,301 km of seismic lines. In June 1963, the Ministry of Petroleum's Maoming Oil Shale Company formed a geological division to take charge of petroleum exploration. In December of the same year, they rented two 500-ton pontoons from the Salvage Bureau of Guangdong Province and drilled three 300 to 400 m deep exploratory wells 14 to 15 m from the surface, at points 4 to 12 km from the shoreline. This was the first offshore petroleum project using modern rotary rigs in the new China. Even though there were oil and gas indications in these three wells, they did not have any commercial value. Afterward, as the war in Vietnam intensified, Yinggehai petroleum exploration was halted.

3 Hai-1: the Springtime of Offshore Oil

Starting in 1957, the Huabei Petroleum Exploration Division of the Ministry of Petroleum and the Huabei Petroleum Reconnaissance Crew of the Ministry of Geology were surveying oil and gas along the Bohai shore. In the following years, the Ministry of Geology, the Ministry of Petroleum, the Chinese Academy of Sciences, and the Chinese National Oceanic Bureau all included Bohai oil exploration as a high-priority project.

In 1963, the Ministry of Petroleum, after much research, made the decision to "Go up to the mountains, go down to the seas, and fight big battles on the plains." "Go up to the mountains" meant carrying out Mao Zedong's order to "find oil and gas in the interior" and entailed conducting another campaign in the Sichuan Basin. "Go down to the seas" meant looking for oil and gas in the Bohai and South China Sea (Nanhai). "Fight big battles on the plains" meant conducting big oil campaigns on the Huabei plain, i.e., the Bohai Bay Basin. Considering that Shengli Oilfield and Dagang Oilfield had been discovered near the Bohai, it was logical to conclude that the focus of offshore oil should be placed in the Bohai. The Ministry of Petroleum decided to form the Offshore Petroleum Exploration Headquarters (later known as the Offshore Oil Exploration Bureau) in Tanggu District, Tianjin, and to transfer workers, purchase necessary boats, and build platforms for offshore drilling. In 1964, the first Bohai Seismic Crew was formed. They installed equipment used on land onto a wooden motorboat and three motorized wooden sailboats. They used simple sextant positioning and manually placed geophones and cable lines. Oil workers who had no ocean experience thus had to work on small boats that were easily tossed on the waves, and despite suffering from motion sickness, they persisted with their seismic work. People made fun of it afterward: these oil workers were like dry ducks trying to swim, using on-land oil exploration methods to do difficult offshore exploration. In March 1965, a geophysical exploration crew consisting of hundreds of people was formed and large-scale seismic prospecting began in the Bohai. In May, the PLA's North Sea Fleet and the Ministry of Transportation dispatched a fleet of boats, including the warship Yinghai (营海号), formerly called Chongqing, and Libra (天秤号), which was originally the Empress Dowager Cixi's yacht.

In August 1966, the Deputy Director of Dagang Oilfield, Zhong Yiming 钟一鸣, was assigned concurrent duty as the Commander of the Offshore Petroleum Exploration Headquarters. On December 31, 1966 the 3026 Drilling Crew of this organization, using a self-made offshore drilling platform started drilling Hai-1 in the Bohai in waters 25 m deep. On June 14, 1967, after the well had reached a depth of 2,441 m, it began producing 35.2 tons of oil daily, as well as 1,941 m^3 of natural gas. This was China's first offshore well that had commercial value. The good news spread. The State Council sent a congratulatory telegram on June 21. Zhong Yiming called Hai-1 "the first spring flower" for offshore petroleum in China.

By 1980, that is to say within a span of twenty years, Bohai oil exploration workers had developed some of their own ships, equipment, and advanced technology, besides importing some others. Over those twenty years they had carried out very difficult oil exploration work and had discovered 15 oil-containing structures, nine gas-containing structures, and six oilfields. In 1982, Bohai produced more than 90,000 tons of oil. This marked the true advent of Chinese offshore oil production. However, due to the low yield during this twenty-year span, Bohai was still not a major boon to Chinese petroleum.

4 Drilling in the Xisha Archipelago: the Determination to Find Oil in the South China Sea

Starting from around the beginning of the Cultural Revolution, Premier Zhou Enlai paid much attention to the offshore petroleum industry. In September 1972, Zou Jiazhi, who was working in the Ministry of Petroleum's May Seventh Cadre School in Hubei, was summoned back to Beijing. Kang Shi'en had a meeting with Zou. Kang said, "At a recent State Council meeting, Premier Zhou asked whether or not we could start offshore drilling. I replied yes. Countries bordering on the South China Sea are all looking for oil there nowadays. We should resume looking in South China Sea, starting with oil exploration on the continental shelf of Nanhai. Once we start, we won't back down no matter what kind of difficulty we encounter! You go to Guangdong and tell the provincial leadership of the Ministry of Petroleum's resolve." Soon, Zou Jiazhi and Zhang Zhiyou 张志友 were appointed to be in charge of the headquarters of Nanhai oil exploration.

Drilling an exploratory well on Yongxing Island (永兴岛; also known as Woody Island) at the southern tip of the Xisha (Paracel) Islands was going to be a difficult job. As soon as they were appointed, Zou Jiazhi and Zhang Zhiyou, both PLA veterans, started contacting relevant army units in Beijing as well as relevant units of the State Council. Once they went to Guangdong, they received great support from Guangdong Province and Nanhai Fleet. All kinds of preparatory work were carried out in an orderly and timely way.

In November 1973, the 32554 Drilling Crew from Tanggu, Tianjin were transferred to Zhanjiang Harbor. As soon as they got off the train, they boarded the ocean-going cargo vessel Haiyun 092 and went to Xisha. While crossing Guangzhou Bay, colloquially called the "Gate of Hell (鬼门关)," the young men all threw up due to the choppy waters. Even though there was no wind, there were three-foot waves. But no one complained. Zou Jiazhi took the second group of personnel and equipment, traveling on the naval landing ship of the Nanhai Fleet. They set off from Zhanjiang Harbor on November 27, passing through Yulin 榆林 military port at the southern tip of Hainan Island. The Chinese Navy at Yulin military port specially dispatched four convoy gunboats to accompany them. When they reached Yongxing Island, the largest of the Xisha islands as well as the site of Xisha Archipelago government (now Sansha government), they were greeted by government employees and some local people. Without the benefit of any large-scale hoisting equipment, the crew unloaded a 3,200 m long, multi-hundred-ton rig and auxiliary equipment from the two boats by manually dragging, carrying, and rolling it.

They began installing the rig immediately. On December 16, 1973, drilling began on Xiyong-1 (西永一井); it was the first exploratory well in the South China Sea. During this time, American military planes often carried out reconnaissance missions over the area. South Vietnamese naval ships often shadowed them. During the Spring Festival in 1974, when South Vietnam initiated the Battle of the Xisha Islands, the oil-drilling crew took on an important support role for the Chinese Navy at Yongxing Island, carrying ammunition, as well as transporting and caring for injured soldiers. After the battle, the crew received commendations from the Navy. On April 5, 1974, drilling was completed at Xiyong-1 at a depth of 1,384.6 m; it provided important geological information on Nanhai and satisfactorily fulfilled its purpose as a regional exploratory well.

Starting with Xiyong-1, oil exploration gradually spread to the Yinggehai Basin, Zhujiangkou Basin, and Beibu Gulf. Oil structures and oil flow were found in all three. In addition, Wei 洄 11-4 Oilfield, which was of commercial value, was found in Beibuwan. In general, however, there were no major breakthroughs with regard to oil and gas in the South China Sea before 1980.

Meanwhile, from the early 1950s to 1979, many entities such as the Ministry of Geology, the State Science and Technology Commission, the State Bureau of Oceanic Administration, the Chinese Academy of Sciences, the Geology Bureau of the State Planning Commission, Shandong Oceanography Institute, and the Ministry of Petroleum, also conducted extensive geological investigations in the East China Sea and the southern and northern parts of the Yellow Sea.

5 Troops, Bases, and Equipment

In 1973, Kang Shi'en said, "To look for oil in the sea, we should first build a port and a nest—even birds need to go back to their nests! We should also have some advanced equipment. At the very least, seismic ships, drilling ships, and operation ships should be purchased!"

The development of offshore oil workers, bases, and equipment was integrated into a system over a series of stages. In the twenty years leading up to 1979, it went through the following stages:

During the first stage, from 1957 to 1965, there was the Geology Division of Maoming Shale Oil Company, a subsidiary of the Ministry of Petroleum. This Division mainly looked for oil on land in Guangdong Province and did little work offshore. Based on sightings by local people, three shallow wells were drilled just offshore from Yinggehai village, but no offshore oil-working teams were formed or bases built.

The second stage started in 1965, when the Ministry of Petroleum decided on Bohai Bay as the focal point for offshore oil exploration and Tanggu Port in Tianjin as the base of operation. With large-scale support from the city of Tianjin, the Ministry of Transportation, the PLA Beihai Fleet, and Dagang Oilfield, the Ministry of Petroleum started amassing offshore oil workers, constructing the base, and developing equipment for the project.

The Ministry of Petroleum established its Offshore Oil Exploration Bureau in Tanggu District in Tianjin. A total of 14,435 managers on all levels, technical personnel, and specialized workers were transferred from oilfields in Dagang, Daqing, Shengli, Yumen, and so on, thus forming a complete complement of geophysical, drilling, oil recovery, shipping, port, and offshore platform construction teams.

Meanwhile, the Ministry of Petroleum spent large sums of money building 11 fixed drilling platforms (six were later changed into oil recovery platforms). In 1972 in Bohai alone, the Ministry of Petroleum, with the permission of the State Council, spent US$250 million of foreign reserves. They imported large jackup rigs Bo-2, Bo-4, Bo-6, Bo-8, and Bo-10 and a group of advanced digital seismic ships, and various other types of specialized ships. They built a specialized port and dock dedicated to offshore oil.

On the barren salt beaches on the outskirts of Tanggu District, employee living quarters and on-land production facilities were built. Along the historic Tanggu Port and the Tanggu Fortress, where China suffered humiliation in the 1894–1895 Sino-Japanese naval war, factory buildings and office buildings sprang up. Brightly lit, it turned into a popular gathering place at night.

The third stage started in September 1972, when Kang Shi'en met with Zou Jiazhi and ordered him to "resume exploration in the South China Sea. Once we start, we won't back down no matter what kind of difficulty we encounter." Oil exploration started in the South China Sea soon afterward. By 1982, 12,000 people were employed in various capacities with the South China Sea teams. Meanwhile, at Zhanjiang, Guangdong, large-scale base construction had begun. Later on, at increments, the Ministry of Petroleum built logistics bases in Guangzhou, in Shenzhen's Shekou District, and in Haikou and Sanya, Hainan.

Around 1975, Nanhai bought six advanced drill ships (Nanhai 1, 2, 3, 4, 5, and 6), of which three were semi-submersible, capable of drilling hundreds of meters underwater. They also bought three advanced digital seismic vessels (Nanhai 501, 502, 503), as well as a large quantity of other advanced special equipment. Later, this specialized workforce, using the advanced equipment, cooperated with more than thirty foreign companies, providing them with complete technical services. Not only were they well received by these foreign companies, they earned a lot of US dollars doing contracting work in Chinese waters. This was also good preparation for offshore oil development in the South China Sea.

The fourth stage began on February 8, 1982, when the State Council approved the establishment of China National Offshore Oil Corporation (CNOOC). As an economic entity, it is a state-owned enterprise that cooperates with foreign companies, and it is headquartered in Beijing. On February 15, Kang Shi'en, on behalf of the State Council, announced CNOOC's founding.

It had approximately 25,000 employees, specializing in different aspects of the oil business. Under the aegis of CNOOC were four regional subsidiaries—Bohai (in Tanggu District, Tianjin), Donghai (in Shanghai), Nanhai East (in Guangzhou), and Nanhai West (in Zhanjiang, Guangdong)—as well as several specialized subsidiaries. After more than twenty years of effort, two offshore oil bases were built in Tanggu Port, Tianjin and Zhanjiang Port, Guangdong.

With the founding of CNOOC, the Chinese offshore oil industry had completed preparations to open up to the outside world and cooperate with foreign companies.

6 Three Accidents, Hard Lessons

The first accident occurred in early 1969, at the height of the Cultural Revolution. The sea water around Hai-2's drilling platform was frozen. Huge drifting chunks of ice toppled and sank the newly-installed giant steel platform which had a draught of 40 m, gravely impacting the oil-producing Hai-1. At this critical moment, Premier Zhou Enlai held a nighttime meeting in Zhongnanhai to discuss countermeasures. He asked Yu Qiuli's opinion. Yu proposed sending Kang Shi'en, who was then still "in the cowshed" (a Cultural Revolution term meaning "under detention"). Kang was released and went to the site where he directed PLA naval officers, soldiers, and Bohai oil workers to seal the well in a timely manner. He thus completed the task personally assigned by Premier Zhou.

The second accident occurred in late fall of 1977, when the Offshore Petroleum Exploration Headquarters' Bozhong-17 had a strong blowout during drilling. Large amounts of oil and gas mixed with mud blew out over the platform and surrounding sea surface. It was a crisis. Kang Shi'en sent Qin Wencai, the Ministry of Petroleum Sichuan Bureau's director known as "the Fire Chief," to kill the well. Qin arrived at the scene. Using meager equipment and poor means of transportation, they braved death for seven days and night, but at last brought the blowout under control, and so avoided a well fire and further pollution of the ocean.

The third accident happened on November 25, 1979, when the Bohai No. 2, a drill ship of the Offshore Oil Exploration Bureau, was being tugged in the Bohai to its new well position. It ran into a big storm and sank, resulting in the death of 72 crew members and a tremendous loss of national property. This accident had much to do with our unfamiliarity with the various risks involved at sea and in offshore oil operations. All on board were oil drillers with no prior ocean experience. Meteorological instrumentation, communication technology, and rescue equipment and practices were all subpar. There were basically no ocean disaster rescue mechanisms or emergency guidelines in place, so these oil workers were helpless in the face of these circumstances. This was a heroic and tragic event in Chinese offshore oil history. As the Tang dynasty poet Du Fu lamented, "Passing away [thus] on the campaign, before victory is achieved, / Has made heroes weep ever since."[1]

For nearly thirty years (1953–1979), during the first, difficult exploratory phases of the Chinese offshore petroleum industry, there were many achievements and feats, but there were also many failures and much loss of blood, sweat, and tears—more than can be described here.

1 From Du Fu 杜甫 (712–770), "Shu xiang 蜀相 [The Shu Minister]," which remembers the Shu prime minister Zhuge Liang 诸葛亮 (181–234).

CHAPTER 37

Leading the Reform and Opening-Up

1 An Important Decision to Open Up the Offshore Petroleum Industry

In October 1975, Zhang Wenbin, the Deputy Petroleum Minister, led the first Chinese petroleum delegation to visit the North Sea oilfields in Europe, near Norway, where the weather conditions are extremely volatile. There are 265 bad days annually; the normal wind velocity is 72 m/s, and waves can reach 19 m high. Under water 300 m deep, the oilfield is technically complex and risky. A dozen international oil companies had invested heavily and had been working there for years, but at last in 1974, they achieved breakthrough results. After construction, the North Sea oilfields became one of the most important oil-producing areas in Europe.

In August 1977, the United States government invited a governmental petroleum delegation to visit the US. With the permission of the Chinese State Council, a delegation headed by Deputy Minister of Petrochemistry Sun Jingwen and Deputy Director of the State Planning Commission Li Renjun (he was originally Deputy Minister of Petroleum) went to visit the US. On their way back to China, they also visited petrochemical facilities in Japan. Upon their return, they proposed using the technologies and capital of international oil companies to speed up development of offshore petroleum in our country. On March 26, Hua Guofeng, Ye Jianying, Li Xiannian, et al., after hearing reports from the delegation, completely agreed with the proposal. This is how the important strategic decision of opening offshore petroleum development to the outside world was made.

From May to July 1980, with the permission of the State Council, Kang Shi'en, in his capacity as Vice Councilor of the State Council, led a Chinese governmental delegation to Norway, the United Kingdom, and the Netherlands, and they visited the North Sea oilfields. This trip was undertaken to investigate the possibility of international cooperation and foreign investment. The Chinese delegation was very impressed with what they saw. Kang Shi'en became even more resolute that China should completely open up offshore oil and cooperate with foreign companies.

Since offshore oil exploration involves advanced technology, large investment, high risk, a skilled workforce, a long development period, and international cooperation, Chinese offshore oil needed to depart from many ways of

 | DOI:10.1163/9789004402737_038

thinking and practices used on land. To take on the new challenge, it needed a fresh outlook, advanced technology, fine equipment, and a scientific management system, as well as an overall strategy. It boiled down to one point: Chinese offshore oil had to boldly open its doors, quickly integrate internationally, and fully cooperate with foreign companies. We did not have an alternative. After Kang Shi'en returned from the trip, in July 1980, he reported the findings of his visit to the Financial and Economic Leadership Group of the CCP Central Committee. He proposed an overall plan for international cooperation for offshore oil. The national leaders at the time happily granted his proposal. A new chapter of offshore oil's international cooperation had thus truly begun.

The reason for documenting this process is that the significance of the early 1980s decision to open up offshore oil for international cooperation is on par with the Daqing Oil Campaign and with the Jianghan Oil Campaign during the Cultural Revolution. The decision was far-sighted, groundbreaking, careful, and feasible. Its economic benefits were tremendous.

2 Studying Hard: the First Step of Reform and Opening

The cooperation between CNOOC and foreign companies required our people to study hard and master advanced science, technology, and management methodologies. CNOOC's breakthroughs came by dint of hard labor and study on the part of the entire workforce involved. Since the mid-1970s, CNOOC has been sending large groups of managers, technical personnel, management personnel, and technical workers to Japan, the US, and the big oil-producing countries in Europe and South America. Some of these have visited, some trained with foreign oil companies long-term, some have gone to pick up equipment and ships and receive on-site instruction on their operation. At the same time, the China Council for the Promotion of International Trade and the Ministry of Petroleum, over the years, have invited more than thirty international oil companies and government officials from many countries, such as the US, United Kingdom, France, Japan, Germany, Norway, Holland, and Canada, to discuss business. The whole process of cooperation has been a good learning opportunity.

To ensure that international cooperation on offshore oil made a good start, the Ministry of Petroleum in 1978 put Deputy Minister Zhang Wenbin in full charge of international cooperation on offshore oil. The first thing he did was to expand personnel training. The Ministry of Petroleum established an offshore oil training center in the city of Zhuozhou, Hebei Province. To organize the offshore oil workforce, managers and technical personnel were transferred, without exception, from different on-land oilfields and sent to the offshore oil training center in Zhuozhou to take classes and seminars on international cooperation. They were then sent to the Beijing Petroleum Institute to take English courses. Only then were they allowed to go to work.

Over the years, great efforts have been made to train the offshore oil workforce. For example, in the first fifteen years after the Nanhai East Corporation was founded, it spent a cumulative 91.4 million yuan on training, of which 33.5 million yuan were from joint-venture foreign partners. The training included English, advanced technologies, modern management, technical and managerial methodology, as well as on-the-job training at oilfield operations. Altogether, 11,034 people took part in the training. Of these, 304 people received overseas training. This kind of heavy investment reaped great returns. It was well worth it! The company employs only 1,000-plus people, but it produced 12 million tons of crude oil per year, so that Nanhai (South China Sea) ranked as the fourth largest oilfield nationally, after Daqing, Shengli, and Liaohe Oilfields.

Liu Jianan 刘家南, of CNOOC's Southern Drilling Company, became a driller in 1965 in Sichuan. In 1976, he was transferred from Sichuan Oilfield to Nanhai Oil Exploration Headquarters and designated a deputy leader of a drilling crew. He was young, energetic, and studious. In 1978, while receiving the newly-imported advanced drilling platform at Nanhai-2, he almost completely severed the middle finger on his right hand while learning how to examine equipment. He was sent to a simple clinic in Hainan's Sanya for treatment. The doctor told him if he were to go to a bigger hospital, his finger could be sewn on. To save time, he asked the doctor to cut off the severed part of his finger. After his finger was wrapped up, he immediately went back to work on the drilling platform. His bravery and gallantry were reminiscent of Guan Yu 关羽 (d. 220), the legendary Three Kingdoms general. Once when Guan Yu was injured in battle by a poisonous arrow, his doctor scraped the poison off his arm bone while he dined with his officers.

Liu Jianan studied English assiduously. Even a foreign professor working on the drilling platform called him a model student. After close to ten years of working on the drilling platform and mingling with foreign workers, he was appointed as CNOOC Southern Drilling Company's CEO, in which capacity he managed more than 10 billion yuan of assets. When he represented the company and competed with foreign companies for bids, he was good at repartee. In 1999, the company paid the state 60 million yuan in profits and taxes. Liu Jianan has become a manager well-versed in offshore oil drilling and management.

The offshore oil training has brought about great efficiency. On March 8, 2000, Phillips Petroleum Company's Global Operations Manager Carl Lawson, after visiting Bohai Oilfield, praised the Chinese crew they employed for the

Bohai-7 platform, even though the platform was made in China and its equipment was not advanced. He said that Bohai-7's operation crew was the best and that the management personnel was first-rate. He commended the service, safety awareness, technical capability, and management shown by the Chinese workers when they worked on Bohai Penglai 19-3-8. He made a decision on the spot that his company would no longer send a safety supervisor for Bohai-7.

3 Prerequisite for Reform and Opening Up: Coming to an Understanding

The Ministry of Petroleum decided to enter into some small joint ventures in order to gain the experience it would need for larger-scale international bidding and to become familiar with legislative work in this regard. So in the early eighties, after nearly two years of negotiations, the Ministry signed contracts with Japanese and French companies, agreeing to jointly explore and develop a few small regions in Bohai and Beibu Gulf in the South China Sea.

There were different reactions to the signing of these contracts inside and outside of the country. In the United States, someone wrote an article in the New York-based *China Daily News* citing false data. The author of the article alleged that China had suffered losses by signing these contracts. In reaction to this, a letter of complaint was written to China's leaders, which attached the article from the overseas *China Daily News*. It claimed that the cooperation with foreigners was "selling off China's resources," that "the Ministry of Petroleum's butt was sitting on the side of the foreigners," that "in the Bohai contract we signed with Japan, we were at a big loss," and so on. China's leaders took it seriously. Deng Xiaoping wrote to Yu Qiuli, Commissioner of the National Energy Commission, and Gu Mu, Commissioner of the National Import and Export Regulatory Commission, asking them to invite some experts for a hearing.

On March 29, 1981, a hearing involving more than 60 managers and experts from 24 different government agencies was held in Beijing. The meeting was presided over by Yu Qiuli and Jiang Zemin, who was then a Deputy Commissioner of the National Import and Export Regulatory Commission. Participants came from the Chinese Academy of Sciences, the Chinese Academy of Social Sciences, the State Planning Commission, the State Construction Commission, the State Economic Commission, the State Science and Technology Commission, the Legislative Affairs Commission of the Standing Committee of National People's Congress, the Research Office of the Secretariat of the Communist Party of China, the Ministry of Geology, the Ministry of Petroleum, the Ministry of Finance, the Ministry of Foreign Trade, the Ministry of Foreign Affairs, the National Bureau of Oceanography, the Bank of China, the China Council for the Promotion of International Trade, the Geological Society of China, the Chinese Geophysical Society, *People's Daily*, *Guangming Daily*, Xinhua News Agency, etc. The hearing lasted nine and half days, with more than 30 speakers. The atmosphere of it was scientific and democratic. Speakers debated from different perspectives. The majority of the speakers advocated foreign cooperation. Naysayers could provide no good reasons for their arguments. Truth became clearer through debate. The meeting concluded that, "The Bohai contract signed with Japan was generally sound, mutually beneficial, and not unfair to China. We should cooperate with other countries on offshore oil development. This policy must be decided and adhered to." Soon after the hearing, on April 7, Premier Zhao Ziyang, at a meeting of the Leading Group for Financial and Economic Affairs of the CCP Central Committee, affirmed the conclusions of the hearing. General Secretary Hu Yaobang on many occasions affirmed the policy of international cooperation regarding offshore oil.

In retrospect, it was a good thing to have an opposing opinion. Agreement between a thousand men is not as good as one man's honest opposition. Opposition allows one to do things with more caution and to consider things more thoroughly and carefully.

4 Offshore Oil Legislation: Rules and Prerequisites of Reform and Opening-up

To ensure that China would suffer fewer disadvantages in international cooperation and to have a set of feasible, effective rules for international cooperation, the Ministry of Petroleum from the beginning underscored the importance of legislation on China's cooperation with overseas firms on offshore oil. In 1978, Deputy Minister of Petroleum Zhang Wenbin gathered relevant experts and set up a special work group to review more than 120 legal contracts from different countries, study their "fit" with realities in China, and engage in discussion. After a few years of effort, this group gradually formed the guiding ideology and legislative model for China's offshore oil legislation.

Legislation was key to the success of international cooperation on offshore oil. This systematic policy-related project touched upon many areas. It would need to be universally feasible and practical, accepted not only by the Chinese but also by foreigners so that they would actively participate and benefit. Under the auspices of the State Council, in October 1980, Kang Shi'en held a coordination meeting attended by managers from nine government agencies to discuss principles and policies for international cooperation on offshore oil that would pave the way for this project. Representatives from the Ministries of Petroleum, Foreign Affairs, Foreign Trade, Public Security, and Finance took part. To gain support of the Standing Committee of the National People's Congress, the highest legislative body, they invited the Chairman of the committee Peng Zhen to listen to the report on offshore oil legislation. The charismatic Peng Zhen was enthusiastic. He proposed that, since offshore oil legislation was a matter of urgency, they save time by devising a set of regulations for the State Council to ratify and put to use.

Under Kang Shi'en's charge, the basic offshore oil model was established after thorough research into the subject. The State would issue regulations (*tiaoli* 条例) regarding international cooperation on offshore oil, including adjustments to relevant tax laws and tax agencies; it would also issue a set of standardized contracts (*biaozhun hetong* 标准合同) to serve as models. On January 30, 1982, the State Council formally issued the *Regulations of the People's Republic of China on the Exploitation of Offshore Petroleum Resources in Cooperation with Foreign Enterprises* (《中华人民共和国对外合作开采海洋石油资源条例》). Subsequently all kinds of rules, regulations, and standard contracts were issued, forming a complete legislative opus governing international offshore petroleum cooperation. This established our nation's first set of offshore oil regulations. It protected our own national interests and was generally accepted by foreign companies. A transparent and attractive environment was thus created for international cooperation.

On February 15, 1982, when Kang Shi'en announced the State Council's formal ratification of the founding of CNOOC, he also proposed two principles for international cooperation on offshore oil exploration. The first was, "Do everything according to the law—do not do things that might seem reasonable but are not in keeping with the law, but [instead] strive to have the legal basis to do things well." The second was, "Do everything according to the regulations—do not finalize matters until all procedures have been executed, but follow all rules with high efficiency and do things well according to the regulations." In this way, international cooperation on offshore oil was given a regulatory foundation.

CHAPTER 38

International Cooperation Increases, an Offshore Daqing Emerges

1 Large-Scale Offshore Geophysical Exploration

From March to July 1979, with the permission of the State Council, the Ministry of Petroleum signed eight offshore seismic exploration agreements with 48 oil companies from 13 countries, including the United States, United Kingdom, France, Italy, and Japan. The agreements covered a total area of 420,000 km^2 in the South Yellow Sea, the mouth of the Pearl River, Beibu Gulf, and Yinggehai Basin. Following internationally accepted norms of risk contracts, the overseas parties assumed all costs for seismic exploration and provided China with the complete findings free of charge; China promised to designate certain areas for international competitive bidding after geophysical exploration, and the foreign companies that participated in geophysical exploration were entitled to participate in the bidding. After signing the agreements, those 48 foreign companies, within a year, had spent US$109.6 million, rented 14 advanced seismic ships to conduct geophysical exploration, and finished 107,170 km of seismic survey lines. They brought in more than 700 seismic and geological personnel from around the world and used 18 advanced mainframe computers to process and analyze the data. Eighty experts from the Ministry of Petroleum and the Ministry of Geology took part in the project. In 1980, after the work was completed, 466 people, in 60 delegations from 31 companies from 11 countries, came to China to report their findings. The seismic data weighed 350 tons. All of these petroleum geophysical and geological experts gathered in Guangzhou. Exxon alone sent more than 200 people, who took up a whole floor in the five-star White Swan Hotel. BP took up a few floors of the China Hotel. One experienced foreign expert commented, "There has never been this kind of large-scale offshore seismic exploration. The Chinese people were bold to have come up with this kind of big campaign in offshore petroleum geophysical exploration."

Through this initiative, a large number of local structures were discovered in the designated cooperation area. These included 169 structures in the Pearl River Mouth Basin, 47 in Yinggehai Basin, and 74 in the South Yellow Sea. Foreign companies that participated in the project gave positive assessments regarding these structures, saying that oil and gas fields of commercial value

 | DOI:10.1163/9789004402737_039

could be found. They considered the Pearl River Mouth Basin to be the most promising. The results encouraged and attracted all those foreign oil companies that were willing to get involved in the exploration. It spurred on everyone who had cared about developing China's offshore petroleum.

If there is competition for offshore oil, there will certainly be big rewards!

2 Four Rounds of International Bidding

On January 22, 1982, with the permission of the State Council, a notice for the first round of international bidding was issued. Forty-one companies from 11 countries made bids, and 19 contracts were signed, with such companies as BP, Exxon, Shell, Chevron, Phillips Petroleum Company, Amoco, ARCO, Tesco, Occidental Petroleum, Agip, and Japan National Oil Corporation. Per the first round of contracts, the foreign companies spent US$1.08 billion, finished 85,000 km of seismic line, and drilled 100 wells.

The second round of international bidding started in November 1984. Since the results of first round of exploration were not satisfactory, and international oil prices were declining, Kang Shi'en directed that conditions for the second round of bidding should be somewhat looser and more flexible in order to continue attracting foreign investment and make it potentially profitable for them. Seven contracts and one agreement were signed for this round.

FIGURE 38.1 Nanhai Oilfield (2002)

In January 1989 and June 1992, the third and fourth rounds of offshore petroleum international bidding were conducted. In 1996, another effort was made in the United States to attract investment.

By the end of 2000, 70 foreign companies from 18 countries and regions had signed 146 petroleum contracts and agreements with China; they had invested a total of US$6.64 billion, of which half was venture capital investment for exploration, and developed 25 offshore oil and gas fields in China.[1]

3 Remarkable Achievements of International Cooperation

A CNOOC company history records that in 1982, the year of its founding, it produced 95,000 tons of oil. As international cooperation quickly expanded, from the mid-eighties up to 2000, its production of offshore oil dramatically increased by one million tons each year on average. In 1993, production was 4.63 million tons, of which 3.819 million tons came from joint-venture oilfields. In 1997, production was 16.285 million tons, of which 13.3535 million tons were produced by joint-venture oilfields; the latter also produced 4.05 billion m^3 of natural gas. In other words, these joint-venture oilfields respectively produced 83% of the CNOOC's oil and 91% of its natural gas.[2]

Looking back, international offshore oil cooperation was not all smooth sailing or an overnight success. Many factors figured in arousing the enthusiasm and determination of experienced foreign oil companies that were good at turning a profit. In the span of 20 years, foreign companies sank US$1.5 billion of venture capital into Chinese offshore oil exploration, and the oil and gas fields that were eventually discovered came to fruition as a result of the foreign companies' persistence. This is the challenge and allure of offshore oil exploration. Specifically, there were three big offshore oil accomplishments, each with large returns.

The first accomplishment was the construction of Ya 13-1 Gas Field in Yinggehai Basin, in the western part of the South China Sea. After China signed a contract with ARCO in September 1982, ARCO completed more than 70,000 km of seismic line and drilled 39 wells. In 1983, 98.4 billion m^3 of natural gas reserves were proven. In 1996, the gas field was developed and put into operation. The terms of the twenty-year contract stipulated that it would supply CLP

1 Bainian shiyou bianxie zu, ed., *Bainian shiyou*, 323.

2 Zhongguo haiyang shiyou zong gongsi bianzuan weiyuanhui 中国海洋石油总公司编纂委员会, ed., *Zhongguo haiyang shiyou zong gongsi zhi*《中国海洋石油总公司志》[*The Record of China National Offshore Oil Corporation*] (Beijing: Gaige chubanshe, 1999), 2.

Power Hong Kong Ltd 2.95 billion m^3 of gas and Hainan's Sanya 0.5 billion m^3 of gas annually. This was the largest offshore gas field in China found in the twentieth century, and in terms of production, it is secondary only to the gas fields in Sichuan Basin.

The second accomplishment was the development of ten oilfields, including Huizhou 惠州, Xijiang 西江, and Liuhua 流花, among others, in the Pearl River Mouth Basin in the eastern part of the South China Sea. In July 1985, a consortium of Agip, Chevron, Texaco, and other well-known companies first discovered the Huizhou 21-1 Oilfield; in February 1987, Amoco discovered the Liuhua 11-1 Oilfield; and in August 1988, Phillips discovered Xijiang 30-2 Oilfield. During the 1990s, CNOOC's Nanhai East Petroleum Corporation successively built in the Pearl River Mouth Basin five sets of oil recovery installations for the oil from these ten oilfields. In 1996 and 1997, respectively, 10 million and just under 13 million tons of oil were produced in this area, which amounted to more than 80% of CNOOC's total production. It thus became CNOOC's premier force.

The third accomplishment was the discovery of Penglai Oilfield in the southwestern tip of Bohai. On July 7, 1999, Phillips Petroleum Company, which had been cooperating with China in Bohai, announced in New York that they had discovered a large oilfield there. On February 16, 2000, CCTV announced that the biggest oilfield since the discovery of Daqing had been found in Bohai. This immediately drew a lot of national and international attention. Phillips shares increased by 3% on the New York Stock Exchange. It had been more than 20 years since Bohai offshore oil exploration had been opened up for foreign cooperation. Many foreign companies had explored there, without any kind of breakthrough. Only Phillips persevered and signed a contract with CNOOC on December 7, 1994. After years of exploring, it finally proved, between May and July 1999, that this oilfield was 50 km^2 in area. The sea in the area was around 20 m deep, and the petroleum was buried 900 to 1,400 m beneath the ocean floor. Some of the oil reservoirs were as thick as 150 m, and the oil reserves were 0.6 billion tons (in New York, Phillips claimed it to be 1 billion tons). Thus an oilfield producing 20 million tons of oil per year could be built.

4 Autonomous Management as Part of Walking on Two Legs

Autonomous management usually means running things on one's own. As mentioned earlier, China started offshore oil exploration in the 1950s, using on-land methods for offshore work. This practice continued into the mid-seventies. Here "autonomous management" refers to management by the

FIGURE 38.2 Bohai Oilfield (2005)

Chinese, when they began taking control of managing operations after the large scale international cooperation on offshore exploration. By 1983 and 1984, this international cooperation in the South China Sea reached its climax. By comparison, in Bohai, where international cooperation had started earlier, the French company terminated its contract, the Japanese company was preparing to withdraw, and the other foreign companies had become cautious about geologically complex areas. Meanwhile, international oil prices had dropped, which presented a potential crisis for international offshore oil cooperation.

In February 1983, during Kang Shi'en's visit to the South China Sea (Nanhai) petroleum oil base in the port of Zhanjiang, Guangdong Province, he proposed that future offshore oil exploration should adopt the policy of "walking on two legs," i.e., international cooperation as well as autonomous management. His proposal met with approval from CNOOC.

In terms of feasibility, from 1978 to 1983, while international cooperation was being fully promoted, offshore oil workers accumulated capital for the country by doing contractual work for foreign companies. This enabled them to update a lot of equipment and improve both technology and management. As a result, we had become better equipped to explore, develop, and manage offshore oil and gas fields on our own. No matter how complex those areas were geologically, they still were China's treasured areas. Chinese offshore oil workers said: "If the foreign devils don't want to do it, we'll do it ourselves."

In 1984, by doing it ourselves, we made breakthroughs in the Liaodong Bay area of Bohai and discovered the Jinzhou 20-2 condensate oilfield. In 1986, we discovered Suizhong 绥中 36-1 Oilfield in Liaodong Bay. In 1992, we discovered the Qikou 歧口 and Qinhuangdao 秦皇岛 32-6 Oilfields. An article in *China Trade News* (*Zhongguo maoyi bao* 《中国贸易报》; Beijing) on August 2, 2001, reported that in the previous two years, seven oilfields each with hundreds of millions tons of oil reserves had been found in Bohai. CNOOC's Nanhai West Oil Corporation soon followed suit. In 1992, they explored and discovered in Yinggehai Basin extensive shallow gas fields such as Dongfang 东方 1-1 and Ledong 乐东 22-1 with reserves close to 10 billion m^3. In 1995, they discovered the Wenchang 文昌 9-1 and Wei 涠 12-1 Oilfields. By 2000, production from autonomous fields accounted for 17% of CNOOC's oil and 9% of its gas production.

5 Exploring for Oil and Gas in the East China Sea—Pinghu Oil and Gas Field

Pinghu 平湖 Oil and Gas Field is situated in the East China Sea continental shelf approximately 375 km from Shanghai and 374 km from the landing spot of the pipeline at Luxiang Port 芦湘港 in Nanhui County, Shanghai. The ocean depth is between 84 to 94 m. The Ministry of Geology started exploration in 1982. After having struggled through many hardships, they proved this oil and gas field in 1990. In June 1990, the National Commission of Mineral Reserves ratified the Ministry's report regarding the reserves of the oil and gas field, affirming that its oil-containing area was 5.7 km^2 and had proven oil reserves of 59.73 million tons; the gas-containing area was 12 km^2 in size, had proven natural gas reserves of 17.05 billion m^3 and proven condensate oil reserves of 3.063 million tons. To develop this oil and gas field, Shanghai Municipality, the Ministry of Geology and Mineral Resources (now Sinopec Group's Star Company), and CNOOC on July 26, 1992, jointly formed Shanghai Petroleum Co., Ltd. Shanghai had a 40% share in this venture, and the Ministry of Geology and Mineral Resources and CNOOC each had a 30% share.

In September 1995, the State Planning Commission approved the development plan for Pinghu Oil and Gas Field. The main facility at Pinghu would have a multifunctional platform with the capabilities of drilling and oil production. It also had ample living quarters. They drilled 13 wells (six for oil and seven for gas), laid a 385 km 14 in. submarine gas pipeline and a 306 km 10 in. submarine oil pipeline, and built a gas-processing plant in Nanhui County, Shanghai and a crude oil transfer station in Daishan County, Zhejiang Province. The project was completed and operations started in 1998. In 2000, Pinghu Oil and Gas Field produced 3,557 barrels of crude oil per day and 7.8 million in^3 of natural gas per day. According to normal calculating methods used in our country, Pinghu Oil and Gas Field's annual hydrocarbon equivalent totaled approximately 430,000 tons.

6 Another Daqing Realized Offshore

Ever since the founding of the new China, offshore oil development has had the full attention and support from the entire country. After the fall of the Gang of Four, the implementation of the reform and opening-up policies ushered in the new epoch of China's offshore oil development. Under the leadership of Hua Guofeng, Ye Jianying, Deng Xiaoping, Li Xiannian, Chen Yun, Peng Zhen, Hu Yaobang, Zhao Ziyang, et al., China's offshore oil industry moved quickly and steadily forward.

The following passage from Kang Shi'en's biography offers a glimpse into how these leaders discussed these matters among themselves:

> In August 1979, Kang Shi'en wrote a letter to Deng Xiaoping, Li Xiannian, Chen Yun, and others, in which he pointed out, "Since a relatively large gap indeed exists between the level of our nation's petroleum geophysical exploration technology and that of advanced foreign technology, it will be necessary and beneficial for us to use the methods of compensation trade, [that is,] sign risk contracts with foreign oil companies that have advanced technologies and conduct joint explorations." A few days later, Deng Xiaoping wrote his instruction: "I agree, and advocate carrying it out fast."[3]

Another source reports:

> In his report delivered at the fourth Plenary Session of the fifth People's Congress in November of 1981, Premier Zhao Ziyang, when talking about cooperating with foreign countries on offshore oil exploration and development, declared, "Our government, in accordance with the principle of

3 Kang Shi'en zhuan bianxie zu, ed., *Kang Shi'en zhuan*, 440.

> mutual benefit, has determined to start the bidding process in the near future, to intensify exploration, development, and construction soon.[4]

On October 21, 1984, General Secretary Hu Yaobang visited Bohai Oil Corporation at Tanggu Port, Tianjin. At the time, offshore oil joint-venture exploration was in full swing in the western and eastern parts of the South China Sea. In Bohai, where joint-venture projects had started relatively early, such projects as well as autonomous (exclusively Chinese) projects were unfolding. Picking up a calligraphy brush, Hu Yaobang gladly wrote, "Create an Offshore Daqing, Forge a New Generation."

CNOOC's achievements through reform and opening-up in the 1980s and 1990s were remarkable. CNOOC's annual production increased from 95,000 tons when it was first established to 18.1 million tons of oil and 4.25 billion m³ of natural gas in 2000. With its 2014 production level hitting more than 50 million tons, another Daqing was indeed realized offshore.

4 Dangdai Zhongguo congshu bianjibu 当代中国丛书编辑部, ed., *Dangdai Zhongguo shiyou gongye* 《当代中国石油工业》 [*The Modern Chinese Petroleum Industry*] (Beijing: Zhongguo shehui kexue chubanshe, 1988), 176.

CHAPTER 39

Integrating Internationally, Creating a New Company

1 What Do We Mean by "a New Company"?

As mentioned above, offshore petroleum in 2000 produced 20 million tons of crude oil. The profitability of the offshore enterprise, however, was even more remarkable than that figure alone might show. CNOOC was producing 1,000 tons of oil per worker, whereas on-land oil production was only 100 tons per worker. Offshore profitability was 10 times better than on land. From 1983 until 2000, thanks to large-scale international cooperation, offshore oil production increased by a factor of 200, while the workforce increased by only 0.4%, meaning that labor productivity rose by a factor of close to 70.

The focus of this chapter is on how CNOOC, in its effort to reform the Chinese oil industry, integrate it internationally, and learn from foreign oil companies, built a whole new petroleum company. Their success in reforming China's oil industry and integrating internationally was more meaningful than the material achievements.

2 Eighty People Manage a Big Gas Field

One morning in April 2000, we visited Ya 崖 13-1 Offshore Gas Field's Nanshan terminal in the outskirts of Sanya, Hainan Province. Ya 13-1 Gas Field is situated in Yinggehai, approximately 100 km from the southeast shore of Sanya. Discovered in August 1983 by ARCO, its construction was completed in 1996. It produced 3.45 billion m^3 of natural gas annually, with a hydrocarbon equivalent of 3.45 million tons. Of the natural gas produced, 2.9 billion m^3 were transported by a 778 km submarine gas pipeline to its terminal in Hong Kong, to supply CLP Power Hong Kong Ltd. for electricity generation. The remaining 550 million m^3 of natural gas and condensate was transferred by a 70 km pipeline to Nanshan 南山, in Sanya, for electricity generation. The condensate was shipped away from Sanya.

Compared with gas fields on land, this gas field was a big one. However, it only had three managing stops. The first point was at the gigantic gas platform

 | DOI:10.1163/9789004402737_040

itself. The sea water around the platform was 40 m deep. Six gas wells were drilled. The 40 to 50 m^2 platform had an advanced gas collecting, pressurizing, and transport unit and a highly efficient oil and gas separation unit. Once the oil and gas was separated, most of the dry gas went to Hong Kong, and the rest of the oil and gas mixture was transported to Nanshan in Sanya. On the platform, there were living quarters, environmental protection and safety facilities, and a helicopter landing pad. The gas platform was managed by some 30 people who worked on two shifts. Every 28 days (and there would be no delays), they would be transported either via boat or helicopter on shore to rest for 28 days. Whenever there was a typhoon alert that signaled a dangerous level, depending on the typhoon's strength and distance, everyone would evacuate the platform. This showed that it was a truly modern enterprise, following the principle of safety first.

I visited Nanshan Gas Transportation Terminal, the pipeline terminus, accompanied by Li Li 李力, its Chinese Deputy Supervisor. He was only 26 years old and had majored in English in college. Starting out as an operator, he eventually was promoted to foreman. Since he had worked several years with foreigners in oil exploration and was familiar with offshore oil and gas production, both the Chinese and the Americans agreed to hire him as Deputy Supervisor. According to the contract, after the American supervisor would leave in 2001, the remaining 30-plus people at the terminal would be all Chinese.

This heavily guarded terminal was situated at a scenic spot on the shore, occupying 50 acres. There were two gas transportation units (one as spare), one oil and gas separator, a few oil tanks and gas tanks, a port, and so on. It was very orderly, with the pedestrian paths sparkling clean and the site kept like a garden. This young deputy supervisor and his 30 or so workers worked systematically. The American supervisor was not there. Li Li told me that the Hong Kong terminal had fewer people, a dozen or so. That is to say, an offshore gas field that produced 3.45 billion m^3 of gas annually, in three locations, including the production platform, Nanshan and Hong Kong terminals, was manned by only 80 to 90 people. By comparison, Huabei Oilfield in the mid-1980s had an annual production of 3 million tons while its workforce was 120,000 people. Eighty people compared to 120,000. This is the difference.

At the terminal's cafeteria, I saw a foreigner. I asked Li if he was one of his workers. He replied, "No, the workers here are all Chinese. He's an Indian chef who works for the Hong Kong dining services company that serves us. Our employees only do gas production. We contract out the port management and services, hiring the best."

3 A New Type of Offshore Drilling Company

In the evening of April 6, 2000, I had a nice conversation with a few financial managers of China Offshore Oil Nanhai West Corporation (Nanhai West). They all used to work for oilfields on land before working for Nanhai West. They told me about their offshore experience.

They said that the company had five imported drilling vessels, two of which were jack-up rigs capable of drilling at 90 m depth; three were semisubmersible vessels capable of drilling at 300 to 500 m depth offshore. The vessels had been worth 500 to 600 million yuan but were now worth 1.5 billion yuan. The company had more than 800 employees, and each vessel had a fixed crew of around 70 people, determined by the number of beds on each. Even if the captain of the vessel (the drilling crew leader) wanted to add more people, they would have no place to sleep. This foreign setup's best feature was that everyone had a clear responsibility—when to do something, and how to do it. The boss's drilling supervisor gave very clear instructions. Workers did everything according to orders, each performing his duty. In addition, after 28 days, the workers would go for a rest on land and be rotated with another crew. Their offshore salary was more than 5,000 yuan per month, as opposed to 1,200 yuan per month for their stay on land, so that their salary averaged out to 2,000 to 3,000 yuan. When they did drilling in joint-venture oilfields, they had often won the bid due to their cheap price and good service. One vessel could make US$20,000 per day. If they were drilling in Chinese oilfields, CNOOC only gave them 730 yuan per month, not even half that given by Sino-foreign joint-venture oilfields. Even so, the previous year, they paid 60 million yuan for revenue and taxes. They maintained and repaired their own vessels, saving the company millions of yuan, most of which they had to surrender to CNOOC. They were allowed to keep only 20% for themselves.

4 Much to Be Done to Truly Integrate Internationally

Despite the two positive examples discussed above, China's offshore oil industry is still far from being truly integrated internationally in terms of their management. On April 5, 2000, my good friend Zou Jiazhi told me the following:

In 1981, Zou was the Secretary of the Party Committee of Nanhai Oil Exploration Headquarters. They encountered opposition to allowing Chinese drilling vessels to drill for foreign companies. He had proposed that our vessels should be rented out, otherwise "the vessels will get rusty, people will become lazy, and we will fall behind technologically." Soon, one American oil company was offering to rent a jack-up drilling vessel for US$30,000 per day, but the person in charge in the Ministry would not agree to it. He reported the incident to Kang Shi'en. Kang instructed him to go ahead. The American company proposed to rent the vessel for two years, but the Ministry of Petroleum would only rent it out for one year. The Americans responded that with a one-year lease, as soon as the vessel arrived, it would need to go back. In the end, there was no deal, and US$30,000 of daily income was lost right under their eyes!

In 1994, Nanhai Drilling Company learned that there was a used vessel in good condition for sale in Singapore for an affordable US$3 million. Another foreign oil company also wanted to buy it. However, by the time Nanhai had gone through the lengthy process of getting the project examined and approved, funds pooled, and visas granted, months had passed. When their purchasing group was finally ready to buy the boat, they learned that the boat had already been bought long before.

In 1997, in an oilfield that Nanhai East Oil Corporation had jointly developed with a foreign oil company, a piece of equipment on the submarine oil transportation pipeline was damaged by a passing fishing boat. This no doubt was an accident. Nanhai East Corporation reported it to CNOOC. The response from CNOOC's production safety division was: Who was involved? When did it happen? Who was responsible? Meanwhile, our foreign joint-venture partner also reported the accident to its company headquarters. Its Safety and Environmental Protection Department's response consisted of concrete interim measures to take in order to deal with the accident and reduce losses; its Supply Department's response was to tell them that the spare parts to replace the damaged parts could be found in such storage, in such a company, at such a place, in such a country, one could purchase in such a way; its Legal Department responded with directions on how much compensation one should seek with the insurance company, how much was definite and how much one could get if one tried ... Afterward, Nanhai East Corporation's Deputy General Manager relayed the above story to CNOOC. Although sympathetic, a manager in CNOOC said helplessly, "Indeed! We at CNOOC had no such personnel. It would be a long time for us to do things like foreign oil companies."

CHAPTER 40

Engaging in International Cooperation, Exploring Foreign Oil and Gas Supplies

1 Building Pipelines along Borders to Import Foreign Oil and Gas

Since the beginning of the reform and opening up period, China has been building pipelines along its borders in order to import foreign oil and gas, due to increasing demand. In the northwest, in 1997, CNPC started investing in Kazakhstan, and it built the Sino-Kazakhstan oil and gas pipelines. As of 2017, CNPC had invested US$42 billion there, making it Kazakhstan's leading Chinese investor. CNPC had cumulatively paid US$41 billion in business income taxes to Kazakhstan and employed more than 30,000 Kazakhs. During the twenty years from 1997 to 2017, the pipelines transported 100 million tons of oil to our country. Meanwhile, China was also cooperating with other central Asian countries such as Turkmenistan and Kyrgyzstan on the development and utilization of oil and gas resources. More recently, three auxiliary lines have been added to the pipelines, and half of China's imported gas is transported by these pipelines.

In the northeast, starting in the 1980s, CNPC built oil and gas pipelines in the Heilongjiang and Inner Mongolia area to import oil and gas from Russia. Since then, these pipelines have been steadily expanded. Currently, we import more than 38 million tons of oil and gas annually from Russia.

In the southwest, in 2010 after many years of planning and negotiating, construction of the Sino-Myanmar pipelines started. Completed in 2013, these 2,400 km-long pipelines (1,634 km in China) now transport tens of millions of tons of oil and tens of billions of m^3 of gas each year.

These measures now guarantee a steady supply of foreign oil and gas to China through multiple channels.

2 Building Large-Scale Petrochemical Enterprises on the Southeastern Coast

More than half of China's imported oil and gas is shipped from Africa (especially Nigeria), the Middle East (Saudi Arabia and Iran), and South America (Venezuela and Brazil). From Saudi Arabia alone, we import 20 million tons of

 | DOI:10.1163/9789004402737_041

oil annually. Because of this, China started building petrochemical enterprises as early as the 1950s in areas such as Guangdong's Maoming and Huizhou, Fujian, Zhejiang's Zhenhai, Shanghai and the Yangtze River Delta, in order to refine the imported oil and gas.

3 Going Abroad to Search for Oil and Gas Resources

After China became a net petroleum importing country in 1993, its oil imports increased rapidly. By 2017, China had become the world's largest oil importer, and 72% of our oil consumption now depends on imports. As a result, since the 1980s, the three big Chinese petrochemical companies (CNPC, Sinopec, CNOOC) have been gathering capital, forming oil teams, transferring equipment, and going abroad to Africa, the Middle East, South America, and Russia, cooperating with other nations on oil and gas exploration projects, and looking for oil and gas resources overseas. They have been very successful in this regard. In 2017 alone, they imported 150 million tons of oil and gas through these projects.

CHAPTER 41

The Oil Industry in Taiwan

1 Adopting New Technology, Exploring New Oil and Gas Areas

Taiwan Province lies off the southeastern coast of the Chinese mainland across the Taiwan Strait. The province includes the island of Taiwan itself, the Penghu Islands, Orchid Island, Green Island, Pengjia Islet, Diaoyu Islands, and Chiwei Island. Taiwan is the largest island of the province and China's largest island, with an area of 35,883 km^2. In the central part of Taiwan, Yushan Mountain, at 3,952 m in elevation, is the tallest mountain in the eastern part of our country. Taiwan in the past has been reported to be rich in petroleum, coal, gold, copper and sulfur resources, but in fact, much of the coal, gold, and copper has already been mined out, and its petroleum has never been able to meet local demand.

Taiwan has been part of China since antiquity. Taiwan's oil industry is an important component of the Chinese oil industry.

Taiwan's oil industry is one of the oldest in the history of Chinese petroleum. Petroleum was extracted using modern technology at Chuhuangkeng in Miaoli County from 1877 until the end of the twentieth century. Taiwan's oil industry is around 140 years old, predating both Yanchang Oilfield (1905) in Shaanxi Province and Yumen Oilfield (1939) in Gansu Province. During the Japanese occupation of Taiwan, the oil industry underwent some development, but after the Japanese surrendered unconditionally in 1945, the Nationalist government took over the oil industry in Taiwan.

On June 1, 1946, the Nationalist government's National Resources Commission established Chinese Petroleum Corporation (CPC) in Shanghai, to oversee a dozen petroleum enterprises such as Ganqing Company (Yumen Oilfield), a few refineries in the northeast, and the Taiwan Oil Exploration Division, Kaohsiung Refinery.

At that time, oil production in Taiwan had been depressed for a long time. Due to financial problems, oil exploration could not be carried out. After the Nationalist government took over Taiwan, CPC sent a gravity geological crew to survey. Besides managing and repairing the five wells left behind by the Japanese, only two 1,200 to 1,300 m exploratory wells were drilled, with unsatisfactory results.

In 1949, CPC moved from Shanghai to Taipei, in Taiwan. Under the leadership of its Chairman Chin Kai-ying, with the full support of the Nationalist government, oil exploration and oil refining developed and surpassed other industries in Taiwan. In 1953, in an effort to promote the petroleum industry,

 | DOI:10.1163/9789004402737_042

Chiang Kai-shek visited the Taiwan Petroleum Exploration Division in Miaoli County, encouraging them to adopt new technologies and explore to find new oil and gas fields.

In 1959, after careful reconnaissance and research, they chose Jinshui Structure, rich in natural gas. They dug deeper into the existing Jinshui-38 well. During the Japanese occupation period, this 3,583 m deep well was producing 9,000 m^3 of gas and 1.3 m^3 of oil daily. After eighty days of hard work, they deepened the well to 4,063 m, drilling through five hydrocarbon zones. When they finished drilling, the well produced 1 million m^3 of gas and 10 m^3 of oil per day. This was Taiwan's first deep well. It also held the Taiwan record for the quantity of oil and natural gas produced. The crew received four medals and ten certificates of merit from Chiang Kai-shek.

As a result of Jinshui-38's success in obtaining high-yield oil and gas flow, CPC became more active in oil exploration and discovered a group of new oil and gas fields. In 1978, Taiwan's oil production reached its zenith, producing 2.5 million m^3 of crude oil and 19.6 billion m^3 of natural gas.

2 Building Oil Refineries, Actively Seeking Foreign Oil Supplies

During the Japanese occupation period, Taiwan's Kaohsiung Refinery was a large, modern, world-renowned refinery, but when the Japanese Imperial Army retreated, it was severely damaged. After CPC took over, they worked hard in "the spirit of Laojunmiao" to rebuild and expand the plant out of the rubble to the point that it was again providing high-grade aviation fuel and liquified gas for civilian use. As such it could satisfy much of Taiwan's domestic oil and gas needs. To stabilize the refinery's crude oil supply, Kaohsiung's oil port was also built. Meanwhile, in 1970, in accordance with Chiang Kai-shek's instructions to "enrich the energy supply, spread out the oil refineries," construction on a much larger refinery started in Taoyuan. As a result, Taiwan's petrochemical industry, using oil and gas as raw materials, expanded.

Since Taiwan is small, with limited resources, the government has been looking for oil and gas overseas from the beginning. Since 1970, the CPC has been cooperating with foreign oil companies in oil and gas exploration and enjoyed remarkable results.

3 A Key Player in Taiwan's Economic Success

The oil industry's development in Taiwan is closely linked to Taiwan's economic development. To satisfy increasing market demand, Taiwan's petroleum industry, including relevant downstream industries, gradually increased refining capability to 600,000 barrels (85,714 tons) per day. During the 20 years when Taiwan's economy was experiencing rapid growth, the petroleum and petrochemical industries provided many people with employment. In the 1960s and 1970s, thirty percent of the jobs in the manufacturing sector were related to the petroleum and petrochemical industries, and output value reached US$52 billion. As Ye Jinlong, Vice Chairman of CPC Corporation, put it, "In Taiwan, the oil industry has the power to influence finance, economics and politics. It has made undeniable contributions to the Taiwan economic miracle ..."[1]

According to a September 4, 2001 report in *Global Times* (*Huanqiu shibao* 《环球时报》; Beijing), CPC was about to cooperate with CNOOC to jointly explore and develop the oil and gas resources in the Taiwan Strait. While mainland China's CNOOC is stronger in oil extraction, Taiwan's CPC excels in oil refining. Taiwan's Central News Agency (Taipei) reported that on May 16, 2002, in Taipei, CNOOC and CPC formally signed a cooperation agreement to explore and develop oil and gas resources in the southern part of the Taiwan Strait. This was mutually beneficial for both Taiwan and mainland China.

1 Ye Jinlong, *Shiyou gongye: Taiwan jingji qiji de zhujiao*, 6.

Epilogue

In closing this volume, I would like to quote from two astute observers of industrial history. One is Daniel Yergin. In his 1991 book *The Prize: The Epic Quest for Oil, Money, and Power,* he claims that "the twentieth century rightly deserves the title 'the century of oil.'" Other observations that he made resonated with my own experience:

> In war and peace, oil would achieve the capacity to make or break nations ...
>
> Oil has brought us the best and the worst in our civilization. It has been both boon and burden ...
>
> This history has been a panorama of triumphs and a litany of tragic and costly mistakes. It has been a theater for the noble and the base in the human character. Creativity, dedication, entrepreneurship, ingenuity, and technical innovation have coexisted with avarice, corruption, blind political ambition, and brute force.[1]

The other comment comes from Karl Marx. He concluded his *Critique of the Gotha Programme* with a Latin phrase meaning, "I have spoken and saved my soul."[2]

---

1 Daniel Yergin, *The Prize: The Epic Quest for Oil, Money, and Power,* rev. ed. (New York: Free Press, 2009), 762.

2 Karl Marx, *Critique of the Gotha Programme* (New York: International Publishers, 1938), 23. The Latin phrase: *Dixi et salvavi animam meam.*

 | DOI:10.1163/9789004402737_043

Index